THE
TENSORFLOW
WORKSHOP

A hands-on guide to building deep learning models
from scratch using real-world datasets

Matthew Moocarme, Anthony So, and Anthony Maddalone

THE TENSORFLOW WORKSHOP

Authors: Matthew Moocarme, Anthony So, and Anthony Maddalone

Reviewer: Abhranshu Bagchi

Managing Editor: Prachi Jain

Acquisitions Editors: Royluis Rodrigues, Kunal Sawant, and Sneha Shinde

Production Editor: Salma Patel

Editorial Board: Megan Carlisle, Heather Gopsill, Manasa Kumar, Alex Mazonowicz, Monesh Mirpuri, Bridget Neale, Abhishek Rane, Brendan Rodrigues, Ankita Thakur, Nitesh Thakur, and Jonathan Wray

First published: December 2021

Production reference: 1141221

ISBN: 978-1-80020-525-3

Published by Packt Publishing Ltd.

Livery Place, 35 Livery Street

Birmingham B3 2PB, UK

Table of Contents

Chapter 2: Loading and Processing Data   49

Chapter 5: Classification Models   159

Chapter 6: Regularization and Hyperparameter Tuning

Chapter 7: Convolutional Neural Networks   239

Appendix   453

Index   559

PREFACE

ABOUT THE BOOK

If you want to learn to build deep learning models in TensorFlow to solve real-world problems, then this is the book for you.

Beginning with an introduction to TensorFlow, this book gives you a tour of the basic mathematical operations of tensors, as well as various methods of data-preparation for modeling and time-saving model-development using TensorFlow resources. You will build regression and classification models, use regularization to prevent models from overfitting training data, and create convolutional neural networks to solve classification tasks on image datasets. Finally, you'll learn to implement pre-trained, recurrent, and generative models and create your own custom TensorFlow components to use within your models.

By the end of this book, you'll have the practical skills to build, train, and evaluate deep learning models using the TensorFlow framework.

ABOUT THE AUTHORS

Matthew Moocarme is an accomplished data scientist with more than eight years of experience in creating and utilizing machine learning models. He comes from a background in the physical sciences, in which he holds a Ph.D. in physics from the Graduate Center of CUNY. Currently, he leads a team of data scientists and engineers in the media and advertising space to build and integrate machine learning models for a variety of applications. In his spare time, Matthew enjoys sharing his knowledge with the data science community through published works, conference presentations, and workshops.

Anthony So is a renowned leader in data science. He has extensive experience in solving complex business problems using advanced analytics and AI in different industries including financial services, media, and telecommunications. He is currently the chief data officer of one of the most innovative fintech start-ups. He is also the author of several best-selling books on data science, machine learning, and deep learning. He has won multiple prizes at several hackathon competitions, such as Unearthed, GovHack, and Pepper Money. Anthony holds two master's degrees, one in computer science and the other in data science and innovation.

Anthony Maddalone is a research engineer at TieSet, a Silicon Valley-based leader in distributed artificial intelligence and federated learning. He is a former founder and CEO of a successful start-up. Anthony lives with his wife and two children in Colorado, where they enjoy spending time outdoors. He is also a master's candidate in analytics with a focus on industrial engineering at the Georgia Institute of Technology.

WHO THIS BOOK IS FOR

This TensorFlow book is for anyone who wants to develop their understanding of deep learning and get started building neural networks with TensorFlow. Basic knowledge of Python programming and its libraries, as well as a general understanding of the fundamentals of data science and machine learning, will help you grasp the topics covered in this book more easily.

ABOUT THE CHAPTERS

Chapter 1, Introduction to Machine Learning with TensorFlow, introduces you to the mathematical concepts that underly TensorFlow and machine learning model development, which include tensors and linear algebra.

Chapter 2, Loading and Processing Data, teaches you how to load and process a variety of different data types including tabular, images, audio, and text so that they can be input into machine learning models.

Chapter 3, TensorFlow Development, introduces you to a variety of development tools that TensorFlow offers to aid your model building, including TensorBoard, TensorFlow Hub, and Google Colab. These tools can help speed up development as well as aiding your understanding of the architecture and performance of your models.

Chapter 4, Regression and Classification Models, guides you through building models using TensorFlow for regression and classification tasks. You will learn how to build simple models, which layers to use, and the appropriate loss functions to use for each.

Chapter 5, Classification Models, demonstrates how to build classification models using TensorFlow. You will learn how to customize the architecture of neural networks for binary, multi-class, or multi-label classification.

Chapter 6, Regularization and Hyperparameter Tuning, discusses the different methods that can help prevent models from overfitting, such as regularization, dropout, or early stopping. You will also learn how to perform automatic hyperparameter tuning.

Chapter 7, Convolutional Neural Networks, demonstrates how to build neural networks with convolutional layers. These networks are popular due to their good performance when working with images because of the convolutional layers they contain.

Chapter 8, Pre-Trained Networks, teaches you how to leverage pre-trained models in order to achieve better performance without having to train a model from scratch.

Chapter 9, Recurrent Neural Networks, introduces a different type of deep learning architecture known as recurrent neural networks, which are best suited for sequential data such as time-series or text.

Chapter 10, Custom TensorFlow Components, expands your repertoire by teaching you how to build your own custom TensorFlow components such as loss functions and neural network layers.

Chapter 11, Generative Models, shows you how you can generate new and novel data by training models on a dataset to discover the underlying patterns and representations. The trained model will then be able to generate convincingly real examples for itself that are completely novel.

CONVENTIONS

Code words in text, database table names, folder names, filenames, file extensions, pathnames, dummy URLs, and user input are shown as follows:

"TensorFlow can be used in Python by importing certain libraries. You can import libraries in Python using the **import** statement."

Words that you see on the screen, for example, in menus or dialog boxes, also appear in the same format.

A block of code is set as follows:

```
int_variable = tf.Variable(4113, tf.int16)
int_variable
```

New important words are shown like this: "**Backpropagation** is the process of determining the derivative of the loss with respect to the model parameter."

Key parts of code snippets are emboldened as follows:

```
df = pd.read_csv('Bias_correction_ucl.csv')
```

CODE PRESENTATION

Lines of code that span multiple lines are split using a backslash (\). When the code is executed, Python will ignore the backslash, and treat the code on the next line as a direct continuation of the current line.

For example,

```
year_dummies = pd.get_dummies(df['Date'].dt.year, \
                              prefix='year')
year_dummies
```

Comments are added into code to help explain specific bits of logic. Single-line comments are denoted using the # symbol, as follows:

```
# Importing the matplotlib library
import matplotlib.pyplot as plt
```

MINIMUM HARDWARE REQUIREMENTS

For an optimal experience, we recommend the following hardware configuration:

- Processor: Dual-core or better
- Memory: 4 GB RAM
- Storage: 10 GB available space

DOWNLOADING THE CODE BUNDLE

Download the code files from GitHub at https://packt.link/Z7pcq. Refer to these code files for the complete code bundle. The files here contain the exercises, activities, and some intermediate code for each chapter. This can be a useful reference when you become stuck.

On the GitHub repo's page, you can click the green **Code** button and then click the **Download ZIP** option to download the complete code as a ZIP file to your disk (refer to *Figure 0.1*). You can then extract these code files to a folder of your choice, for example, **C:\Code**.

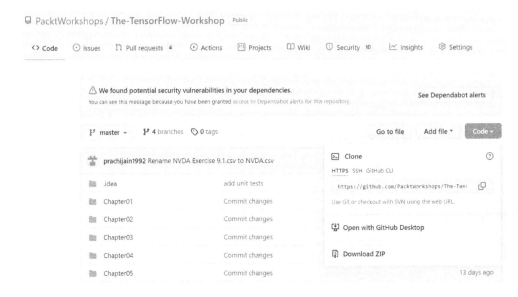

Figure 0.1: Download ZIP option

On your system, the extracted ZIP file should contain all the files present in the GitHub repository:

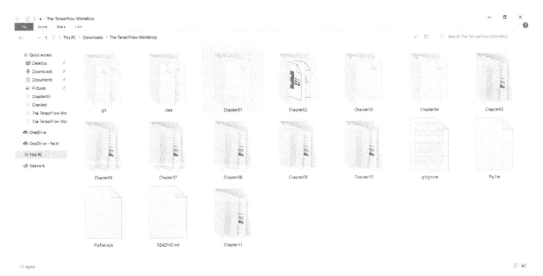

Figure 0.2: GitHub code directory structure

SETTING UP YOUR ENVIRONMENT

Before you explore the book in detail, you need to set up specific software and tools. In the following section, you will see how to do that.

INSTALLING ANACONDA ON YOUR SYSTEM

The code for all the exercises and activities in this book can be executed using the Jupyter Notebook. You'll first need to install Anaconda Navigator, which is an interface through which you can access your Jupyter notebooks. Anaconda Navigator will be installed as part of Anaconda Individual Edition, which is an open source Python distribution platform available for Windows, macOS, and Linux. Installing Anaconda will also install Python. Head to https://www.anaconda.com/distribution/:

1. From the page that opens, click the **Download** button (annotated by *1*). Make sure you are downloading the **Individual Edition**.

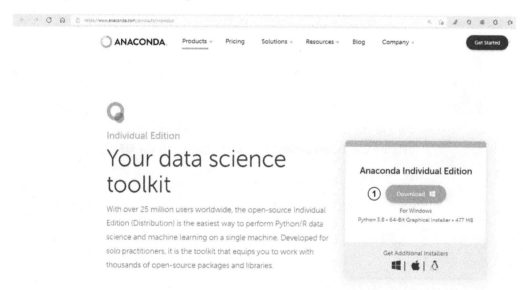

Figure 0.3: Anaconda home page

2. The installer should start downloading immediately. The website will, by default, choose an installer based on your system configuration. If you prefer downloading Anaconda for a different operating system (Windows, macOS, or Linux) and system configuration (32- or 64-bit), click the `Get Additional Installers` link at the bottom of the box (refer to *Figure 0.3*). The page should scroll down to a section (refer to *Figure 0.4*) that lets you choose from various options based on the operating system and configuration you desire. For this book, it is recommended that you use the latest version of Python (3.8 or higher).

Figure 0.4: Downloading Anaconda based on the OS

3. Follow the installation steps presented on the screen.

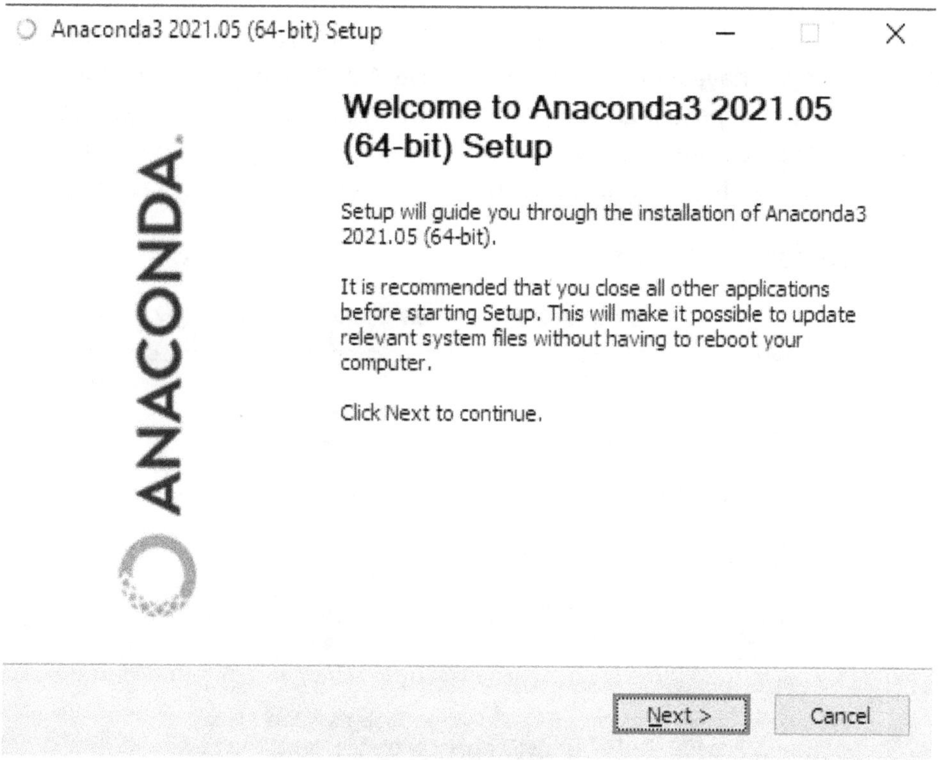

Figure 0.5: Anaconda setup

4. On Windows, if you've never installed Python on your system before, you can select the checkbox that prompts you to add Anaconda to your **PATH**. This will let you run Anaconda-specific commands (like **conda**) from the default command prompt. If you have Python installed or have installed an earlier version of Anaconda in the past, it is recommended that you leave it unchecked (you may run Anaconda commands from the Anaconda Prompt application instead). The installation may take a while depending on your system configuration.

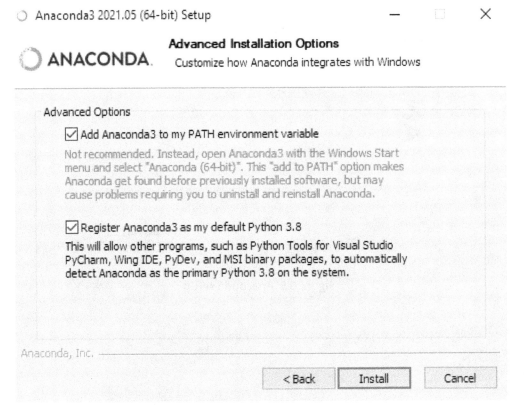

Figure 0.6: Anaconda installation steps

For more detailed instructions, you may refer to the official documentation for Linux by clicking this link (https://docs.anaconda.com/anaconda/install/linux/); for macOS using this link (https://docs.anaconda.com/anaconda/install/mac-os/); for Windows using this link (https://docs.anaconda.com/anaconda/install/windows/).

5. To check if Anaconda Navigator is correctly installed, look for **Anaconda Navigator** in your applications. Look for an application that has the following icon. Depending on your operating system, the icon's aesthetics may vary slightly.

Anaconda Navigator (Anaconda3)

App

Figure 0.7: Anaconda Navigator icon

You can also search for the application using your operating system's search functionality. For example, on Windows 10, you can use the *Windows Key + S* combination and type in *Anaconda Navigator*. On macOS, you can use Spotlight search. On Linux, you can open the terminal and type the `anaconda-navigator` command and press the *Return* key.

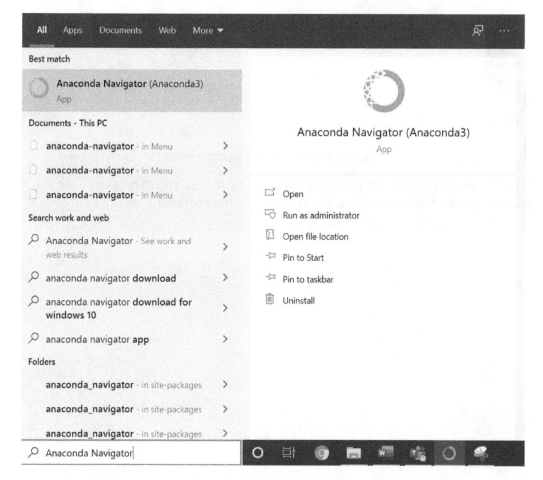

Figure 0.8: Searching for Anaconda Navigator on Windows 10

For detailed steps on how to verify if Anaconda Navigator is installed, refer to the following link: https://docs.anaconda.com/anaconda/install/verify-install/.

6. Click the icon to open Anaconda Navigator. It may take a while to load for the first time, but upon successful installation, you should see a similar screen:

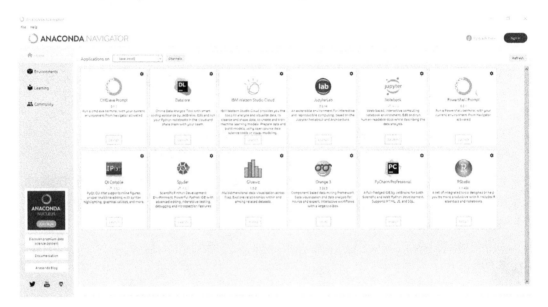

Figure 0.9: Anaconda Navigator screen

If you have more questions about the installation process, you may refer to the list of frequently asked questions from the Anaconda documentation: https://docs.anaconda.com/anaconda/user-guide/faq/.

LAUNCHING JUPYTER NOTEBOOK

Once Anaconda Navigator is open, you can launch the Jupyter Notebook interface from this screen. The following steps will show you how to do that:

1. Open Anaconda Navigator. You should see the following screen:

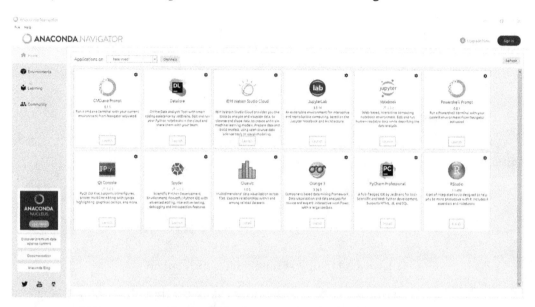

Figure 0.10: Anaconda Navigator screen

2. Now, click **Launch** under the **Jupyter Notebook** panel to start the notebook interface on your local system:

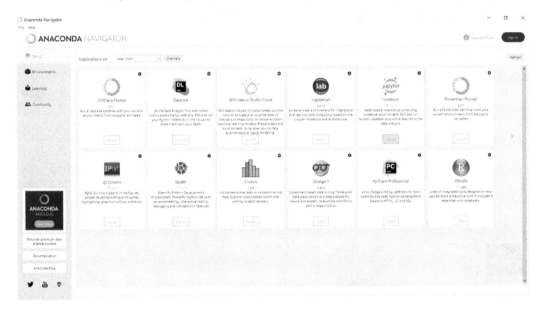

Figure 0.11: Jupyter notebook launch option

3. On clicking the **Launch** button, you'll notice that even though nothing changes in the window shown in the preceding screenshot, a new tab opens up in your default browser. This is known as the *Notebook Dashboard*. It will, by default, open to your root folder. For Windows users, this path would be something similar to `C:\Users\<username>`. On macOS and Linux, it will be `/home/<username>/`.

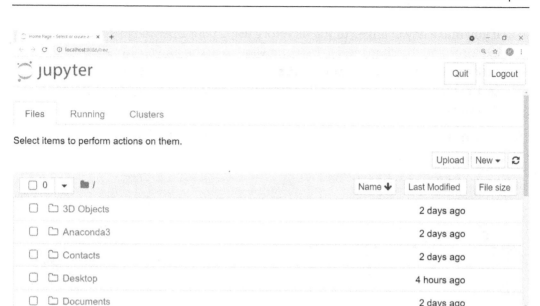

Figure 0.12: Notebook Dashboard

Note that you can also open a Jupyter notebook by simply running the command **`jupyter notebook`** in the terminal or command prompt. Or, you can search for **Jupyter Notebook** in your applications just like you did in *Figure 0.8*.

4. You can use this dashboard as a file explorer to navigate to the directory where you have downloaded or stored the code files for the book (refer to the *Downloading the Code Bundle* section on how to download the files from GitHub). Once you have navigated to your desired directory, you can start by creating a new notebook. Alternatively, if you've downloaded the code from our repository, you can open an existing notebook as well (notebook files will have a `.inpyb` extension). The menus here are quite simple to use:

Figure 0.13: Jupyter notebook navigator menu options walk-through

If you make any changes to the directory using your operating system's file explorer and the changed file isn't showing up in the Jupyter Notebook navigator, click the **Refresh Notebook List** button (annotated as *1*). To quit, click the **Quit button** (annotated as *2*). To create a new file (a new Jupyter notebook), you can click the **New** button (annotated as *3*).

5. Clicking the **New** button will open a dropdown menu as follows:

Figure 0.14: Creating a new Jupyter notebook

You can get started and create your first notebook by selecting **Python 3**; however, it's recommended that you also install the virtual environment we've provided to help you install all the packages required for the title. The following section will show you how to install it.

> **NOTE**
>
> A detailed tutorial on the interface and the keyboard shortcuts for Jupyter notebooks can be found here: https://jupyter-notebook.readthedocs.io/en/stable/notebook.html#the-jupyter-notebook.

INSTALLING THE TENSORFLOW VIRTUAL ENVIRONMENT

As you run the code for the exercises and activities, you'll notice that even after installing Anaconda, there are certain libraries that you'll need to install separately as you progress through the book. Then again, you may already have these libraries installed, but their versions may be different from the ones we've used, which may lead to varying results. That's why we've provided an **environment.yml** file with this book that will:

1. Install all the packages and libraries required for this book at once.

2. Make sure that the version numbers of your libraries match the ones we've used to write the code for this book.

3. Make sure that the code you write based on this course remains separate from any other coding environment you may have.

You can download the **environment.yml** file by clicking the following link: https://packt.link/Z7pcq.

Save this file, ideally in the same folder where you'll be running the code for this book. If you've downloaded the code from GitHub as detailed in the *Downloading the Code Bundle* section, this file should already be present in the parent directory, and you won't need to download it separately.

To set up the environment, follow these steps:

1. On macOS, open Terminal from Launchpad (you can find more information about Terminal here: https://support.apple.com/en-in/guide/terminal/apd5265185d-f365-44cb-8b09-71a064a42125/mac). On Linux, open the Terminal application that's native to your distribution. On Windows, you can open Anaconda Prompt instead by simply searching for the application. You can do this by opening the **Start** menu and searching for **Anaconda Prompt**.

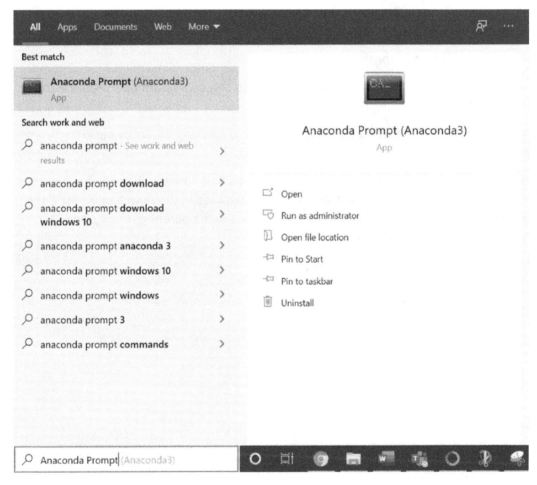

Figure 0.15: Searching for Anaconda Prompt on Windows

A new terminal like the following should open. By default, it will start in your home directory:

Figure 0.16: Anaconda terminal prompt

In the case of Linux, it will look like the following:

Figure 0.17: Terminal in Linux

2. In the terminal, navigate to the directory where you've saved the **environment.yml** file on your computer using the **cd** command. Say you've saved the file in **Documents\The-TensorFlow-Workshop**. In that case, you'll type the following command in the prompt and press *Enter*:

```
cd Documents\The-TensorFlow-Workshop
```

Note that the command may vary slightly based on your directory structure and your operating system.

3. Now that you've navigated to the correct folder, create a new **conda** environment by typing or pasting the following command in the terminal. Press *Enter* to run the command:

```
conda env create -f environment.yml
```

This will install the **tensorflow** virtual environment along with the libraries that are required to run the code in this book. If you see a prompt asking you to confirm before proceeding, type **y** and press *Enter* to continue creating the environment. Depending on your system configuration, it may take a while for the process to complete.

> **NOTE**
>
> For a complete list of **conda** commands, visit the following link: https://conda.io/projects/conda/en/latest/index.html.
>
> For a detailed guide on how to manage **conda** environments, please visit the following link: https://conda.io/projects/conda/en/latest/user-guide/tasks/manage-environments.html.

4. Once complete, type or paste the following command in the shell to activate the newly installed environment – **tensorflow**:

```
conda activate tensorflow
```

If the installation is successful, you'll see the environment name in brackets change from **base** to **tensorflow**:

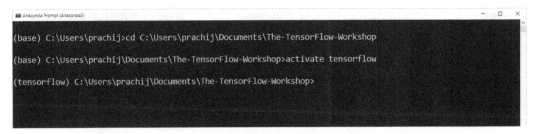

Figure 0.18: Environment name showing up in the shell

5. Run the following command to install **ipykernel** in the newly activated **conda** environment:

```
pip install ipykernel
```

> **NOTE**
>
> On macOS and Linux, you'll need to specify **pip3** instead of **pip**.

6. In the same environment, run the following command to add **ipykernel** as a Jupyter kernel:

```
python -m ipykernel install --user --name=tensorflow
```

7. **Windows only:** If you're on Windows, type or paste the following command. Otherwise, you may skip this step and exit the terminal:

```
conda install pywin32
```

8. Select the created **tensorflow** kernel when you start your Jupyter notebook.

Figure 0.19: Selecting the tensorflow kernel

A new tab will open with a fresh, untitled Jupyter notebook where you can start writing your code:

Figure 0.20: A new Jupyter notebook

GET IN TOUCH

Feedback from our readers is always welcome.

General feedback: If you have any questions about this book, please mention the book title in the subject of your message and email us at `customercare@packtpub.com`.

Errata: Although we have taken every care to ensure the accuracy of our content, mistakes do happen. If you have found a mistake in this book, we would be grateful if you could report this to us. Please visit www.packtpub.com/support/errata and complete the form.

Piracy: If you come across any illegal copies of our works in any form on the internet, we would be grateful if you could provide us with the location address or website name. Please contact us at `copyright@packt.com` with a link to the material.

If you are interested in becoming an author: If there is a topic that you have expertise in and you are interested in either writing or contributing to a book, please visit authors.packtpub.com.

PLEASE LEAVE A REVIEW

Let us know what you think by leaving a detailed, impartial review on Amazon. We appreciate all feedback – it helps us continue to make great products and help aspiring developers build their skills. Please spare a few minutes to give your thoughts – it makes a big difference to us. You can leave a review by clicking the following link: https://packt.link/r/1800205252.

1

INTRODUCTION TO MACHINE LEARNING WITH TENSORFLOW

OVERVIEW

In this chapter, you will learn how to create, utilize, and apply linear transformations to the fundamental building blocks of programming with TensorFlow: tensors. You will then utilize tensors to understand the complex concepts associated with neural networks, including tensor reshaping, transposition, and multiplication.

INTRODUCTION

Machine learning (ML) has permeated various aspects of daily life that are unknown to many. From the recommendations of your daily social feeds to the results of your online searches, they are all powered by machine learning algorithms. These algorithms began in research environments solving niche problems, but as their accessibility broadened, so too have their applications for broader use cases. Researchers and businesses of all types recognize the value of using models to optimize every aspect of their respective operations. Doctors can use machine learning to decide diagnosis and treatment options, retailers can use ML to get the right products to their stores at the right time, and entertainment companies can use ML to provide personalized recommendations to their customers.

In the age of data, machine learning models have proven to be valuable assets to any data-driven company. The large quantities of data available allow powerful and accurate models to be created to complete a variety of tasks, from regression to classification, recommendations to time series analysis, and even generative art, many of which will be covered in this workshop. And all can be built, trained, and deployed with TensorFlow.

The TensorFlow API has a huge amount of functionality that has made it popular among all machine learning practitioners building machine learning models or working with tensors, which are multidimensional numerical arrays. For researchers, TensorFlow is an appropriate choice to create new machine learning applications due to its advanced customization and flexibility. For developers, TensorFlow is an excellent choice of machine learning library due to its ease in terms of deploying models from development to production environments. Combined, TensorFlow's flexibility and ease of deployment make the library a smart choice for many practitioners looking to build performant machine learning models using a variety of different data sources and to replicate the results of that learning in production environments.

This chapter provides a practical introduction to TensorFlow's API. You will learn how to perform mathematical operations pertinent to machine learning that will give you a firm foundation for building performant ML models using TensorFlow. You will first learn basic operations such as how to create variables with the API. Following that, you will learn how to perform linear transformations such as addition before moving on to more advanced tasks, including tensor multiplication.

IMPLEMENTING ARTIFICIAL NEURAL NETWORKS IN TENSORFLOW

The advanced flexibility that TensorFlow offers lends itself well to creating **artificial neural networks (ANNs)**. ANNs are algorithms that are inspired by the connectivity of neurons in the brain and are intended to replicate the process in which humans learn. They consist of layers through which information propagates from the input to the output.

Figure 1.1 shows a visual representation of an ANN. An input layer is on the left-hand side, which, in this example, has two features (**X**$_1$ and **X**$_2$). The input layer is connected to the first hidden layer, which has three units. All the data from the previous layer gets passed to each unit in the first hidden layer. The data is then passed to the second hidden layer, which also has three units. Again, the information from each unit of the prior layer is passed to each unit of the second hidden layer. Finally, all the information from the second hidden layer is passed to the output layer, which has one unit, representing a single number for each set of input features.

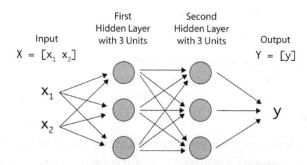

Figure 1.1: A visual representation of an ANN with two hidden layers

ANNs have proven to be successful in learning complex and nonlinear relationships with large, unstructured datasets, such as audio, images, and text data. While the results can be impressive, there is a lot of variability in how ANNs can be configured. For example, the number of layers, the size of each layer, and which nonlinear function should be used are some of the factors that determine the configuration of ANNs. Not only are the classes and functions that TensorFlow provides well-suited to building and training ANNs, but the library also supplies a suite of tools to help visualize and debug ANNs during the training process.

Compared with traditional machine learning algorithms, such as linear and logistic regression, ANNs can outperform them when provided with large amounts of data. ANNs are advantageous since they can be fed unstructured data and feature engineering is not necessarily required. Data pre-processing can be a time-intensive process. Therefore, many practitioners prefer ANNs if there is a large amount of data.

Many companies from all sectors utilize TensorFlow to build ANNs for their applications. Since TensorFlow is backed by Google, the company utilizes the library for much of its research, development, and production of machine learning applications. However, there are many other companies that also use the library. Companies such as Airbnb, Coca-Cola, Uber, and GE Healthcare all utilize the library for a variety of tasks. The use of ANNs is particularly appealing since they can achieve remarkable accuracy if provided with sufficient data and trained appropriately. For example, GE Healthcare uses TensorFlow to build ANNs to identify specific anatomy regardless of orientation from magnetic resonance images to improve speed and accuracy. By using ANNs, they can achieve over 99% accuracy in identifying anatomy in seconds, regardless of head rotation, which would otherwise take a trained professional much more time.

While the number of companies utilizing ANNs is vast, ANNs may not be the most appropriate choice for solving all business problems. In such an environment, you must answer the following questions to determine whether ANNs are the most appropriate choice:

- **Does the problem have a numerical solution?** Machine learning algorithms, ANNs included, generate predicted numerical results based on input data. For example, machine learning algorithms may predict a given number, such as the temperature of a city given the location and previous weather conditions, or the stock price given previous stock prices, or label images into a given number of categories. In each of these examples, a numerical output is generated based on the data provided and, given enough labeled data, models can perform well. However, when the desired result is more abstract, or creativity is needed, such as creating a new song, then machine learning algorithms may not be the most appropriate choice, since a well-defined numerical solution may not be available.

- **Is there enough appropriately labeled data to train a model?** For a supervised learning task, you must have at least some labeled data to train a model. For example, if you want to build a model to predict financial stock data for a given company, you will first need historical training data. If the company in question has not been public for very long, there may not be adequate training data. ANNs can often require a lot of data. When working with images, ANNs often need millions of training examples to develop accurate, robust models. This may be a determining factor for consideration when deciding which algorithm is appropriate for a given task.

Now that you are aware of what TensorFlow is, consider the following advantages and disadvantages of TensorFlow.

ADVANTAGES OF TENSORFLOW

The following are a few of the main advantages of using TensorFlow that many practitioners consider when deciding whether to pursue the library for machine learning purposes:

- **Library Management**: There is a large community of practitioners that maintain the TensorFlow library to keep it up to date with frequent new releases to help fix bugs, add new functions and classes to reflect current advances in the field, and add support for multiple programming languages.

- **Pipelining**: TensorFlow supports end-to-end model production, from model development in highly parallelizable environments that support GPU processing to a suite of model deployment tools. Also, there are lightweight libraries in TensorFlow that are used for deploying trained TensorFlow models on mobile and embedded devices, such as **Internet of Things (IoT)** devices.

- **Community Support**: The community of practitioners that use and support the library is vast and they support each other, because of which those practitioners who are new to the library achieve the results they are looking for easily.

- **Open Source**: TensorFlow is an open source library, and its code base is available for anyone to use and modify for their own applications.

- **Works with Multiple Languages**: While the library is natively designed for Python, models can now be trained and deployed in JavaScript.

DISADVANTAGES OF TENSORFLOW

The following are a few of the disadvantages of using TensorFlow:

- **Computational Speed**: Since the primary programming language of TensorFlow is Python, the library is not as computationally fast as it could be if it were native to other languages, such as C++.

- **Steep Learning Curve**: Compared to other machine learning libraries, such as Keras, the learning curve is steeper, and this can make it challenging for new practitioners to create their own models outside of given example code.

Now that you have understood what TensorFlow is, the next section will demonstrate how to use the TensorFlow library using Python.

THE TENSORFLOW LIBRARY IN PYTHON

TensorFlow can be used in Python by importing certain libraries. You can import libraries in Python using the **import** statement:

```
import tensorflow as tf
```

In the preceding command, you have imported the TensorFlow library and used the shorthand **tf**.

In the next exercise, you will learn how to import the TensorFlow library and check its version so that you can utilize the classes and functions supplied by the library, which is an important and necessary first step when utilizing the library.

EXERCISE 1.01: VERIFYING YOUR VERSION OF TENSORFLOW

In this exercise, you will load TensorFlow and check which version is installed on your system.

Perform the following steps:

1. Open a Jupyter notebook to implement this exercise by typing **jupyter notebook** in the terminal.

2. Import the TensorFlow library by entering the following code in the Jupyter cell:

```
import tensorflow as tf
```

3. Verify the version of TensorFlow using the following command:

```
tf.__version__
```

This will result in the following output:

```
'2.6.0'
```

As you can see from the preceding output, the version of TensorFlow is **2.6.0**.

> **NOTE**
>
> The version may vary on your system if you have not set up the environment using the steps provided in *Preface*.

In this exercise, you successfully imported TensorFlow. You have also checked which version of TensorFlow is installed on your system.

This task can be done for any imported library in Python and is useful for debugging and referencing documentation.

The potential applications of using TensorFlow are numerous, and it has already achieved impressive results, as evidenced by the results from companies such as Airbnb, which uses TensorFlow to classify images on their platform, to GE Healthcare, which uses TensorFlow to identify anatomy on MRIs of the brain. To learn how to create powerful models for your own applications, you first must learn the basic mathematical principles and operations that make up the machine learning models that can be achieved in TensorFlow. The mathematical operations can be intimidating to new users, but a comprehensive understanding of how they operate is key to making performant models.

INTRODUCTION TO TENSORS

Tensors can be thought of as the core components of ANNs—the input data, output predictions, and weights that are learned throughout the training process are all tensors. Information propagates through a series of linear and nonlinear transformations to turn the input data into predictions. This section demonstrates how to apply linear transformations such as additions, transpositions, and multiplications to tensors. Other linear transformations, such as rotations, reflections, and shears, also exist. However, their applications as they pertain to ANNs are less common.

SCALARS, VECTORS, MATRICES, AND TENSORS

Tensors can be represented as multi-dimensional arrays. The number of dimensions a tensor spans is known as the tensor's rank. Tensors with ranks **0**, **1**, and **2** are used often and have their own names, which are **scalars**, **vectors**, and **matrices**, respectively, although the term *tensors* can be used to describe each of them. *Figure 1.2* shows some examples of tensors of various ranks. From left to right are a scalar, vector, matrix, and a 3-dimensional tensor, where each element represents a different number, and the subscript represents the location of the element in the tensor:

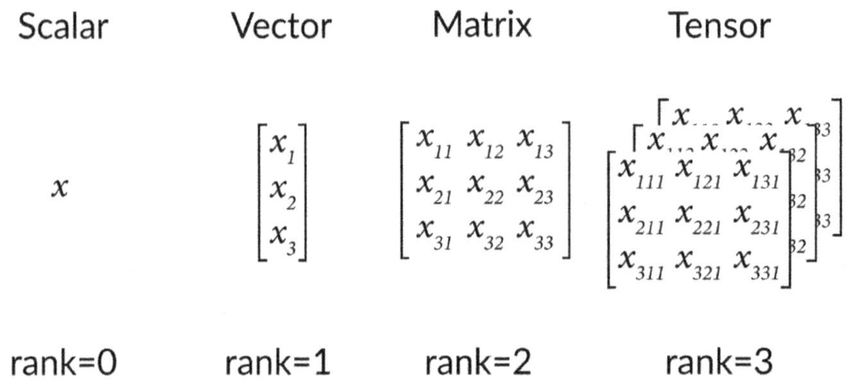

Scalar　　　　Vector　　　Matrix　　　　Tensor

$$x \qquad \begin{bmatrix} x_1 \\ x_2 \\ x_3 \end{bmatrix} \qquad \begin{bmatrix} x_{11} & x_{12} & x_{13} \\ x_{21} & x_{22} & x_{23} \\ x_{31} & x_{32} & x_{33} \end{bmatrix} \qquad \begin{bmatrix} x_{111} & x_{121} & x_{131} \\ x_{211} & x_{221} & x_{231} \\ x_{311} & x_{321} & x_{331} \end{bmatrix}$$

rank=0　　　　rank=1　　　rank=2　　　　rank=3

Figure 1.2: A visual representation of a scalar, vector, matrix, and tensor

The formal definitions of a scalar, vector, matrix, and tensor are as follows:

- **Scalar**: A scalar consists of a single number, making it a zero-dimensional array. It is an example of zero-order tensors. Scalars do not have any axes. For instance, the width of an object is a scalar.

- **Vector**: Vectors are one-dimensional arrays and are an example of first-order tensors. They can be considered lists of values. Vectors have one axis. The size of a given object denoted by the width, height, and depth is an example of a vector field.

- **Matrix**: Matrices are two-dimensional arrays with two axes. They are an example of second-order tensors. Matrices might be used to store the size of several objects. Each dimension of the matrix comprises the size of each object (width, height, depth) and the other matrix dimension is used to differentiate between objects.

- **Tensor**: Tensors are the general entities that encapsulate scalars, vectors, and matrices, although the name is generally reserved for tensors of rank **3** or more. A tensor can be used to store the size of many objects and their locations over time. The first dimension of the matrix comprises the size of each object (width, height, depth), the second dimension is used to differentiate between the objects, and the third dimension describes the location of these objects over time.

Tensors can be created using the **Variable** class present in the TensorFlow library and passing in a value representing the tensor. A float or integer can be passed for scalars, a list of floats or integers can be passed for vectors, a nested list of floats or integers for matrices, and so on. The following command demonstrates the use of the **Variable** class where a list of the intended values for the tensor as well as any other attributes that are required to be explicitly defined are passed:

```
tensor1 = tf.Variable([1,2,3], dtype=tf.int32, \
                    name='my_tensor', trainable=True)
```

The resultant **Variable** object has several attributes that may be commonly called, and these are as follows:

- **dtype**: The datatype of the **Variable** object (for the tensor defined above, the datatype is **tf.int32**). The default value for this attribute is determined from the values passed.

- **shape**: The number of dimensions and length of each dimension of the **Variable** object (for the tensor defined above, the shape is **[3]**). The default value for this attribute is also determined from the values passed.

- **name**: The name of the **Variable** object (for the tensor defined above, the name of the tensor is defined as **'my_tensor'**). The default for this attribute is **Variable**.

- **trainable**: This attribute indicates whether the **Variable** object can be updated during model training (for the tensor defined above, the **trainable** parameter is set to **true**). The default for this attribute is **true**.

> **NOTE**
>
> You can read more about the attributes of the **Variable** object here:
> https://www.tensorflow.org/api_docs/python/tf/Variable.

The **shape** attribute of the **Variable** object can be called as follows:

```
tensor1.shape
```

The **shape** attribute gives the shape of the tensor, that is, is it a scalar, vector, matrix, and so on. The output of the preceding command will be **[3]** since the tensor has a single dimension with three values along that dimension.

The rank of a tensor can be determined in TensorFlow using the **rank** function. It can be used by passing the tensor as the single argument to the function and the result will be an integer value:

```
tf.rank(tensor1)
```

The output of the following command will be a zero-dimensional integer tensor representing the rank of the input. In this case, the rank of **tensor1** will be **1** as the tensor has only one dimension.

In the following exercise, you will learn how to create tensors of various ranks using TensorFlow's **Variable** class.

EXERCISE 1.02: CREATING SCALARS, VECTORS, MATRICES, AND TENSORS IN TENSORFLOW

The votes cast for different candidates of three different political parties in districts A and B are as follows:

		Political Party X	Political Party Y	Political Party Z
District A	Candidate 1	4,113	3,870	5,102
	Candidate 2	7,511	6,725	7,038
	Candidate 3	6,529	6,962	6,591
District B	Candidate 1	3,611	951	870
	Candidate 2	5,901	1,208	645
	Candidate 3	6,235	1,098	948

Figure 1.3: Votes cast for different candidates of three different political
parties in districts A and B

You are required to do the following:

- Create a scalar to store the votes cast for **Candidate 1** of political party **X** in district **A**, that is, **4113**, and check its shape and rank.

- Create a vector to represent the proportion of votes cast for three different candidates of political party **X** in district **A** and check its shape and rank.

- Create a matrix to represent the votes cast for three different candidates of political parties **X** and **Y** and check its shape and rank.

- Create a tensor to represent the votes cast for three different candidates in two different districts, for three political parties, and check its shape and rank.

Perform the following steps to complete this exercise:

1. Import the TensorFlow library:

    ```
    import tensorflow as tf
    ```

2. Create an integer variable using TensorFlow's **Variable** class and pass **4113** to represent the number of votes cast for a particular candidate. Also, pass **tf.int16** as a second argument to ensure that the input number is an integer datatype. Print the result:

    > **NOTE**
    >
    > The datatype does not have to be explicitly defined. If one is not defined, the datatype will be determined by TensorFlow's **convert_to_tensor** function.

    ```
    int_variable = tf.Variable(4113, tf.int16)
    int_variable
    ```

    This will result in the following output:

    ```
    <tf.Variable 'Variable:0' shape=() dtype=int32, numpy=4113>
    ```

    Here, you can see the attributes of the variable created, including the name, **Variable:0**, the shape, datatype, and the NumPy representation of the tensor.

3. Use TensorFlow's **rank** function to print the rank of the variable created:

    ```
    tf.rank(int_variable)
    ```

    This will result in the following output:

    ```
    <tf.Tensor: shape=(), dtype=int32, numpy=0>
    ```

    You can see that the rank of the integer variable that was created is **0** from the NumPy representation of the tensor.

4. Access the integer variable of the rank by calling the **numpy** attribute:

```
tf.rank(int_variable).numpy()
```

This will result in the following output:

```
0
```

The rank of the scalar is **0**.

> **NOTE**
>
> All attributes of the result of the **rank** function can be called, including the **shape** and **dtype** attributes.

5. Call the **shape** attribute of the integer to find the shape of the tensor:

```
int_variable.shape
```

This will result in the following output:

```
TensorShape([])
```

The preceding output signifies that the shape of the tensor has no size, which is representative of a scalar.

6. Print the **shape** of the scalar variable as a Python list:

```
int_variable.shape.as_list()
```

This will result in the following output:

```
[]
```

7. Create a **vector** variable using TensorFlow's **Variable** class. Pass a list for the vector to represent the proportion of votes cast for three different candidates, and pass in a second argument for the datatype as **tf.float32** to ensure that it is a **float** datatype. Print the result:

```
vector_variable = tf.Variable([0.23, 0.42, 0.35], \
                              tf.float32)
vector_variable
```

This will result in the following output:

```
<tf.Variable 'Variable:0' shape(3,) dtype=float32,
numpy=array([0.23, 0.42, 0.35], dtype=float32)>
```

You can see that the shape and NumPy attributes are different from the scalar variable created earlier. The shape is now **(3,)**, indicating that the tensor is one-dimensional with three elements along that dimension.

8. Print the rank of the **vector** variable using TensorFlow's **rank** function as a NumPy variable:

```
tf.rank(vector_variable).numpy()
```

This will result in the following output:

```
1
```

Here, you can see that the rank of the vector variable is **1**, confirming that this variable is one-dimensional.

9. Print the shape of the **vector** variable as a Python list:

```
vector_variable.shape.as_list()
```

This will result in the following output:

```
[3]
```

10. Create a matrix variable using TensorFlow's **Variable** class. Pass a list of lists of integers for the matrix to represent the votes cast for three different candidates in two different districts. This matrix will have three columns representing the candidates, and two rows representing the districts. Pass in a second argument for the datatype as **tf.int32** to ensure that it is an integer datatype. Print the result:

```
matrix_variable = tf.Variable([[4113, 7511, 6259], \
                               [3870, 6725, 6962]], \
                              tf.int32)
matrix_variable
```

This will result in the following output:

```
<tf.Variable 'Variable:0' shape=(2, 3) dtype=int32, numpy=
array([[4113, 7511, 6259],
       [3870, 6725, 6962]])>
```

Figure 1.4: The output of the TensorFlow variable

11. Print the rank of the matrix variable as a NumPy variable:

```
tf.rank(matrix_variable).numpy()
```

This will result in the following output:

```
2
```

Here, you can see that the rank of the matrix variable is **2**, confirming that this variable is two-dimensional.

12. Print the shape of the matrix variable as a Python list:

```
matrix_variable.shape.as_list()
```

This will result in the following output:

```
[2, 3]
```

13. Create a tensor variable using TensorFlow's **Variable** class. Pass in a triple nested list of integers for the tensor to represent the votes cast for three different candidates in two different districts, for three political parties. Print the result:

```
tensor_variable = tf.Variable([[[4113, 7511, 6259], \
                                [3870, 6725, 6962]], \
                               [[5102, 7038, 6591], \
                                [3661, 5901, 6235]], \
                               [[951, 1208, 1098], \
                                [870, 645, 948]]])
tensor_variable
```

This will result in the following output:

```
<tf.Variable 'Variable:0' shape=(3, 2, 3) dtype=int32, numpy=
array([[[4113, 7511, 6259],
        [3870, 6725, 6962]],

       [[5102, 7038, 6591],
        [3661, 5901, 6235]],

       [[ 951, 1208, 1098],
        [ 870,  645,  948]]])>
```

Figure 1.5: The output of the TensorFlow variable

14. Print the rank of the tensor variable as a NumPy variable:

```
tf.rank(tensor_variable).numpy()
```

This will result in the following output:

```
3
```

Here, you can see that the rank of the tensor variable is **3**, confirming that this variable is three-dimensional.

15. Print the shape of the tensor variable as a Python list:

```
tensor_variable.shape.as_list()
```

This will result in the following output:

```
[3, 2, 3]
```

The result shows that the shape of the resulting tensor is a list object.

In this exercise, you have successfully created tensors of various ranks from political voting data using TensorFlow's **Variable** class. First, you created scalars, which are tensors that have a rank of **0**. Next, you created vectors, which are tensors with a rank of **1**. Matrices were then created, which are tensors of rank **2**. Finally, tensors were created that have rank **3** or more. You confirmed the rank of the tensors you created by using TensorFlow's **rank** function and verified their shape by calling the tensor's **shape** attribute.

In the next section, you will combine tensors to create new tensors using tensor addition.

TENSOR ADDITION

Tensors can be added together to create new tensors. You will use the example of matrices in this chapter, but the concept can be extended to tensors with any rank. Matrices may be added to scalars, vectors, and other matrices under certain conditions in a process known as broadcasting. Broadcasting refers to the process of array arithmetic on tensors of different shapes.

Two matrices may be added (or subtracted) together if they have the same shape. For such matrix-matrix addition, the resultant matrix is determined by the element-wise addition of the input matrices. The resultant matrix will therefore have the same shape as the two input matrices. You can define the matrix $Z = [Z_{ij}]$ as the matrix sum $Z = X + Y$, where $z_{ij} = x_{ij} + y_{ij}$ and each element in Z is the sum of the same element in X and Y.

Matrix addition is commutative, which means that the order of **X** and **Y** does not matter, that is, **X + Y = Y + X**. Matrix addition is also associative, which means that the same result is achieved even when the order of additions is different or even if the operation is applied more than once, that is, **X + (Y + Z) = (X + Y) + Z**.

The same matrix addition principles apply to scalars, vectors, and tensors. An example is shown in the following figure:

$$\begin{bmatrix} x_{11} & x_{12} & x_{13} \\ x_{21} & x_{22} & x_{23} \\ x_{31} & x_{32} & x_{33} \end{bmatrix} + \begin{bmatrix} y_{11} & y_{12} & y_{13} \\ y_{21} & y_{22} & y_{23} \\ y_{31} & y_{32} & y_{33} \end{bmatrix} = \begin{bmatrix} x_{11}+y_{11} & x_{12}+y_{12} & x_{13}+y_{13} \\ x_{21}+y_{21} & x_{22}+y_{22} & x_{23}+y_{23} \\ x_{31}+y_{31} & x_{32}+y_{32} & x_{33}+y_{33} \end{bmatrix}$$

Figure 1.6: A visual example of matrix-matrix addition

Scalars can also be added to matrices. Here, each element of the matrix is added to the scalar individually, as shown in *Figure 1.7*:

$$\begin{bmatrix} x_{11} & x_{12} & x_{13} \\ x_{21} & x_{22} & x_{23} \\ x_{31} & x_{32} & x_{33} \end{bmatrix} + y = \begin{bmatrix} x_{11}+y & x_{12}+y & x_{13}+y \\ x_{21}+y & x_{22}+y & x_{23}+y \\ x_{31}+y & x_{32}+y & x_{33}+y \end{bmatrix}$$

Figure 1.7: A visual example of matrix-scalar addition

Addition is an important transformation that can be applied to tensors since the transformation occurs so frequently. For example, a common transformation in developing ANNs is to add a bias to a layer. This is when a constant tensor array of the same size of the ANN layer is added to that layer. Therefore, it is important to know how and when this seemingly simple transformation can be applied to tensors.

Tensor addition can be performed in TensorFlow by using the **add** function and passing in the tensors as arguments, or simply by using the **+** operator as follows:

```
tensor1 = tf.Variable([1,2,3])
tensor2 = tf.Variable([4,5,6])
tensor_add1 = tf.add(tensor1, tensor2)
tensor_add2 = tensor1 + tensor2
```

In the following exercise, you will perform tensor addition on scalars, vectors, and matrices in TensorFlow.

EXERCISE 1.03: PERFORMING TENSOR ADDITION IN TENSORFLOW

The votes cast for different candidates of three different political parties in districts A and B are as follows:

		Political Party X	Political Party Y	Political Party Z
District A	Candidate 1	4,113	3,870	5,102
	Candidate 2	7,511	6,725	7,038
	Candidate 3	6,529	6,962	6,591
District B	Candidate 1	3,611	951	870
	Candidate 2	5,901	1,208	645
	Candidate 3	6,235	1,098	948

Figure 1.8: Votes cast for different candidates of three different political
parties in districts A and B

Your requisite tasks are as follows:

- Store the total number of votes cast for political party X in district A.

- Store the total number of votes cast for each political party in district A.

- Store the total number of votes cast for each political party in both districts.

Perform the following steps to complete the exercise:

1. Import the TensorFlow library:

```
import tensorflow as tf
```

2. Create three scalar variables using TensorFlow's **Variable** class to represent the votes cast for three candidates of political party X in district A:

```
int1 = tf.Variable(4113, tf.int32)
int2 = tf.Variable(7511, tf.int32)
int3 = tf.Variable(6529, tf.int32)
```

3. Create a new variable to store the total number of votes cast for political party X in district A:

```
int_sum = int1+int2+int3
```

4. Print the result of the sum of the two variables as a NumPy variable:

```
int_sum.numpy()
```

This will result in the following output:

```
18153
```

5. Create three vectors to represent the number of votes cast for different political parties in district A, each with one row and three columns:

```
vec1 = tf.Variable([4113, 3870, 5102], tf.int32)
vec2 = tf.Variable([7511, 6725, 7038], tf.int32)
vec3 = tf.Variable([6529, 6962, 6591], tf.int32)
```

6. Create a new variable to store the total number of votes for each political party in district A:

```
vec_sum = vec1 + vec2 + vec3
```

7. Print the result of the sum of the two variables as a NumPy array:

```
vec_sum.numpy()
```

This will result in the following output:

```
array([18153, 17557, 18731])
```

8. Verify that the vector addition is as expected by performing the addition of each element of the vector:

```
print((vec1[0] + vec2[0] + vec3[0]).numpy())
print((vec1[1] + vec2[1] + vec3[1]).numpy())
print((vec1[2] + vec2[2] + vec3[2]).numpy())
```

This will result in the following output:

```
18153
17557
18731
```

You can see that the **+** operation on three vectors is simply element-wise addition of the vectors.

9. Create three matrices to store the votes cast for candidates of each political party in each district:

```
matrix1 = tf.Variable([[4113, 3870, 5102], \
                       [3611, 951, 870]], tf.int32)
matrix2 = tf.Variable([[7511, 6725, 7038], \
                       [5901, 1208, 645]], tf.int32)
matrix3 = tf.Variable([[6529, 6962, 6591], \
                       [6235, 1098, 948]], tf.int32)
```

10. Verify that the three tensors have the same shape:

```
matrix1.shape == matrix2.shape == matrix3.shape
```

This will result in the following output:

```
True
```

11. Create a new variable to store the total number of votes cast for each political party in both districts:

```
matrix_sum = matrix1 + matrix2 + matrix3
```

12. Print the result of the sum of the two variables as a NumPy array:

```
matrix_sum.numpy()
```

This will result in the following output representing the total votes for each candidate and each party across districts:

```
array([[18153, 17557, 18731],
       [15747,  3257,  2463]])
```

Figure 1.9: The output of the matrix summation as a NumPy variable

13. Verify that the tensor addition is as expected by performing the addition of each element of the vector:

```
print((matrix1[0][0] + matrix2[0][0] + matrix3[0][0]).numpy())
print((matrix1[0][1] + matrix2[0][1] + matrix3[0][1]).numpy())
print((matrix1[0][2] + matrix2[0][2] + matrix3[0][2]).numpy())
print((matrix1[1][0] + matrix2[1][0] + matrix3[1][0]).numpy())
print((matrix1[1][1] + matrix2[1][1] + matrix3[1][1]).numpy())
print((matrix1[1][2] + matrix2[1][2] + matrix3[1][2]).numpy())
```

This will result in the following output:

```
18153
17557
18731
15747
3257
2463
```

You can see that the **+** operation is equivalent to the element-wise addition of the three matrices created.

In this exercise, you successfully performed tensor addition on data representing votes cast for political candidates. The transformation can be applied by using the **+** operation. You also verified that addition is performed element by element, and that one way to ensure that the transformation is valid is for the tensors to have the same rank and shape.

In the following activity, you will further practice tensor addition in TensorFlow.

ACTIVITY 1.01: PERFORMING TENSOR ADDITION IN TENSORFLOW

You work in a company that has three locations, each with two salespersons and each location sells three products. You are required to sum the tensors to represent the total revenue for each product across locations.

		Product A	Product B	Product C
Location X	Salesperson 1	2706	2799	5102
	Salesperson 2	2386	4089	5932
Location Y	Salesperson 3	5901	1208	645
	Salesperson 4	6235	1098	948
Location Z	Salesperson 5	3908	2339	5520
	Salesperson 6	4544	1978	4729

Figure 1.10: Number of different products sold by each salesperson at different locations

The steps you will take are as follows:

1. Import the TensorFlow library.

2. Create two scalars to represent the total revenue for **Product A** by all salespeople at **Location X** using TensorFlow's **Variable** class. The first variable will have a value of **2706** and the second will have a value of **2386**.

3. Create a new variable as the sum of the scalars and print the result.

   You should get the following output:

   ```
   5092
   ```

4. Create a vector with values **[2706, 2799, 5102]** and a scalar with the value **95** using TensorFlow's **Variable** class.

5. Create a new variable as the sum of the scalar with the vector to represent the sales goal for **Salesperson 1** at **Location X** and print the result.

   You should get the following output:

   array([2801, 2894, 5197])

   Figure 1.11: The output of the integer-vector summation as a NumPy variable

6. Create three tensors with a rank of 2 representing the revenue for each salesperson, product, and location using TensorFlow's **Variable** class. The first tensor will have the value **[[2706, 2799, 5102], [2386, 4089, 5932]]**, the second will have the value **[[5901, 1208, 645], [6235, 1098, 948]]**, and the third will have **[[3908, 2339, 5520], [4544, 1978, 4729]]**.

7. Create a new variable as the sum of the matrices and print the result:

   array([[12515, 6346, 11267],
          [13165, 7165, 11609]])

   Figure 1.12: The output of the matrix summation as a NumPy variable

> **NOTE**
>
> The solution to this activity can be found on page 454.

In the following section, you will learn how to change a tensor's shape and rank.

RESHAPING

Some operations, such as addition, can only be applied to tensors if they meet certain conditions. Reshaping is one method for modifying the shape of tensors so that such operations can be performed. Reshaping takes the elements of a tensor and rearranges them into a tensor of a different size. A tensor of any size can be reshaped so long as the number of total elements remains the same.

For example, a **(4x3)** matrix can be reshaped into a **(6x2)** matrix since they both have a total of **12** elements. The rank, or number, of dimensions, can also be changed in the reshaping process. For instance, a **(4x3)** matrix that has a rank equal to **2** can be reshaped into a **(3x2x2)** tensor that has a rank equal to 3. The **(4x3)** matrix can also be reshaped into a **(12x1)** vector in which the rank has changed from **2** to **1**.

Figure 1.13 illustrates tensor reshaping. On the left is a tensor with shape **(3x2)**, which can be reshaped to a tensor of shape equal to either **(2x3)**, **(6)**, or **(6x1)**. Here, the number of elements, that is, six, has remained constant, though the shape and rank of the tensor have changed:

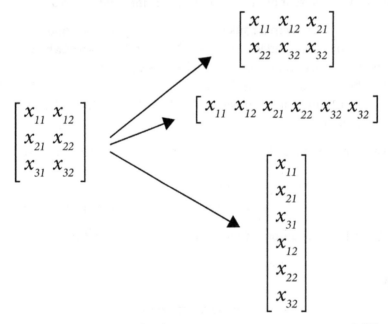

Figure 1.13: Visual representation of reshaping a (3x2) tensor to tensors of different shapes

Tensor reshaping can be performed in TensorFlow by using the **reshape** function and passing in the tensor and the desired shape of the new tensor as the arguments:

```
tensor1 = tf.Variable([1,2,3,4,5,6])
tensor_reshape = tf.reshape(tensor1, shape=[3,2])
```

Here, a new tensor is created that has the same elements as the original; however, the shape is **[3,2]** instead of **[6]**.

The next section introduces tensor transposition, which is another method for modifying the shape of a tensor.

TENSOR TRANSPOSITION

When a tensor is transposed, the elements in the tensor are rearranged in a specific order. The transpose operation is usually denoted as a **T** superscript on the tensor. The new position of each element in the tensor can be determined by $(\mathbf{x}_{12...k})^T = \mathbf{x}_{k...21}$. For a matrix or tensor of rank equal to **2**, the rows become the columns and vice versa. An example of matrix transposition is shown in *Figure 1.14*. Tensors of any rank can be transposed, and often the shape changes as a result:

$$\begin{bmatrix} x_{11} & x_{12} \\ x_{21} & x_{22} \\ x_{31} & x_{32} \end{bmatrix}^T = \begin{bmatrix} x_{11} & x_{21} & x_{31} \\ x_{12} & x_{22} & x_{32} \end{bmatrix}$$

Figure 1.14: A visual representation of tensor transposition on a (3x2) matrix

The following diagram shows the matrix transposition properties of matrices **A** and **B**:

$$(X^T)^T = X$$
$$(X+Y)^T = X^T + Y^T$$
$$(XY)^T = Y^T X^T$$
$$(X_1 X_2 ... X_k)^T = X_k ... X_2 X_1$$
$$(X^{-1})^T = (X^T)^{-1}$$

Figure 1.15: Tensor transposition properties where X and Y are tensors

A tensor is said to be symmetrical if the transpose of a tensor is equivalent to the original tensor.

Tensor transposition can be performed in TensorFlow by using its **transpose** function and passing in the tensor as the only argument:

```
tensor1 = tf.Variable([1,2,3,4,5,6])
tensor_transpose = tf.transpose(tensor1)
```

When transposing a tensor, there is only one possible result; however, reshaping a tensor has multiple possible results depending on the desired shape of the output.

In the following exercise, reshaping and transposition are demonstrated on tensors using TensorFlow.

EXERCISE 1.04: PERFORMING TENSOR RESHAPING AND TRANSPOSITION IN TENSORFLOW

In this exercise, you will learn how to perform tensor reshaping and transposition using the TensorFlow library.

Perform the following steps:

1. Import the TensorFlow library and create a matrix with two rows and four columns using TensorFlow's **Variable** class:

```
import tensorflow as tf
matrix1 = tf.Variable([[1,2,3,4], [5,6,7,8]])
```

2. Verify the shape of the matrix by calling the **shape** attribute of the matrix as a Python list:

```
matrix1.shape.as_list()
```

This will result in the following output:

```
[2, 4]
```

You see that the shape of the matrix is **[2,4]**.

3. Use TensorFlow's **reshape** function to change the matrix to a matrix with four rows and two columns by passing in the matrix and the desired new shape:

```
reshape1 = tf.reshape(matrix1, shape=[4, 2])
reshape1
```

You should get the following output:

```
<tf.Tensor: shape=(4, 2), dtype=int32, numpy=
array([[1, 2],
       [3, 4],
       [5, 6],
       [7, 8]])>
```

Figure 1.16: The reshaped matrix

4. Verify the shape of the reshaped matrix by calling the **shape** attribute as a Python list:

```
reshape1.shape.as_list()
```

This will result in the following output:

```
[4, 2]
```

Here, you can see that the shape of the matrix has changed to your desired shape, **[4,2]**.

5. Use TensorFlow's **reshape** function to convert the matrix into a matrix with one row and eight columns. Pass the matrix and the desired new shape as parameters to the **reshape** function:

```
reshape2 = tf.reshape(matrix1, shape=[1, 8])
reshape2
```

You should get the following output:

```
<tf.Tensor: shape=(1, 8), dtype=int32, numpy=array([[1, 2, 3, 4, 5,
6, 7, 8]])>
```

6. Verify the shape of the reshaped matrix by calling the **shape** attribute as a Python list:

```
reshape2.shape.as_list()
```

This will result in the following output:

```
[1, 8]
```

The preceding output confirms the shape of the reshaped matrix as **[1, 8]**.

7. Use TensorFlow's **reshape** function to convert the matrix into a matrix with eight rows and one column, passing the matrix and the desired new shape as parameters to the **reshape** function:

```
reshape3 = tf.reshape(matrix1, shape=[8, 1])
reshape3
```

You should get the following output:

```
<tf.Tensor: shape=(8, 1), dtype=int32, numpy=
array([[1],
       [2],
       [3],
       [4],
       [5],
       [6],
       [7],
       [8]])>
```

Figure 1.17: Reshaped matrix of shape (8, 1)

8. Verify the shape of the reshaped matrix by calling the **shape** attribute as a Python list:

```
reshape3.shape.as_list()
```

This will result in the following output:

```
[8, 1]
```

The preceding output confirms the shape of the reshaped matrix as **[8, 1]**.

9. Use TensorFlow's **reshape** function to convert the matrix to a tensor of size **2x2x2**. Pass the matrix and the desired new shape as parameters to the reshape function:

```
reshape4 = tf.reshape(matrix1, shape=[2, 2, 2])
reshape4
```

You should get the following output:

```
<tf.Tensor: shape=(2, 2, 2), dtype=int32, numpy=
array([[[1, 2],
        [3, 4]],

       [[5, 6],
        [7, 8]]])>
```

Figure 1.18: Reshaped matrix of shape (2, 2, 2)

10. Verify the shape of the reshaped matrix by calling the **shape** attribute as a Python list:

```
reshape4.shape.as_list()
```

This will result in the following output:

```
[2, 2, 2]
```

The preceding output confirms the shape of the reshaped matrix as **[2, 2, 2]**.

11. Verify that the rank has changed using TensorFlow's **rank** function and print the result as a NumPy variable:

```
tf.rank(reshape4).numpy()
```

This will result in the following output:

```
3
```

12. Use TensorFlow's **transpose** function to convert the matrix of size **2X4** to a matrix of size **4x2**:

```
transpose1 = tf.transpose(matrix1)
transpose1
```

You should get the following output:

```
<tf.Tensor: shape=(4, 2), dtype=int32, numpy=
array([[1, 5],
       [2, 6],
       [3, 7],
       [4, 8]])>
```

Figure 1.19: Transposed matrix

13. Verify that the **reshape** function and the **transpose** function create different resulting matrices when applied to the given matrix:

```
transpose1 == reshape1
```

```
<tf.Tensor: shape=(4, 2), dtype=bool, numpy=
array([[ True, False],
       [False, False],
       [False, False],
       [False,  True]])>
```

Figure 1.20: Verification that transposition and reshaping produce different results

14. Use TensorFlow's **transpose** function to transpose the reshaped matrix in *step 9*:

```
transpose2 = tf.transpose(reshape4)
transpose2
```

This will result in the following output:

```
<tf.Tensor: shape=(2, 2, 2), dtype=int32, numpy=
array([[[1, 5],
        [3, 7]],

       [[2, 6],
        [4, 8]]])>
```

Figure 1.21: The output of the transposition of the reshaped tensor

This result shows how the resulting tensor appears after reshaping and transposing a tensor.

In this exercise, you have successfully modified the shape of a tensor either through reshaping or transposition. You studied how the shape and rank of the tensor changes following the reshaping and transposition operation.

In the following activity, you will test your knowledge on how to reshape and transpose tensors using TensorFlow.

ACTIVITY 1.02: PERFORMING TENSOR RESHAPING AND TRANSPOSITION IN TENSORFLOW

In this activity, you are required to simulate the grouping of 24 school children for class projects. The dimensions of each resulting reshaped or transposed tensor will represent the size of each group.

Perform the following steps:

1. Import the TensorFlow library.

2. Create a one-dimensional tensor with 24 monotonically increasing elements using the **Variable** class to represent the IDs of the school children. Verify the shape of the matrix.

    You should get the following output:

    ```
    [24]
    ```

3. Reshape the matrix so that it has 12 rows and 2 columns using TensorFlow's **reshape** function representing 12 pairs of school children. Verify the shape of the new matrix.

    You should get the following output:

    ```
    [12, 2]
    ```

4. Reshape the original matrix so that it has a shape of **3x4x2** using TensorFlow's **reshape** function representing 3 groups of 4 sets of pairs of school children. Verify the shape of the new tensor.

    You should get the following output:

    ```
    [3, 4, 2]
    ```

5. Verify that the rank of this new tensor is **3**.

6. Transpose the tensor created in *step 3* to represent 2 groups of 12 students using TensorFlow's **transpose** function. Verify the shape of the new tensor.

You should get the following output:

```
[2, 12]
```

> **NOTE**
>
> The solution to this activity can be found on page 455.

In this section, you were introduced to some of the basic components of ANNs—tensors. You also learned about some basic manipulation of tensors, such as addition, transposition, and reshaping. You implemented these concepts by using functions in the TensorFlow library.

In the next topic, you will extend your understanding of linear transformations by covering another important transformation related to ANNs—tensor multiplication.

TENSOR MULTIPLICATION

Tensor multiplication is another fundamental operation that is used frequently in the process of building and training ANNs since information propagates through the network from the inputs to the result via a series of additions and multiplications. While the rules for addition are simple and intuitive, the rules for tensors are more complex. Tensor multiplication involves more than simple element-wise multiplication of the elements. Rather, a more complicated procedure is implemented that involves the dot product between the entire rows/columns of each of the tensors to calculate each element of the resulting tensor. This section will explain how multiplication works for two-dimensional tensors or matrices. However, tensors of higher orders can also be multiplied.

Given a matrix, $\mathbf{X} = [\mathbf{x}_{ij}]_{m \times n}$, and another matrix, $\mathbf{Y} = [\mathbf{y}_{ij}]_{n \times p}$, the product of the two matrices is $\mathbf{Z} = \mathbf{XY} = [\mathbf{z}_{ij}]_{m \times p}$, and each element, $\mathbf{z}_{ij}$, is defined element-wise as $z_{ij} = \sum_{k=1}^{n} x_{ik} y_{kj}$. The shape of the resultant matrix is the same as the outer dimensions of the matrix product, or the number of rows of the first matrix and the number of columns of the second matrix. For the multiplication to work, the inner dimensions of the matrix product must match, or the number of columns in the first matrix and the number of columns in the second matrix must correspond.

The concept of inner and outer dimensions of matrix multiplication is shown in the following diagram, where **X** represents the first matrix and **Y** represents the second matrix:

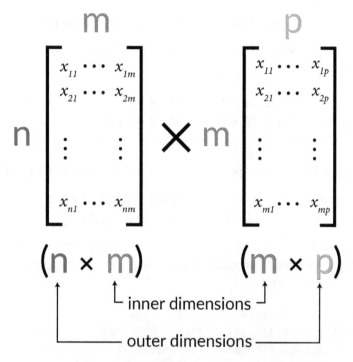

Figure 1.22: A visual representation of inner and outer dimensions in matrix multiplication

Unlike matrix addition, matrix multiplication is not commutative, which means that the order of the matrices in the product matters:

$$XY \neq XY$$

Figure 1.23: Matrix multiplication is non-commutative

For example, say you have the following two matrices:

$$X = \begin{bmatrix} 5 & 4 & -1 \\ -2 & 0 & 6 \end{bmatrix}, \quad Y = \begin{bmatrix} 4 & 7 \\ 3 & 9 \\ -2 & 1 \end{bmatrix}$$

Figure 1.24: Two matrices, X and Y

One way to construct the product is to have matrix **X** first, multiplied by **Y**:

$$XY = \begin{bmatrix} 5 \times 4 + 4 \times 3 + -1 \times -2 & 5 \times 7 + 4 \times 9 + -1 \times 1 \\ -2 \times 4 + 0 \times 3 + 6 \times -2 & -2 \times 7 + 0 \times 9 + 6 \times 1 \end{bmatrix} = \begin{bmatrix} 34 & 70 \\ -20 & 20 \end{bmatrix}$$

Figure 1.25: Visual representation of matrix X multiplied by Y, X•Y

This results in a **2x2** matrix. Another way to construct the product is to have **Y** first, multiplied by **X**:

$$YX = \begin{bmatrix} 4 \times 5 + 7 \times -2 & 4 \times 4 + 7 \times 0 & 4 \times -1 + 7 \times 6 \\ 3 \times 5 + 9 \times -2 & 3 \times 4 + 9 \times 0 & 3 \times -1 + 9 \times 6 \\ -2 \times 5 + 1 \times -2 & -2 \times 4 + 1 \times 0 & -2 \times -1 + 1 \times 6 \end{bmatrix} = \begin{bmatrix} 6 & 16 & 38 \\ -3 & 12 & 51 \\ -12 & -8 & 8 \end{bmatrix}$$

Figure 1.26: Visual representation of matrix Y multiplied by X, Y•X

Here you can see that the matrix formed from the product **YX** is a **3x3** matrix and is very different from the matrix formed from the product **XY**.

Tensor multiplication can be performed in TensorFlow by using the **matmul** function and passing in the tensors to be multiplied in the order in which they are to be multiplied as the arguments:

```
tensor1 = tf.Variable([[1,2,3]])
tensor2 = tf.Variable([[1],[2],[3]])
tensor_mult = tf.matmul(tensor1, tensor2)
```

Tensor multiplication can also be achieved by using the @ operator as follows:

```
tensor_mult = tensor1 @ tensor2
```

Scalar-tensor multiplication is much more straightforward and is simply the product of every element in the tensor multiplied by the scalar so that $\lambda X = [\lambda x_{ij...k}]$, where λ is a scalar and **X** is a tensor.

Scalar multiplication can be achieved in TensorFlow either by using the **matmul** function or by using the * operator:

```
tensor1 = tf.Variable([[1,2,3]])
scalar_mult = 5 * tensor1
```

In the following exercise, you will perform tensor multiplication using the TensorFlow library.

EXERCISE 1.05: PERFORMING TENSOR MULTIPLICATION IN TENSORFLOW

In this exercise, you will perform tensor multiplication in TensorFlow using TensorFlow's `matmul` function and the @ operator. In this exercise, you will use the example of data from a sandwich retailer representing the ingredients of various sandwiches and the costs of different ingredients. You will use matrix multiplication to determine the costs of each sandwich.

Sandwich recipe:

	Ingredient A	Ingredient B	Ingredient C	Ingredient D	Ingredient E
Sandwich 1	1.0	0	3.0	1.0	2.0
Sandwich 2	0	1.0	1.0	1.0	1.0
Sandwich 3	2.0	1.0	0	2.0	0

Figure 1.27: Sandwich recipe

Ingredient details:

	Cost	Weight
Ingredient A	0.49	103
Ingredient B	0.18	38
Ingredient C	0.24	69
Ingredient D	1.02	75
Ingredient E	0.68	78

Figure 1.28: Ingredient details

Sales projections:

	Sandwich 1	Sandwich 2	Sandwich 3
Location 1	120	100	90
Location 2	30	15	20
Location 3	220	240	185
Location 4	145	160	155
Location 5	330	295	290

Figure 1.29: Sales projections

Perform the following steps:

1. Import the TensorFlow library:

```
import tensorflow as tf
```

2. Create a matrix representing the different sandwich recipes, with the rows representing the three different sandwich offerings and the columns representing the combination and number of the five different ingredients using the **Variable** class:

```
matrix1 = tf.Variable([[1.0,0.0,3.0,1.0,2.0], \
                       [0.0,1.0,1.0,1.0,1.0], \
                       [2.0,1.0,0.0,2.0,0.0]], \
                       tf.float32)
matrix1
```

You should get the following output:

```
<tf.Variable 'Variable:0' shape=(3, 5) dtype=float32, numpy=
array([[1., 0., 3., 1., 2.],
       [0., 1., 1., 1., 1.],
       [2., 1., 0., 2., 0.]], dtype=float32)>
```

Figure 1.30: Matrix representing the number of ingredients needed to make sandwiches

3. Verify the shape of the matrix by calling the **shape** attribute of the matrix as a Python list:

```
matrix1.shape.as_list()
```

This will result in the following output:

```
[3, 5]
```

4. Create a second matrix representing the cost and weight of each individual ingredient in which the rows represent the five ingredients, and the columns represent the cost and weight of the ingredients in grams:

```
matrix2 = tf.Variable([[0.49, 103], \
                       [0.18, 38], \
                       [0.24, 69], \
                       [1.02, 75], \
                       [0.68, 78]])
matrix2
```

You should get the following result:

```
<tf.Variable 'Variable:0' shape=(5, 2) dtype=float32, numpy=
array([[  0.49, 103.  ],
       [  0.18,  38.  ],
       [  0.24,  69.  ],
       [  1.02,  75.  ],
       [  0.68,  78.  ]], dtype=float32)>
```

Figure 1.31: A matrix representing the cost and weight of each ingredient

5. Use TensorFlow's **matmul** function to perform the matrix multiplication of **matrix1** and **matrix2**:

```
matmul1 = tf.matmul(matrix1, matrix2)
matmul1
```

This will result in the following output:

```
<tf.Tensor: shape=(3, 2), dtype=float32, numpy=
array([[  3.5900002, 541.      ],
       [  2.1200001, 260.      ],
       [  3.2      , 394.      ]], dtype=float32)>
```

Figure 1.32: The output of the matrix multiplication

6. Create a matrix to represent the sales projections of five different stores for each of the three sandwiches:

```
matrix3 = tf.Variable([[120.0, 100.0, 90.0], \
                       [30.0, 15.0, 20.0], \
                       [220.0, 240.0, 185.0], \
                       [145.0, 160.0, 155.0], \
                       [330.0, 295.0, 290.0]])
```

7. Multiply **matrix3** by the result of the matrix multiplication of **matrix1** and **matrix2** to give the expected cost and weight for each of the five stores:

```
matmul3 = matrix3 @ matmul1
matmul3
```

This will result in the following output:

```
<tf.Tensor: shape=(5, 2), dtype=float32, numpy=
array([[9.3080005e+02, 1.2638000e+05],
       [2.0350000e+02, 2.8010000e+04],
       [1.8906001e+03, 2.5431000e+05],
       [1.3557501e+03, 1.8111500e+05],
       [2.7381001e+03, 3.6949000e+05]], dtype=float32)>
```

Figure 1.33: The output of matrix multiplication

The resulting tensor from the multiplication shows the expected cost of sandwiches and the expected weight of the total ingredients for each of the stores.

In this exercise, you have successfully learned how to perform matrix multiplication in TensorFlow using several operators. You used TensorFlow's **matmul** function, as well as the shorthand @ operator. Each will perform the multiplication; however, the **matmul** function has several different arguments that can be passed into the function that make it more flexible.

> **NOTE**
>
> You can read more about the **matmul** function here:
> https://www.tensorflow.org/api_docs/python/tf/linalg/matmul.

In the next section, you will explore some other mathematical concepts that are related to ANNs. You will explore forward and backpropagation, as well as activation functions.

OPTIMIZATION

In this section, you will learn about some optimization approaches that are fundamental to training machine learning models. Optimization is the process by which the weights of the layers of an ANN are updated such that the error between the predicted values of the ANN and the true values of the training data is minimized.

FORWARD PROPAGATION

Forward propagation is the process by which information propagates through ANNs. Operations such as a series of tensor multiplications and additions occur at each layer of the network until the final output. Forward propagation is explained in *Figure 1.37*, showing a single hidden layer ANN. The input data has two features, while the output layer has a single value for each input record.

The weights and biases for the hidden layer and output are shown as matrices and vectors with the appropriate indexes. For the hidden layer, the number of rows in the weight matrix is equal to the number of features of the input, and the number of columns is equal to the number of units in the hidden layer. Therefore, **W1** has two rows and three columns because the input, **X**, has two features. Likewise, **W2** has three rows and one column, the hidden layer has three units, and the output has the size one. The bias, however, is always a vector with a size equal to the number of nodes in that layer and is added to the product of the input and weight matrix.

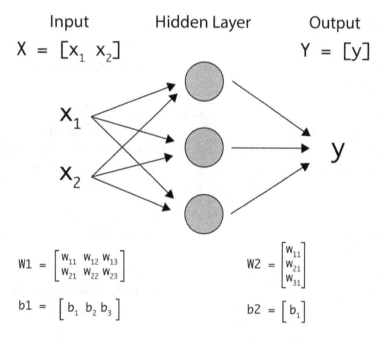

Figure 1.34: A single-layer artificial neural network

The steps to perform forward propagation are as follows:

1. **X** is the input to the network and the input to the hidden layer. First, the input matrix, **X**, is multiplied by the weight matrix for the hidden layer, **W1**, and then the bias, **b1**, is added:

   `z1 = X*W1 + b1`

   Here is an example of what the shape of the resulting tensor will be after the operation. If the input is size **nX2**, where **n** is the number of input examples, **W1** is of size **2X3**, and **b1** is of size **1X3**, the resulting matrix, **z1**, will have a size of **nX3**.

2. **z1** is the output of the hidden layer, which is the **input** for the output layer. Next, the output of the hidden layer is the input matrix multiplied by the weight matrix for the output layer, **W2**, and the bias, **b2**, is added:

   `Y = z1 * W2 + b2`

   To understand the shape of the resulting tensor, consider the following example. If the input to the output layer, **z1**, is of size **nX3**, **W2** is of size **3X1**, and **b1** is of size **1X1**, the resulting matrix, **Y**, will have a size of **nX1**, representing one result for each training example.

The total number of parameters in this model is equal to the sum of the number of elements in **W1**, **W2**, **b1**, and **b2**. Therefore, the number of parameters can be calculated by summing the elements in each of the parameters in weight matrices and biases, which is equal to **6 + 3 + 3 + 1 = 13**. These are the parameters that need to be learned in the process of training the ANN.

Following the forward propagation step, you must evaluate your model and compare it to the real target values. This is achieved using a loss function. Mean squared error, that is, the mean value of the squared difference between true and predicted values, is one of the examples of the loss function of the regression task. Once the loss is calculated, the weights must be updated to reduce the loss, and the amount and direction that the weights should be updated are found using backpropagation.

BACKPROPAGATION

Backpropagation is the process of determining the derivative of the loss with respect to the model parameter. The loss is calculated by applying the `loss` function to the predicted outputs as follows:

```
loss = L(y_predicted)
```

The derivative of the loss with respect to the model parameters will inform you if increasing or decreasing the model parameter will result in increasing or decreasing the loss. The process of backpropagation is achieved by applying the chain rule of calculus from the output layer to the input layer of a neural network, at each layer computing the derivatives of the `loss` function with respect to the model parameters.

The chain rule of calculus is a technique used to compute the derivative of a composite function via intermediate functions. A generalized version of the function can be written as follows:

```
dz/dx = dz/dy * dy/dx
```

Here, **dz/dx** is the composite function and **y** is the intermediate function. In the case of ANNs, the composite function is the loss as a function of the model parameters and the intermediate functions represent the hidden layers. Therefore, the derivative of the loss with respect to the model parameters can be computed by multiplying the derivative of the loss with respect to the predicted output by the derivative of the predicted output with respect to the model parameters.

In the next section, you will learn how the weight parameters are updated given the derivatives of the loss function with respect to each of the weights so that the loss is minimized.

LEARNING OPTIMAL PARAMETERS

In this section, you will see how optimal weights are iteratively chosen. You know that forward propagation transfers information through the network via a series of tensor additions and multiplications, and that backpropagation is the process of understanding the change in loss with respect to each model weight. The next step is to use the results from backpropagation to update the weights so that they reduce the error according to the loss function. This process is known as learning the parameters and is achieved using an optimization algorithm. A common optimization algorithm often utilized is called **gradient descent**.

In learning the optimal parameters, you apply the optimization algorithm until a minimum in the loss function is reached. You usually stop after a given number of steps or when there is a negligible change in the loss function. If you plot the loss as a function of each model parameter, the shape of the loss function resembles a convex shape, having only one minimum, and it is the goal of the optimization function to find this minimum.

The following figure shows the loss function of a particular feature:

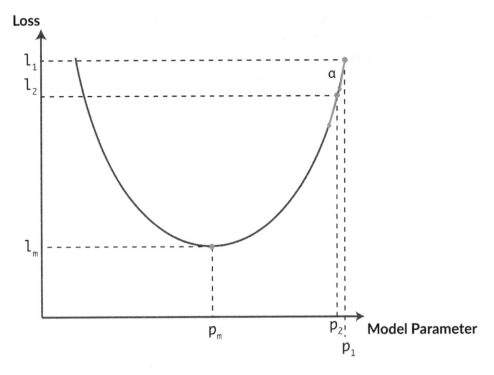

Figure 1.35: A visual representation of the gradient descent algorithm finding the optimal parameter to minimize the loss

This is achieved, first, by randomly setting parameters for each weight, indicated by p_1 in the diagram. The loss is then calculated for that model parameter, l_1. The backpropagation step determines the derivative of the loss with respect to the model parameter and will determine in which direction the model should be updated. The next model parameter, p_2, is equal to the current model parameter minus the learning rate (α) multiplied by the derivative value. The learning rate is a hyperparameter that is set before the model training process. By multiplying by the derivative value, larger steps will be taken when the parameter is far from the minimum where the absolute value for the derivative is larger. The loss, l_2, is then calculated and the process continues until the minimum loss is reached, l_m, with the optimal parameter, p_m.

To summarize, these are the iterative steps that the optimization algorithm performs to find the optimal parameters:

1. Use forward propagation and current parameters to predict the outputs for the entire dataset.

2. Apply the loss function to compute the loss over all the examples from the predicted output.

3. Use backpropagation to compute the derivatives of the loss with respect to the weights and biases at each layer.

4. Update the weights and biases using the derivative values and the learning rate.

OPTIMIZERS IN TENSORFLOW

There are several different optimizers readily available within TensorFlow. Each is based on a different optimization algorithm that aims to reach a global minimum for the loss function. They are all based on the gradient descent algorithm, although they differ slightly in implementation. The available optimizers in TensorFlow include the following:

*   **Stochastic Gradient Descent (SGD)**: The SGD algorithm applies gradient descent to small batches of training data. A momentum parameter is also available when using the optimizer in TensorFlow that applies exponential smoothing to the computed gradient to speed up training.

*   **Adam**: This optimization is an SGD method that is based on the continuous adaptive estimation of first and second-order moments.

- **Root Mean Squared Propagation (RMSProp)**: This is an unpublished, adaptive learning rate optimizer. RMSprop divides the learning rate by an average of the squared gradients when finding the loss minimum after each step, which results in a learning rate that exponentially decays.

- **Adagrad**: This optimizer has parameter-specific learning rates that are updated depending on how frequently the parameter is updated during the training process. As the parameter receives more updates, each subsequent update is smaller in value.

The choice of optimizer will affect training time and model performance. Each optimizer also has hyperparameters, such as the initial learning rate, that must be selected before training, and tuning of these hyperparameters will also affect training time and model performance. While other optimizers available in TensorFlow are not explicitly stated here (and can be found here: https://www.tensorflow.org/api_docs/python/tf/keras/optimizers), those stated above perform well both in terms of training time and model performance and are a safe first choice when selecting an optimizer for your model. The optimizers available in TensorFlow are located in the `tf.optimizers` module; for example, an Adam optimizer with a learning rate equal to `0.001` can be initialized as follows:

```
optimizer = tf.optimizer.adam(learning_rate=0.001)
```

In this topic, you have seen the steps taken in achieving gradient descent to compute the optimal parameters for model training. In gradient descent, every single training example is used to learn the parameters. However, when working with large volume datasets, such as with images and audio, you will often work in batches and make updates after learning from each batch. When using gradient descent on batch data, the algorithm is known as SGD. The SGD optimizer, along with a suite of other performant optimizers, is readily available in TensorFlow, including the Adam, RMSProp, and Adagrad optimizers, and more.

In the next section, you will explore different activation functions, which are generally applied to the output of each layer.

ACTIVATION FUNCTIONS

Activation functions are mathematical functions that are generally applied to the outputs of ANN layers to limit or bound the values of the layer. The reason that values may want to be bounded is that without activation functions, the value and corresponding gradients can either explode or vanish, thereby making the results unusable. This is because the final value is the cumulative product of the values from each subsequent layer. As the number of layers increases, the likelihood of values and gradients exploding to infinity or vanishing to zero increases. This concept is known as the **exploding and vanishing gradient problem**. Deciding whether a node in a layer should be *activated* is another use of activation functions, hence their name. Common activation functions and their visual representation in *Figure 1.36* are as follows:

- **Step** function: The value is non-zero if it is above a certain threshold, otherwise it is zero. This is shown in *Figure 1.36a*.

- **Linear** function: $A(x) = cx$, which is a scalar multiplication of the input value. This is shown in *Figure 1.36b*.

- **Sigmoid** function: $A(x) = \frac{1}{1+e^{-x}}$, like a smoothed-out step function with smooth gradients. This activation function is useful for classification since the values are bound from zero to one. This is shown in *Figure 1.36c*.

- **Tanh** function: $A(x) = \tanh(x) = \frac{2}{1+e^{-2x}} - 1$, which is a scaled version of the sigmoid with steeper gradients around **x=0**. This is shown in *Figure 1.36d*.

- **ReLU (Rectified Linear Unit)** function: $A(x) = x,\ x > 0$, otherwise **0**. This is shown in *Figure 1.36e*.

- **ELU (Exponential Linear Unit)** function: $A(x) = x,\ x > 0$, otherwise $\beta(e^x - 1)$, where β is a constant.

- **SELU (Scaled Exponential Linear Unit)** function: $A(x) = \alpha x$, otherwise $\alpha\beta(e^x - 1)$, where α, β are constants. This is shown in *Figure 1.36f*.

- **Swish** function: $A(x) = \dfrac{x}{1+e^{-x}}$. This is shown in *Figure 1.36g*:

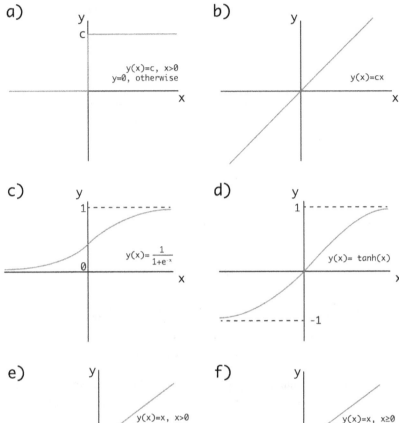

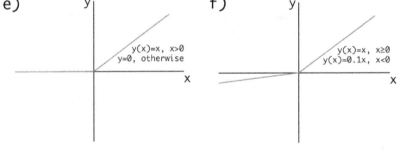

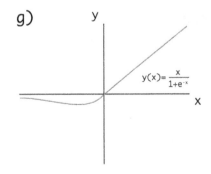

Figure 1.36: A visual representation of the common activation functions

An activation function can be applied to any tensor by utilizing the activation functions in the **tf.keras.activations** module. For example, a sigmoid activation function can be applied to a tensor as follows:

```
y=tf.keras.activations.sigmoid(x)
```

Now, let's test the knowledge that you have gained so far in the following activity.

ACTIVITY 1.03: APPLYING ACTIVATION FUNCTIONS

In this activity, you will recall many of the concepts used throughout the chapter as well as apply activation functions to tensors. You will use example data of car dealership sales, apply these concepts, show the sales records of various salespeople, and highlight those with net positive sales.

Sales records:

	Vehicle W	Vehicle X	Vehicle Y	Vehicle Z
Salesperson 1	-0.013	0.024	0.06	0.022
Salesperson 2	0.001	-0.047	0.039	0.016
Salesperson 3	0.018	0.030	-0.021	-0.028

Figure 1.37: Sales records

Vehicle MSRPs:

	Vehicle W	Vehicle X	Vehicle Y	Vehicle Z
MSRP	1999.95	24995.50	36745.50	29995.95

Figure 1.38: Vehicle MSRPs

Fixed costs:

	Salesperson 1	Salesperson 2	Salesperson 3
Cost	2500	2500	2500

Figure 1.39: Fixed costs

Perform the following steps:

1. Import the TensorFlow library.

2. Create a **3x4** tensor as an input with the values `[[-0.013, 0.024, 0.06, 0.022], [0.001, -0.047, 0.039, 0.016], [0.018, 0.030, -0.021, -0.028]]`. The rows in this tensor represent the sales of various sales representatives, the columns represent various vehicles available at the dealership, and values represent the average percentage difference from MSRP. The values are positive or negative depending on whether the salesperson was able to sell for more or less than the MSRP.

3. Create a **4x1** weights tensor with the shape **4x1** with the values `[[19995.95], [24995.50], [36745.50], [29995.95]]` representing the MSRP of the cars.

4. Create a bias tensor of size **3x1** with the values `[[-2500.0], [-2500.0], [-2500.0]]` representing the fixed costs associated with each salesperson.

5. Matrix multiply the input by the weight to show the average deviation from the MSRP on all cars and add the bias to subtract the fixed costs of the salesperson. Print the result.

    You should get the following result:

    ```
    <tf.Tensor: shape=(3, 1), dtype=float32, numpy=
    array([[  704.58545],
           [-1741.7827 ],
           [-3001.75   ]], dtype=float32)>
    ```

    Figure 1.40: The output of the matrix multiplication

6. Apply a ReLU activation function to highlight the net-positive salespeople and print the result.

    You should get the following result:

    ```
    <tf.Tensor: shape=(3, 1), dtype=float32, numpy=
    array([[704.58545],
           [  0.     ],
           [  0.     ]], dtype=float32)>
    ```

    Figure 1.41: The output after applying the activation function

> **NOTE**
>
> The solution to this activity can be found on page 457.

In subsequent chapters, you will see how to add activation functions to your ANNs, either between layers or applied directly after a layer when layers are defined. You will learn how to choose which activation functions are most appropriate, which is often by hyperparameter optimization techniques. The activation function is one example of a hyperparameter, a parameter set before the learning process begins, that can be tuned to find the optimal values for model performance.

SUMMARY

In this chapter, you were introduced to the TensorFlow library. You learned how to use it in the Python programming language. You created the building blocks of ANNs (tensors) with various ranks and shapes, performed linear transformations on tensors using TensorFlow, and implemented addition, reshaping, transposition, and multiplication on tensors—all of which are fundamental for understanding the underlying mathematics of ANNs.

In the next chapter, you will improve your understanding of tensors and learn how to load data of various types and pre-process it such that it is appropriate for training ANNs in TensorFlow. You will work with tabular, visual, and textual data, all of which must be pre-processed differently. By working with visual data (that is, images), you will also learn how to use training data in which the size of the training data cannot fit into memory.

2

LOADING AND PROCESSING DATA

OVERVIEW

In this chapter, you will learn how to load and process a variety of data types for modeling in TensorFlow. You will implement methods to input data into TensorFlow models so that model training can be optimized.

By the end of this chapter, you will know how to input tabular data, images, text, and audio data and preprocess them so that they are suitable for training TensorFlow models.

INTRODUCTION

In the previous chapter, you learned how to create, utilize, and apply linear transformations to tensors using TensorFlow. The chapter started with the definition of tensors and how they can be created using the **Variable** class in the TensorFlow library. You then created tensors of various ranks and learned how to apply tensor addition, reshaping, transposition, and multiplication using the library. These are all examples of linear transformations. You concluded that chapter by covering optimization methods and activation functions and how they can be accessed in the TensorFlow library.

When training machine learning models in TensorFlow, you must supply the model with training data. The raw data that is available may come in a variety of formats—for example, tabular CSV files, images, audio, or text files. Different data sources are loaded and preprocessed in different ways in order to provide numerical tensors for TensorFlow models. For example, virtual assistants use voice queries as input interaction and then apply machine learning models to decipher input speech and perform specific actions as output. To create the models for this task, the audio data of the speech input must be loaded into memory. A preprocessing step also needs to be involved that converts the audio input into text. Following this, the text is converted into numerical tensors for model training. This is one example that demonstrates the complexity of creating models from non-tabular, non-numerical data such as audio data.

This chapter will explore a few of the common data types that are utilized for building machine learning models. You will load raw data into memory in an efficient manner, and then perform some preprocessing steps to convert the raw data into numerical tensors that are appropriate for training machine learning models. Luckily, machine learning libraries have advanced significantly, which means that training models with data types such as images, text, and audio is extremely accessible to practitioners.

EXPLORING DATA TYPES

Depending on the source, raw data can be of different forms. Common forms of data include tabular data, images, video, audio, and text. For example, the output from a temperature logger (used to record the temperature at a given location over time) is tabular. Tabular data is structured with rows and columns, and, in the example of a temperature logger, each column may represent a characteristic for each record, such as the time, location, and temperature, while each row may represent the values of each record. The following table shows an example of numerical tabular data:

	station	Date	Present_Tmax	Present_Tmin	LDAPS_RHmin	LDAPS_RHmax	LDAPS_Tmax_lapse	LDAPS_Tmin_lapse	LDAPS_WS	LDAPS_LH
0	1.0	2013-06-30	28.7	21.4	58.255688	91.116364	28.074101	23.006936	6.818887	69.451805
1	2.0	2013-06-30	31.9	21.6	52.263397	90.604721	29.850689	24.035009	5.691890	51.937448
2	3.0	2013-06-30	31.6	23.3	48.690479	83.973587	30.091292	24.565633	6.138224	20.573050
3	4.0	2013-06-30	32.0	23.4	58.239788	96.483688	29.704629	23.326177	5.650050	65.727144
4	5.0	2013-06-30	31.4	21.9	56.174095	90.155128	29.113934	23.486480	5.735004	107.965535
...	...	...	...	...	...	...	...	...	...	...
7747	23.0	2017-08-30	23.3	17.1	26.741310	78.869858	26.352081	18.775678	6.148918	72.058294
7748	24.0	2017-08-30	23.3	17.7	24.040634	77.294975	27.010193	18.733519	6.542819	47.241457
7749	25.0	2017-08-30	23.2	17.4	22.933014	77.243744	27.939516	18.522965	7.289264	9.090034
7750	NaN	NaN	20.0	11.3	19.794666	58.936283	17.624954	14.272646	2.882580	-13.603212
7751	NaN	NaN	37.6	29.9	98.524734	100.000153	38.542255	29.619342	21.857621	213.414006

Figure 2.1: An example of 10 rows of tabular data that consists of numerical values

Image data represents another common form of raw data that is popular for building machine learning models. These models are popular due to the large volume of data that's available. With smartphones and security cameras recording all of life's moments, they have generated an enormous amount of data that can be used to train models.

The dimensions of image data for training are different than they are for tabular data. Each image has a height and width dimension, as well as a color channel adding a third dimension, and the quantity of images adding a fourth. As such, the input tensors for image data models are four-dimensional tensors, whereas the input tensors for tabular data are two-dimensional. The following figure shows an example of labeled training examples of boats and airplanes taken from the **Open Images** dataset (https://storage.googleapis.com/openimages/web/index.html); the images have been preprocessed so that they all have the same height and width. This data could be used, for example, to train a binary classification model to classify images as boats or airplanes:

Figure 2.2: A sample of image data that can be used for training machine learning models

Other types of raw data that can be used to build machine learning models include text and audio. Like images, their popularity in the machine learning community is derived from the large amount of data that's available. Both audio and text have the challenge of having indeterminate sizes. You will explore how this challenge can be overcome later in this chapter. The following figure shows an audio sample with a sample rate of 44.1 kHz, which means the audio data is sampled 44,100 times per second. This is an example of the type of raw data that is input into virtual assistants, from which they decipher the request and act accordingly:

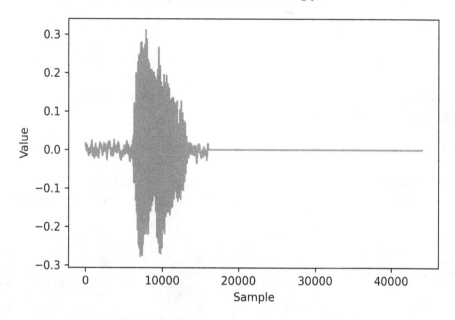

Figure 2.3: A visual representation of audio data

Now that you know about some of the types of data you may encounter when building machine learning models, in the next section, you will uncover ways to preprocess different types of data.

DATA PREPROCESSING

Data preprocessing refers to the process in which raw data is converted into a form that is appropriate for machine learning models to use as input. Each different data type will require different preprocessing steps, with the minimum requirement that the resulting tensor is composed solely of numerical elements, such as integers or decimal numbers. Numerical tensors are required since models rely on linear transformations such as addition and multiplication, which can only be performed on numerical tensors.

While many datasets exist with solely numerical fields, many do not. They may have fields that are of the string, Boolean, categorical, or date data types that must all be converted into numerical fields. Some may be trivial; a Boolean field can be mapped so that **true** values are equal to **1** and **false** values are equal to **0**. Therefore, mapping a Boolean field to a numerical field is simple and all the necessary information is preserved. However, when converting other data types, such as date fields, you may lose information when converting into numerical fields unless it's explicitly stated otherwise.

One example of a possible loss of information occurs when converting a date field into a numerical field by using Unix time. Unix time represents the number of seconds that have elapsed since the Unix epoch; that is, 00:00:00 UTC on January 1, 1970, and leap seconds are ignored. Using Unix time removes the explicit indication of the month, day of the week, hour of the day, and so on, which may act as important features when training a model.

When converting fields into numerical data types, it is important to preserve as much informational context as possible as it will aid any model that is trained to understand the relationship between the features and the target. The following diagram demonstrates how a date field can be converted into a series of numerical fields:

Date		Year	January		December	Weekend
2020-01-01		2020	1		0	0
2020-01-02		2020	1		0	0
2020-01-03	→	2020	1	...	0	0
2020-01-04		2020	1		0	1
2020-01-05		2020	1		0	1
2020-01-06		2020	1		0	0

Figure 2.4: A numerical encoding of a date column

As shown in the preceding diagram, on the left, the date field represents a given date, while on the right, there is a method providing numerical information:

- The year is extracted from the date, which is an integer.

- The month is one-hot encoded. There is a column for each month of the year and the month is binary encoded, if the date's month corresponds with the column's name.

- A column is created indicating whether the date occurs on a weekend.

This is just a method to encode the **date** column here; not all the preceding methods are necessary and there are many more that can be used. Encoding all the fields into numerical fields appropriately is important to create performant machine learning models that can learn the relationships between the features and the target.

Data normalization is another preprocessing technique used to speed up the training process. The normalization process rescales the fields so that they are all of the same scale. This will also help ensure that the weights of the model are of the same scale.

In the preceding diagram, the **year** column has the order of magnitude 10^3, and the other columns have the order 10^0. This implies there are three orders of magnitude between the columns. Fields with values that are very different in scale will result in a less accurate model as the optimal weights to minimize the error function may not be discovered. This may be due to the tolerance limits or the learning rate that are defined as hyperparameters prior to training not being optimal for both scales when the weights are updated. In the preceding example, it may be beneficial to rescale the **year** column so that it has the same order of magnitude as the other columns.

Throughout this chapter, you will explore a variety of methods that can be used to preprocess tabular data, image data, text data, and audio data so that it can be used to train machine learning models.

PROCESSING TABULAR DATA

In this section, you will learn how to load tabular data into a Python development environment so that it can be used for TensorFlow modeling. You will use pandas and scikit-learn to utilize the classes and functions that are useful for processing data. You will also explore methods that can be used to preprocess this data.

Tabular data can be loaded into memory by using the pandas **read_csv** function and passing the path into the dataset. The function is well suited and easy to use for loading in tabular data and can be used as follows:

```
df = pd.read_csv('path/to/dataset')
```

In order to normalize the data, you can use a scaler that is available in scikit-learn. There are multiple scalers that can be applied; **StandardScaler** will normalize the data so that the fields of the dataset have a mean of **0** and a standard deviation of **1**. Another common scaler that is used is **MinMaxScaler**, which will rescale the dataset so that the fields have a minimum value of **0** and a maximum value of **1**.

To use a scaler, it must be initialized and fit to the dataset. By doing this, the dataset can be transformed by the scaler. In fact, the fitting and transformation processes can be performed in one step by using the **fit_transform** method, as follows:

```
scaler = StandardScaler()
transformed_df = scaler.fit_transform(df)
```

In the first exercise, you will learn how to use pandas and scikit-learn to load a dataset and preprocess it so that it is suitable for modeling.

EXERCISE 2.01: LOADING TABULAR DATA AND RESCALING NUMERICAL FIELDS

The dataset, **Bias_correction_ucl.csv**, contains information for bias correction of the next-day maximum and minimum air temperature forecast for Seoul, South Korea. The fields represent temperature measurements of the given date, the weather station at which the metrics were measured, model forecasts of weather-related metrics such as humidity, and projections for the temperature of the following day. You are required to preprocess the data to make all the columns normally distributed with a mean of **0** and a standard deviation of **1**. You will demonstrate the effects with the **Present_Tmax** column, which represents the maximum temperature on the given date at a given weather station.

> **NOTE**
>
> The dataset can be found here: https://packt.link/l83pR.

Perform the following steps to complete this exercise:

1. Open a new Jupyter notebook to implement this exercise. Save the file as **Exercise2-01.ipnyb**.

2. In a new Jupyter Notebook cell, import the pandas library, as follows:

```
import pandas as pd
```

> **NOTE**
>
> You can find the documentation for pandas at the following link: https://pandas.pydata.org/docs/.

3. Create a new pandas DataFrame named **df** and read the
   Bias_correction_ucl.csv file into it. Examine whether your
   data is properly loaded by printing the resultant DataFrame:

```
df = pd.read_csv('Bias_correction_ucl.csv')
df
```

> **NOTE**
>
> Make sure you change the path (highlighted) to the CSV file based on its
> location on your system. If you're running the Jupyter notebook from the
> same directory where the CSV file is stored, you can run the preceding
> code without any modification.

The output will be as follows:

	station	Date	Present_Tmax	Present_Tmin	LDAPS_RHmin	LDAPS_RHmax	LDAPS_Tmax_lapse	LDAPS_Tmin_lapse	LDAPS_WS	LDAPS_LH	...
0	1.0	2013-06-30	28.7	21.4	58.255688	91.116364	28.074101	23.006936	6.818887	69.451805	...
1	2.0	2013-06-30	31.9	21.6	52.263397	90.604721	29.850689	24.035009	5.691890	51.937448	...
2	3.0	2013-06-30	31.6	23.3	48.690479	83.973587	30.091292	24.565633	6.138224	20.573050	...
3	4.0	2013-06-30	32.0	23.4	58.239788	96.483688	29.704629	23.326177	5.650050	65.727144	...
4	5.0	2013-06-30	31.4	21.9	56.174095	90.155128	29.113934	23.486480	5.735004	107.965535	...
...	...	...	...	...	...	...	...	...	...	...	...
7747	23.0	2017-08-30	23.3	17.1	26.741310	78.869858	26.352081	18.775678	6.148918	72.058294	...
7748	24.0	2017-08-30	23.3	17.7	24.040634	77.294975	27.010193	18.733519	6.542819	47.241457	...
7749	25.0	2017-08-30	23.2	17.4	22.933014	77.243744	27.939516	18.522965	7.289264	9.090034	...
7750	NaN	NaN	20.0	11.3	19.794666	58.936283	17.624954	14.272646	2.882580	-13.603212	...
7751	NaN	NaN	37.6	29.9	98.524734	100.000153	38.542255	29.619342	21.857621	213.414006	...

7752 rows × 25 columns

Figure 2.5: The output from printing the DataFrame

4. Drop the **date** column using the **drop** method of the DataFrame and pass
   in the name of the column. The **date** column will be dropped as it is a
   non-numerical field and rescaling will not be possible when non-numerical fields
   exist. Since you are dropping a column, both the **axis=1** argument and the
   inplace=True argument should be passed:

```
df.drop('Date', inplace=True, axis=1)
```

5. Plot a histogram of the **Present_Tmax** column that represents the maximum temperature across dates and weather stations within the dataset:

```
ax = df['Present_Tmax'].hist(color='gray')
ax.set_xlabel("Temperature")
ax.set_ylabel("Frequency")
```

The output will be as follows:

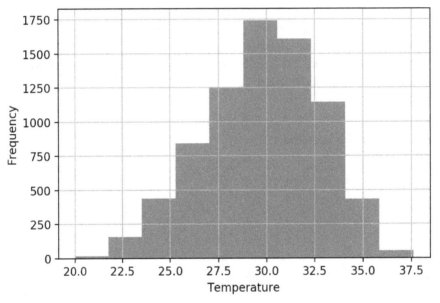

Figure 2.6: A Temperature versus Frequency histogram of the Present_Tmax column

The resultant histogram shows the distribution of values for the **Present_Tmax** column. You can see that the temperature values vary from 20 to 38 degrees Celsius. Plotting a histogram of the feature values is a good way to view the distribution of values to understand whether scaling is required as a preprocessing step.

6. Import the **StandardScaler** class from scikit-learn's preprocessing package. Initialize the scaler, fit the scaler, and transform the DataFrame using the scaler's **fit_transform** method. Create a new DataFrame, **df2**, using the transformed DataFrame since the result of the **fit_transform** method is a NumPy array. The standard scaler will transform the numerical fields so that the mean of the field is **0** and the standard deviation is **1**:

```
from sklearn.preprocessing import StandardScaler
scaler = StandardScaler()
df2 = scaler.fit_transform(df)
df2 = pd.DataFrame(df2, columns=df.columns)
```

> **NOTE**
>
> The values for the mean and standard deviation of the resulting transformed data can be input into the scaler.

7. Plot a histogram of the transformed **Present_Tmax** column:

```
ax = df2['Present_Tmax'].hist(color='gray')
ax.set_xlabel("Normalized Temperature")
ax.set_ylabel("Frequency")
```

The output will be as follows:

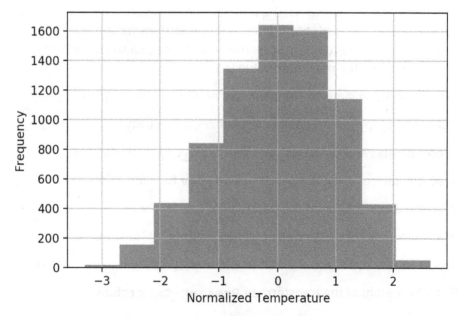

Figure 2.7: A histogram of the rescaled Present_Tmax column

The resulting histogram shows that the temperature values range from around **−3** to **3** degrees Celsius, as evidenced by the range on the *x* axis of the histogram. By using the standard scaler, the values will always have a mean of **0** and a standard deviation of **1**. Having the features normalized can speed up the model training process.

In this exercise, you successfully imported tabular data using the pandas library and performed some preprocessing using the scikit-learn library. The preprocessing of data included dropping the **date** column and scaling the numerical fields so that they have a mean value of **0** and a standard deviation of **1**.

In the following activity, you will load in tabular data using the pandas library and scale that data using the **MinMax** scaler present in scikit-learn. You will do so on the same dataset that you used in the prior exercise, which describes the bias correction of air temperature forecasts for Seoul, South Korea.

ACTIVITY 2.01: LOADING TABULAR DATA AND RESCALING NUMERICAL FIELDS WITH A MINMAX SCALER

In this activity, you are required to load tabular data and rescale the data using a **MinMax** scaler. The dataset, **Bias_correction_ucl.csv**, contains information for bias correction of the next-day maximum and minimum air temperature forecast for Seoul, South Korea. The fields represent temperature measurements of the given date, the weather station at which the metrics were measured, model forecasts of weather-related metrics such as humidity, and projections for the temperature the following day. You are required to scale the columns so that the minimum value of each column is **0** and the maximum value is **1**.

Perform the following steps to complete this activity:

1. Open a new Jupyter notebook to implement this activity.

2. Import pandas and the **Bias_correction_ucl.csv** dataset.

3. Read the dataset using the pandas **read_csv** function.

4. Drop the **date** column of the DataFrame.

5. Plot a histogram of the **Present_Tmax** column.

6. Import **MinMaxScaler** and fit it to and transform the feature DataFrame.

7. Plot a histogram of the transformed **Present_Tmax** column.

You should get an output similar to the following:

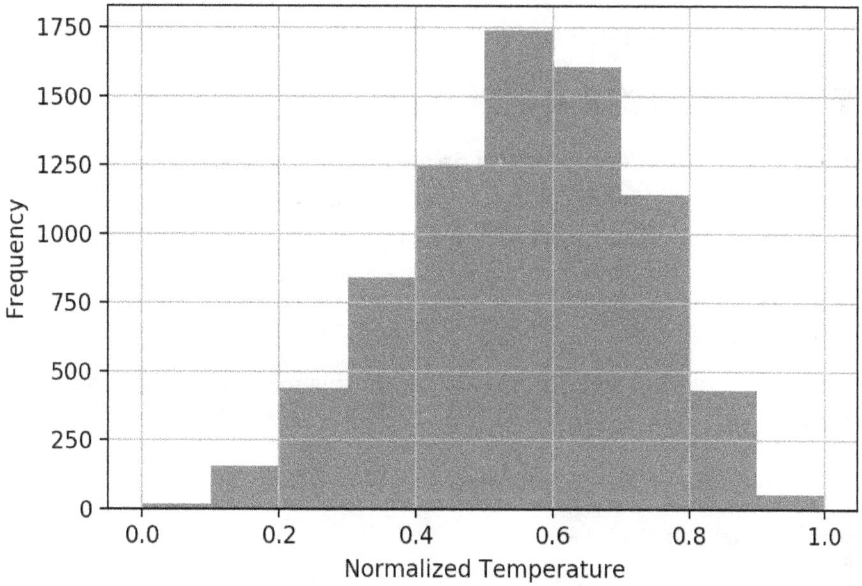

Figure 2.8: Expected output of Activity 2.01

> **NOTE**
>
> The solution to this activity can be found on page 459.

One method of converting non-numerical fields such as categorical or date fields is to one-hot encode them. The **one-hot encoding process** creates a new column for each unique value in the provided column, while each row has a value of **0** except for the one that corresponds to the correct column. The column headers of the newly created dummy columns correspond to the unique values. One-hot encoding can be achieved by using the **get_dummies** function of the pandas library and passing in the column to be encoded. An optional argument is to provide a prefix feature that adds a prefix to the column headers. This can be useful for referencing the columns:

```
dummies = pd.get_dummies(df['feature1'], prefix='feature1')
```

> **NOTE**
>
> When using the **get_dummies** function, **NaN** values are converted into all zeros.

In the following exercise, you'll learn how to preprocess non-numerical fields. You will utilize the same dataset that you used in the previous exercise and activity, which describes the bias correction of air temperature forecasts for Seoul, South Korea.

EXERCISE 2.02: PREPROCESSING NON-NUMERICAL DATA

In this exercise, you will preprocess the **date** column by one-hot encoding the year and the month from the **date** column using the **get_dummies** function. You will join the one-hot-encoded columns with the original DataFrame and ensure that all the fields in the resultant DataFrame are numerical.

Perform the following steps to complete this exercise:

1. Open a new Jupyter notebook to implement this exercise. Save the file as **Exercise2-02.ipnyb**.

2. In a new Jupyter Notebook cell, import the pandas library, as follows:

```
import pandas as pd
```

3. Create a new pandas DataFrame named **df** and read the **Bias_correction_ucl.csv** file into it. Examine whether your data is properly loaded by printing the resultant DataFrame:

```
df = pd.read_csv('Bias_correction_ucl.csv')
```

> **NOTE**
>
> Make sure you change the path (highlighted) to the CSV file based on its location on your system. If you're running the Jupyter notebook from the same directory where the CSV file is stored, you can run the preceding code without any modification.

4. Change the data type of the **date** column to **Date** using the pandas **to_datetime** function:

```
df['Date'] = pd.to_datetime(df['Date'])
```

5. Create dummy columns for **year** using the pandas **get_dummies** function. Pass in the year of the **date** column as the first argument and add a prefix to the columns of the resultant DataFrame. Print out the resultant DataFrame:

```
year_dummies = pd.get_dummies(df['Date'].dt.year, \
                              prefix='year')
year_dummies
```

The output will be as follows:

	year_2013.0	year_2014.0	year_2015.0	year_2016.0	year_2017.0
0	1	0	0	0	0
1	1	0	0	0	0
2	1	0	0	0	0
3	1	0	0	0	0
4	1	0	0	0	0
...	...	...	...	...	...
7747	0	0	0	0	1
7748	0	0	0	0	1
7749	0	0	0	0	1
7750	0	0	0	0	0
7751	0	0	0	0	0

7752 rows × 5 columns

Figure 2.9: Output of the get_dummies function applied to the year of the date column

The resultant DataFrame contains only 0s and 1s. **1** corresponds to the value present in the original **date** column. Null values will have 0s for all columns in the newly created DataFrame.

6. Repeat this for the month by creating dummy columns from the month of the **date** column. Print out the resulting DataFrame:

```
month_dummies = pd.get_dummies(df['Date'].dt.month, \
                               prefix='month')
month_dummies
```

The output will be as follows:

	month_6.0	month_7.0	month_8.0
0	1	0	0
1	1	0	0
2	1	0	0
3	1	0	0
4	1	0	0
...	...	...	...
7747	0	0	1
7748	0	0	1
7749	0	0	1
7750	0	0	0
7751	0	0	0

7752 rows × 3 columns

Figure 2.10: The output of the get_dummies function applied to the month of the date column

The resultant DataFrame now contains only 0s and 1s for the month in the **date** column.

7. Concatenate the original DataFrame and the dummy DataFrames you created in *Steps 5* and *6*:

```
df = pd.concat([df, month_dummies, year_dummies], \
          axis=1)
```

8. Drop the original **date** column since it is now redundant:

```
df.drop('Date', axis=1, inplace=True)
```

9. Verify that all the columns are now of the numerical data type:

```
df.dtypes
```

The output will be as follows:

```
station              float64
Present_Tmax         float64
Present_Tmin         float64
LDAPS_RHmin          float64
LDAPS_RHmax          float64
LDAPS_Tmax_lapse     float64
LDAPS_Tmin_lapse     float64
LDAPS_WS             float64
LDAPS_LH             float64
LDAPS_CC1            float64
LDAPS_CC2            float64
LDAPS_CC3            float64
LDAPS_CC4            float64
LDAPS_PPT1           float64
LDAPS_PPT2           float64
LDAPS_PPT3           float64
LDAPS_PPT4           float64
lat                  float64
lon                  float64
DEM                  float64
Slope                float64
Solar radiation      float64
Next_Tmax            float64
Next_Tmin            float64
month_6.0            uint8
month_7.0            uint8
month_8.0            uint8
year_2013.0          uint8
year_2014.0          uint8
year_2015.0          uint8
year_2016.0          uint8
year_2017.0          uint8
dtype: object
```

Figure 2.11: Output of the dtypes attribute of the resultant DataFrame

Here, you can see that all the data types of the resultant DataFrame are numerical. This means they can now be passed into an ANN for modeling.

In this exercise, you successfully imported tabular data and preprocessed the **date** column using the pandas and scikit-learn libraries. You utilized the **get_dummies** function to convert categorical data into numerical data types.

> **NOTE**
>
> Another method to attain a numerical data type from date data types is by using the **pandas.Series.dt** accessor object. More information about the available options can be found here: https://pandas.pydata.org/docs/reference/api/pandas.Series.dt.html.

Processing non-numerical data is an important step in creating performant models. If possible, any domain knowledge should be imparted to the training data features. For example, when forecasting the temperature using the date, like the dataset used in the prior exercises and activity of this chapter, encoding the month would be helpful since the temperature is likely highly correlated with the month of the year. Encoding the day of the week, however, may not be useful as there is likely no correlation between the day of the week and temperature. Using this domain knowledge can aid the model to learn the underlying relationship between the features and the target.

In the next section, you will learn how to process image data so that it can be input into machine learning models.

PROCESSING IMAGE DATA

A plethora of images is being generated every day by various organizations that can be used to create predictive models for tasks such as object detection, image classification, and object segmentation. When working with image data and some other raw data types, you often need to preprocess the data. Creating models from raw data with minimal preprocessing is one of the biggest benefits of using ANNs for modeling since the feature engineering step is minimal. Feature engineering usually involves using domain knowledge to create features out of the raw data, which is time consuming and has no guarantee of improvements in model performance. Utilizing ANNs with no feature engineering streamlines the training process and has no need for domain knowledge.

For example, locating tumors in medical images requires expert knowledge from those who have been trained for many years, but for ANNs, all that is required is sufficient labeled data for training. There will be a small amount of preprocessing that generally needs to be applied to these images. These steps are optional but helpful for standardizing the training process and creating performant models.

One preprocessing step is rescaling. Since images have color values that are integers that range between **0** and **255**, they are scaled to have values between **0** and **1**, similar to *Activity 2.01, Loading Tabular Data and Rescaling Numerical Fields with a MinMax Scaler*. Another common preprocessing step that you will explore later in this section is image augmentation, which is essentially the act of augmenting images to add a greater number of training examples and build a more robust model.

This section also covers batch processing. Batch processing loads in the training data one batch at a time. This can result in slower training times than if the data was loaded in at once; however, this does allow you to train your models on very large-volume datasets. Training on images or audio are examples that often require large volumes to achieve performant results.

For example, a typical image may be 100 KB in size. For a training dataset of 1 million images, you would need 100 GB of memory, which may be unattainable to most. If the model is trained in batches of 32 images, the memory requirement is orders of magnitude less. Batch training allows you to augment the training data, as you will explore in a later section.

Images can be loaded into memory using a class named `ImageDataGenerator`, which can be imported from Keras' preprocessing package. This is a class originally from Keras that can now be used in TensorFlow. When loading in images, you can rescale them. It is common practice to rescale images by the value of 1/255 pixels. This means that images that have values from 0 to 255 will now have values from 0 to 1.

`ImageDataGenerator` can be initialized with rescaling, as follows:

```
datagenerator = ImageDataGenerator(rescale = 1./255)
```

Once the **ImageDataGenerator** class has been initialized, you can use the **flow_from_directory** method and pass in the directory that the images are located in. The directory should include sub-directories labeled with the class labels, and they should contain the images of the corresponding class. Another argument to be passed in is the desired size for the images, the batch size, and the class mode. The class mode determines the type of label arrays that are produced. Using the **flow_from_directory** method for binary classification with a batch size of 25 and an image size of 64x64 can be done as follows:

```
dataset = datagenerator.flow_from_directory\
         ('path/to/data',\
          target_size = (64, 64),\
          batch_size = 25,\
          class_mode = 'binary')
```

In the following exercise, you will load images into memory by utilizing the **ImageDataGenerator** class.

> **NOTE**
>
> The image data provided comes from the Open Image dataset, a full description of which can be found here: https://storage.googleapis.com/openimages/web/index.html.

Images can be viewed by plotting them using Matplotlib. This is a useful exercise for verifying that the images match their respective labels.

EXERCISE 2.03: LOADING IMAGE DATA FOR BATCH PROCESSING

In this exercise, you'll learn how to load in image data for batch processing. The **image_data** folder contains a set of images of boats and airplanes. You will load the images of boats and airplanes for batch processing and rescale them so that the image values range between **0** and **1**. You are then tasked with printing the labeled images of a batch from the data generator.

> **NOTE**
>
> You can find **image_data** here: https://packt.link/jZ2oc.

Perform the following steps to complete this exercise:

1. Open a new Jupyter notebook to implement this exercise. Save the file as **Exercise2-03.ipnyb**.

2. In a new Jupyter Notebook cell, import the **ImageDataGenerator** class from **tensorflow.keras.preprocessing.image**:

```
from tensorflow.keras.preprocessing.image \
    import ImageDataGenerator
```

3. Instantiate the **ImageDataGenerator** class and pass the **rescale** argument with the value **1./255** to convert image values so that they're between **0** and **1**:

```
train_datagen = ImageDataGenerator(rescale =  1./255)
```

4. Use the data generator's **flow_from_directory** method to direct the data generator to the image data. Pass in the arguments for the target size, the batch size, and the class mode:

```
training_set = train_datagen.flow_from_directory\
                ('image_data',\
                 target_size = (64, 64),\
                 batch_size = 25,\
                 class_mode = 'binary')
```

5. Create a function to display the images in the batch. The function will plot the first 25 images in a 5x5 array with their associated labels:

```
import matplotlib.pyplot as plt

def show_batch(image_batch, label_batch):\
    lookup = {v: k for k, v in \
            training_set.class_indices.items()}
    label_batch = [lookup[label] for label in \
                label_batch]
    plt.figure(figsize=(10,10))
    for n in range(25):
        ax = plt.subplot(5,5,n+1)
        plt.imshow(image_batch[n])
        plt.title(label_batch[n].title())
        plt.axis('off')
```

6. Take a batch from the data generator and pass it to the function to display the images and their labels:

```
image_batch, label_batch = next(training_set)
show_batch(image_batch, label_batch)
```

The output will be as follows:

Figure 2.12: The images from a batch

Here, you can see the output of a batch of images of boats and airplanes that can be input into a model. Note that all the images are the same size, which was achieved by modifying the aspect ratio of the images. This ensures consistency in the images as they are passed into an ANN.

In this exercise, you learned how to import images in batches so they can be used for training ANNs. Images are loaded one batch at a time and by limiting the number of training images per batch, you can ensure that the RAM of the machine is not exceeded.

In the following section, you will see how to augment images as they are loaded in.

IMAGE AUGMENTATION

Image augmentation is the process of modifying images to increase the number of training examples available. This process can include zooming in on the image, rotating the image, or flipping the image vertically or horizontally. This can be performed if the augmentation process does not change the context of the image. For example, an image of a banana, when flipped horizontally, is still recognizable as a banana, and new images of bananas are likely to be of either orientation. In this case, providing a model for both orientations during the training process will help build a robust model.

However, if you have an image of a boat, it may not be appropriate to flip it vertically, as this does not represent how boats commonly exist in images, upside-down. Ultimately the goal of image augmentation is to increase the number of training images that resemble the object in its everyday occurrence, preserving the context. This will help the trained model perform well on new, unseen images. An example of image augmentation can be seen in the following figure, in which an image of a banana has been augmented three times; the left image is the original image, and those on the right are the augmented images.

The top-right image is the original image flipped horizontally, the middle-right image is the original image zoomed in by 15%, and the bottom-right image is the original image rotated by 10 degrees. After this augmentation process, you have four images of a banana, each of which has the banana in different positions and orientations:

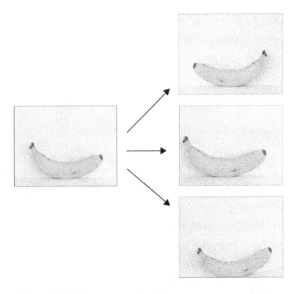

Figure 2.13: An example of image augmentation

Image augmentation can be achieved with TensorFlow's **ImageDataGenerator** class when the images are loaded with each batch. Similar to image rescaling, various image augmentation processes can be applied. The arguments for common augmentation processes include the following:

- **horizontal_flip**: Flips the image horizontally.
- **vertical_flip**: Flips the image vertically.
- **rotation_range**: Rotates the image up to a given number of degrees.
- **width_shift_range**: Shifts the image along its width axis up to a given fraction or pixel amount.
- **height_shift_range**: Shifts the image along its height axis up to a given fraction or pixel amount.
- **brightness_range**: Modifies the brightness of the image up to a given amount.
- **shear_range**: Shears the image up to a given amount.
- **zoom_range**: Zooms in the image up to a given amount.

Image augmentation can be applied when instantiating the **ImageDataGenerator** class, as follows:

```
datagenerator = ImageDataGenerator(rescale = 1./255,\
                          shear_range = 0.2,\
                          rotation_range= 180,\
                          zoom_range = 0.2,\
                          horizontal_flip = True)
```

In the following activity, you perform image augmentation using TensorFlow's **ImageDataGenerator** class. The process is as simple as passing in parameters. You will use the same dataset that you used in *Exercise 2.03, Loading Image Data for Batch Processing*, which contains images of boats and airplanes.

ACTIVITY 2.02: LOADING IMAGE DATA FOR BATCH PROCESSING

In this activity, you will load image data for batch processing and augment the images in the process. The **image_data** folder contains a set of images of boats and airplanes. You are required to load in image data for batch processing and adjust the input data with random perturbations such as rotations, flipping the image horizontally, and adding shear to the images. This will create additional training data from the existing image data and will lead to more accurate and robust machine learning models by increasing the number of different training examples even if only a few are available. You are then tasked with printing the labeled images of a batch from the data generator.

The steps for this activity are as follows:

1. Open a new Jupyter notebook to implement this activity.

2. Import the **ImageDataGenerator** class from **tensorflow.keras.preprocessing.image**.

3. Instantiate **ImageDataGenerator** and set the **rescale=1./255**, **shear_range=0.2**, **rotation_range=180**, **zoom_range=0.2**, and **horizontal_flip=True** arguments.

4. Use the **flow_from_directory** method to direct the data generator to the images while passing in the target size as **64x64**, a batch size of **25**, and the class mode as **binary**.

5. Create a function to display the first 25 images in a 5x5 array with their associated labels.

6. Take a batch from the data generator and pass it to the function to display the images and their labels.

> **NOTE**
>
> The solution to this activity can be found on page 462.

In this activity, you augmented images in batches so they could be used for training ANNs. You've seen that when images are used as input, they can be augmented to generate a larger number of effective training examples.

You learned how to load images in batches, which enables you to train on huge volumes of data that may not fit into the memory of your machine at one time. You also learned how to augment images using the **ImageDataGenerator** class, which essentially generates new training examples from the images in your training set.

In the next section, you will learn how to load and preprocess text data.

TEXT PROCESSING

Text data represents a large class of raw data that is readily available. For example, text data can be from web pages such as Wikipedia, transcribed speech, or social media conversations—all of which are increasing at a massive scale and must be processed before they can be used for training machine learning models.

Working with text data can be challenging for several different reasons, including the following:

- Thousands of different words exist.

- Different languages present challenges.

- Text data often varies in size.

There are many ways to convert text data into a numerical representation. One way is to one-hot encode the words, much like you did with the date field in *Exercise 2.02, Preprocessing Non-Numerical Data*. However, this presents issues when training models since large datasets with many unique words will result in a sparse dataset and can lead to slow training speeds and potentially inaccurate models. Moreover, if a new word is encountered that was not in the training data, the model cannot use that word.

One popular method that's used to represent text data is to convert the entire piece of text into embedding vectors. Pretrained models exist to convert raw text into vectors. These models are usually trained on large volumes of text. Using word embedding vectors from pretrained models has some distinct advantages:

- The resulting vectors have a fixed size.

- The vectors maintain contextual information, so they benefit from transfer learning.

- No further preprocessing of the data needs to be done and the results of the embedding can be fed directly into an ANN.

While TensorFlow Hub will be covered in more depth in the next chapter, the following is an example of how to use pretrained models as a preprocessing step. To load in the pretrained model, you need to import the **tensorflow_hub** library. By doing this, the URL of the model can be loaded. Then, the model can be loaded into the environment by calling the **KerasLayer** class, which wraps the model so that it can be used like any other TensorFlow model. It can be created as follows:

```
import tensorflow_hub as hub
model_url = "url_of_model"
hub_layer = hub.KerasLayer(model_url, \
                           input_shape=[], dtype=tf.string, \
                           trainable=True)
```

The data type of the input data, indicated by the **dtype** parameter, should be used as input for the **KerasLayer** class, as well as a Boolean argument indicating whether the weights are trainable. Once the model has been loaded using the **tensorflow_hub** library, it can be called on text data, as follows:

```
hub_layer(data)
```

This will run the data through the pretrained model. The output will be based on the architecture and weights of the pretrained model.

In the following exercise, you will explore how to load in data that includes a text field, batch the dataset, and apply a pretrained model to the text field to convert the field into embedded vectors.

> **NOTE**
>
> The pretrained model can be found here:
> https://tfhub.dev/google/tf2-preview/gnews-swivel-20dim/1.
>
> The dataset can be found here: https://archive.ics.uci.edu/ml/datasets/
> Drug+Review+Dataset+%28Drugs.com%29.

EXERCISE 2.04: LOADING TEXT DATA FOR TENSORFLOW MODELS

The dataset, **drugsComTrain_raw.tsv**, contains information related to patient reviews on specific drugs, along with their related conditions and a rating indicating the patient's satisfaction with the drug. In this exercise, you will load in text data for batch processing. You will apply a pretrained model from TensorFlow Hub to perform a word embedding on the patient reviews. You are required to work on the **review** field only as that contains text data.

Perform the following steps:

1. Open a new Jupyter notebook to implement this exercise. Save the file as **Exercise2-04.ipnyb**.

2. In a new Jupyter Notebook cell, import the TensorFlow library:

    ```
    import tensorflow as tf
    ```

3. Create a TensorFlow dataset object using the library's **make_csv_dataset** function. Set the **batch_size** argument equal to **1** and the **field_delim** argument to **'\t'** since the dataset is tab-delimited:

    ```
    df = tf.data.experimental.make_csv_dataset\
        ('../Datasets/drugsComTest_raw.tsv', \
         batch_size=1, field_delim='\t')
    ```

4. Create a function that takes a dataset object as input and shuffles, repeats, and batches the dataset:

    ```
    def prep_ds(ds, shuffle_buffer_size=1024, \
                batch_size=32):
        # Shuffle the dataset
    ```

```
        ds = ds.shuffle(buffer_size=shuffle_buffer_size)
        # Repeat the dataset
        ds = ds.repeat()
        # Batch the dataset
        ds = ds.batch(batch_size)


        return ds
```

5. Apply the function to the dataset object you created in *Step 3*, setting **batch_size** equal to **5**:

```
ds = prep_ds(df, batch_size=5)
```

6. Take the first batch and print it out:

```
for x in ds.take(1):\
    print(x)
```

You should get output similar to the following:

```
OrderedDict([('', <tf.Tensor: shape=(5, 1), dtype=int32, numpy=
array([[193905],
       [ 91684],
       [ 38033],
       [108776],
       [ 84770]])>), ('drugName', <tf.Tensor: shape=(5, 1), dtype=string, numpy=
array([[b'Dilaudid'],
       [b'Protonix IV'],
       [b'Adipex-P'],
       [b'Nexplanon'],
       [b'Ethinyl estradiol / norgestimate']], dtype=object)>), ('condition', <tf.Tensor: shape=(5, 1), dtype=string, numpy=
array([[b'Pain'],
       [b"Barrett's Esophagus"],
       [b'Weight Loss'],
       [b'Birth Control'],
       [b'Birth Control']], dtype=object)>), ('review', <tf.Tensor: shape=(5, 1), dtype=string, numpy=
array([[b'"I have been given Dilaudid through an IV on several occasions and on all those occasions I&#039;d rather just dealt
with the pain. Today the hospital gave me Dilaudid, and some nausea medication... On ever occasion that I have been given Dilau
did through IV instantly my head gets too heavy, I begin to get dizzy and lighthead. I always get VERY NAUSEOUS and confused, s
o I usually sleep it off  because I hate the feeling it gives me. This time was different I fell asleep about 40 minutes after
getting the medication and every couple minutes I kept waking up. All of a sudden I became very hot and clamy while feeling ver
y cold.  "'],
       [b'"Have been on Protonix for about 10 yrs. Recently was admitted to the hospital due to pancreatitis. I was vomiting no
n-stop for 4 days and could only get meds via I.V. during my 7 day stay. I found that Protonix I.V. seemed to work almost bette
r at least worked faster and lasted 24hrs until my next dose the following day."'],
       [b'"In Jan of this year I weighed 170.0. I had some gallbladder issues that ended up with it being taken out and caused
some weight gain. It is now May and my last weigh in was 141.4. I love my adipex and I would recommend it for anyone. I have ha
d two children and so my body has been through the ringer. I love the energy it gives me and the appetite suppression. I also l
ike that it makes you thirsty and the only thing that helps is water. I am now a water drinker and I drink well over the daily
recommendation. I love this!!"'],
```

Figure 2.14: A batch from the dataset object

The output represents the input data in tensor format.

7. Import the pretrained word embedding model from TensorFlow Hub and create a Keras layer:

```
import tensorflow_hub as hub
embedding = "https://tfhub.dev/google/tf2-preview"\
            "/gnews-swivel-20dim/1"
hub_layer = hub.KerasLayer(embedding, input_shape=[], \
                           dtype=tf.string, \
                           trainable=True)
```

8. Take one batch from the dataset, flatten the tensor corresponding to the **review** field, apply the pretrained layer, and print it out:

```
for x in ds.take(1):\
    print(hub_layer(tf.reshape(x['review'],[-1])))
```

This will display the following output:

```
tf.Tensor(
[[ 0.85637194 -1.506148    2.0898294   0.76287127 -1.3294315  -3.2016342
  -1.5625719   2.5138297   1.9493035  -0.87237185 -0.71198505  1.2482367
   0.8347453   0.74195063 -3.2654202   1.7437259   3.6106536  -2.433589
  -2.1442554  -1.8830311 ]
 [ 1.3633374  -3.472839    1.6306508   1.3522071  -3.7051663  -5.591705
  -2.5833106   1.9622269   2.148968   -0.5469887  -1.8512614   3.1438973
  -0.81081367  0.9246471  -6.089089    2.6840522   5.7409143  -4.6783543
  -3.0686712  -1.891416  ]
 [-0.38192645 -3.241573    1.20132     0.37701157 -3.3210232  -3.9527822
  -3.1954875   1.5348208   2.1768715   1.4031676  -3.039334    1.8451444
   1.5885328   0.62233907 -4.6009307   3.2810917   2.9764225  -2.7709908
  -1.3856741  -0.16450648]
 [ 1.3386972  -4.202724    2.456265    1.6445506  -4.9147415  -4.971446
  -3.1711445   1.771192    3.425477    0.30268404 -2.1084783   2.555107
   0.74540186  0.77879316 -5.8247786   3.1800213   5.0734744  -2.6060858
  -2.3680193  -1.4511749 ]
 [ 1.1435002  -1.9953871   0.0774065   0.8793874  -1.8057402  -3.9833982
  -2.35677     1.1914377   1.7414727  -0.4663017  -0.9780923   2.705346
   0.29036057  0.60294515 -3.9027228   2.1389863   3.684991   -2.9337153
  -1.1615205  -1.3189584 ]], shape=(5, 20), dtype=float32)
```

Figure 2.15: A batch of the review column after the pretrained model
has been applied to the text

The preceding output represents the embedding vectors for the first batch of drug reviews. The specific values may not mean much at first glance but encoded within the embeddings is contextual information based on the dataset that the embedding model was trained upon. The batch size is equal to **5** and the embedding vector size is **20**, which means the resulting size, after applying the pretrained layer, is **5x20**.

In this exercise, you learned how to import tabular data that might contain a variety of data types. You took the `review` field and applied a pretrained word embedding model to convert the text into a numerical tensor. Ultimately, you preprocessed and batched the text data so that it was appropriate for large-scale training. This is one way to represent text so that it can be input into machine learning models in TensorFlow. In fact, other pretrained word embedding models can be used and are available on TensorFlow Hub. You will learn more about how to utilize TensorFlow Hub in the next chapter.

In this section, you learned about one way to preprocess text data for use in machine learning models. There are a number of different methods you could have used to generate a numerical tensor from the text. For example, you could have one-hot encoded the words, removed the stop words, stemmed and lemmatized the words, or even done something as simple as counting the number of words in each review. The method demonstrated in this section is advantageous as it is simple to implement. Also, the word embedding incorporates contextual information in the text that is difficult to encode in other methods, such as one-hot encoding.

Ultimately, it is up to the practitioner to apply any domain knowledge to the preprocessing step to retain as much contextual information as possible. This will allow any subsequent models to learn the underlying function between the features and the target variable.

In the next section, you will learn how to load and process audio data so that the data can be used for TensorFlow models.

AUDIO PROCESSING

This section will demonstrate how to load audio data in batches, as well as how to process it so that it can be used to train machine learning models. There is some advanced signal processing that takes place to preprocess audio files. Some of these steps are optional, but they are presented to provide a comprehensive approach to processing audio data. Since each audio file can be hundreds of KB, you will utilize batch processing, as you did when processing image data. Batch processing can be achieved by creating a dataset object. A generic method for creating a dataset object from raw data is using TensorFlow's **from_tensor_slice** function. This function generates a dataset object by slicing a tensor along its first dimension. It can be used as follows:

```
dataset = tf.data.Dataset\
            .from_tensor_slices([1, 2, 3, 4, 5])
```

Loading audio data into a Python environment can be achieved using TensorFlow by reading the file into memory using the **read_file** function, then decoding the file using the **decode_wav** function. When using the **decode_wav** function, the sample rate, which represents how many data points comprise 1 second of data, as well as the desired channel to use must be passed in as arguments. For example, if a value of **-1** is passed for the desired channel, then all the audio channels will be decoded. Importing the audio file can be achieved as follows:

```
sample_rate = 44100
audio_data = tf.io.read_file('path/to/file')
audio, sample_rate = tf.audio.decode_wav\
                    (audio_data,\
                     desired_channels=-1,\
                     desired_samples=sample_rate)
```

As with text data, you must preprocess the data so that the resulting numerical tensor has the same size as the data. This is achieved by sampling the audio file after converting the data into the frequency domain. Sampling the audio can be thought of as splitting the audio file into chunks that are always the same size. For example, a 30-second audio file can be split into 30 1-second non-overlapping audio samples, and in the same way, a 15-second audio file can be split into 15 1-second non-overlapping samples. Thus, your result is 45 equally sized audio samples.

Another common preprocessing step that can be performed on audio data is to convert the audio sample from the time domain into the frequency domain. Interpreting the data in the time domain is useful for understanding the intensity or volume of the audio, whereas the frequency domain can help you discover which frequencies are present. This is useful for classifying sounds since different objects have different characteristic sounds that will be present in the frequency domain. Audio data can be converted from the time domain into the frequency domain using the **stft** function.

This function takes the short-time Fourier transform of the input data. The arguments to the function include the frame length, which is an integer value that indicates the window length in samples; the frame step, which is an integer value that describes the number of samples to step; and the **Fast Fourier Transform (FFT)** length, which is an integer value that indicates the length of the FFT to apply. A spectrogram is the absolute value of the short-time Fourier transform as it is useful for visual interpretation. The short-time Fourier transform and spectrogram can be created as follows:

```
stfts = tf.signal.stft(audio, frame_length=1024,\
                       frame_step=256,\
                       fft_length=1024)
spectrograms = tf.abs(stfts)
```

Another optional preprocessing step is to generate the **Mel-Frequency Cepstral Coefficients (MFCCs)**. As the name suggests, the MFCCs are the coefficients of the mel-frequency cepstrum. The cepstrum is a representation of the short-term power spectrum of an audio signal. MFCCs are commonly used in applications for speech recognition and music information retrieval. As such, it may not be important to understand each step of how the MFCCs are generated but understanding that they can be applied as a preprocessing step to increase the information density of the audio data pipeline is beneficial.

MFCCs are generated by creating a matrix to warp the linear scale to the mel scale. This matrix can be created using **linear_to_mel_weight_matrix** and by passing in the number of bands in the resulting mel spectrum, the number of bins in the source spectrogram, the sample rate, and the lower and upper frequencies to be included in the mel spectrum. Once the linear-to-mel weight matrix has been created, a tensor contraction with the spectrograms is applied along the first axis using the **tensordot** function.

Following this, the log of the values is applied to generate the log mel spectrograms. Finally, the **mfccs_from_log_mel_spectrograms** function can be applied to generate the MFCCs that are passing in the log mel spectrograms. These steps can be applied as follows:

```
lower_edge_hertz, upper_edge_hertz, num_mel_bins \
    = 80.0, 7600.0, 80
linear_to_mel_weight_matrix \
    = tf.signal.linear_to_mel_weight_matrix\
        (num_mel_bins, num_spectrogram_bins, sample_rate, \
         lower_edge_hertz, upper_edge_hertz)
mel_spectrograms = tf.tensordot\
                    (spectrograms, \
                     linear_to_mel_weight_matrix, 1)
mel_spectrograms.set_shape\
    (spectrograms.shape[:-1].concatenate\
    (linear_to_mel_weight_matrix.shape[-1:]))

log_mel_spectrograms = tf.math.log(mel_spectrograms + 1e-6)
mfccs = tf.signal.mfccs_from_log_mel_spectrograms\
        (log_mel_spectrograms)[..., :num_mfccs]
```

In the following exercise, you will understand how audio data can be processed. In a similar manner to what you did in *Exercise 2.03*, *Loading Image Data for Batch Processing*, and *Exercise 2.04*, *Loading Text Data for TensorFlow Models*, you will load the data in batches for efficient and scalable training. You will load in the audio files using TensorFlow's generic **read_file** function, then decode the audio data using TensorFlow's **decode_wav** function. You will then create a function that will generate the MFCCs from each audio sample. Finally, a dataset object will be generated that can be passed into a TensorFlow model for training. The dataset that you will be utilizing is Google's speech commands dataset, which consists of 1-second-long utterances of words.

> **NOTE**
>
> The dataset can be found here: https://packt.link/Byurf.

EXERCISE 2.05: LOADING AUDIO DATA FOR TENSORFLOW MODELS

In this exercise, you'll learn how to load in audio data for batch processing. The dataset, **data_speech_commands_v0.02**, contains speech samples of people speaking the word **zero** for exactly 1 second with a sample rate of 44.1 kHz, meaning that for every second, there are 44,100 data points. You will apply some common audio preprocessing techniques, including converting the data into the Fourier domain, sampling the data to ensure the data has the same size as the model, and generating MFCCs for each audio sample. This will generate a preprocessed dataset object that can be input into a TensorFlow model for training.

Perform the following steps:

1. Open a new Jupyter notebook to implement this exercise. Save the file as **Exercise2-05.ipnyb**.

2. In a new Jupyter Notebook cell, import the **tensorflow** and **os** libraries:

```
import tensorflow as tf
import os
```

3. Create a function that will load an audio file using TensorFlow's **read_file** function and **decode_wav** function, respectively. Return the transpose of the resultant tensor:

```
def load_audio(file_path, sample_rate=44100):
    # Load audio at 44.1kHz sample-rate
    audio = tf.io.read_file(file_path)
    audio, sample_rate = tf.audio.decode_wav\
                        (audio,\
                         desired_channels=-1,\
                         desired_samples=sample_rate)
    return tf.transpose(audio)
```

4. Load in the paths to the audio data as a list using **os.list_dir**:

```
prefix = " ../Datasets/data_speech_commands_v0.02"\
        "/zero/"
paths = [os.path.join(prefix, path) for path in \
            os.listdir(prefix)]
```

5. Test the function by loading in the first audio file from the list and plotting it:

```
import matplotlib.pyplot as plt
audio = load_audio(paths[0])
plt.plot(audio.numpy().T)
plt.xlabel('Sample')
plt.ylabel('Value')
```

The output will be as follows:

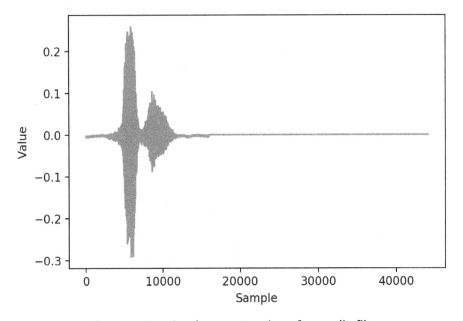

Figure 2.16: A visual representation of an audio file

The figure shows the waveform of the speech sample. The amplitude at a given time corresponds to the volume of the sound; high amplitude relates to high volume.

6. Create a function to generate the MFCCs from the audio data. First, apply the short-time Fourier transform passing in the audio signal as the first argument, the frame length set to **1024** as the second argument, the frame step set to **256** as the third argument, and the FFT length as the fourth parameter. Then, take the absolute value of the result to compute the spectrograms. The number of spectrogram bins is given by the length along the last axis of the short-time Fourier transform. Next, define the upper and lower bounds of the mel weight matrix as **80** and **7600** respectively and the number of mel bins as **80**. Then, compute the mel weight matrix using **linear_to_mel_weight_matrix** from TensorFlow's signal package. Next, compute the mel spectrograms via tensor contraction using TensorFlow's **tensordot** function along axis 1 of the spectrograms with the mel weight matrix. Then, take the log of the mel spectrograms before finally computing the MFCCs using TensorFlow's **mfccs_from_log_mel_spectrograms** function. Then, return the MFCCs from the function:

```
def apply_mfccs(audio, sample_rate=44100, num_mfccs=13):
    stfts = tf.signal.stft(audio, frame_length=1024, \
                           frame_step=256, \
                           fft_length=1024)
    spectrograms = tf.abs(stfts)
    num_spectrogram_bins = stfts.shape[-1]#.value

    lower_edge_hertz, upper_edge_hertz, \
    num_mel_bins = 80.0, 7600.0, 80
    linear_to_mel_weight_matrix = \
      tf.signal.linear_to_mel_weight_matrix\
      (num_mel_bins, num_spectrogram_bins, \
       sample_rate, lower_edge_hertz, upper_edge_hertz)
    mel_spectrograms = tf.tensordot\
                       (spectrograms, \
                        linear_to_mel_weight_matrix, 1)
```

```
    mel_spectrograms.set_shape\
    (spectrograms.shape[:-1].concatenate\
    (linear_to_mel_weight_matrix.shape[-1:]))

    log_mel_spectrograms = tf.math.log\
                        (mel_spectrograms + 1e-6)

    #Compute MFCCs from log_mel_spectrograms
    mfccs = tf.signal.mfccs_from_log_mel_spectrograms\
            (log_mel_spectrograms)[..., :num_mfccs]
    return mfccs
```

7. Apply the function to generate the MFCCs for the audio data you loaded in
   Step 5:

```
mfcc = apply_mfccs(audio)
plt.pcolor(mfcc.numpy()[0])
plt.xlabel('MFCC log coefficient')
plt.ylabel('Sample Value')
```

The output will be as follows:

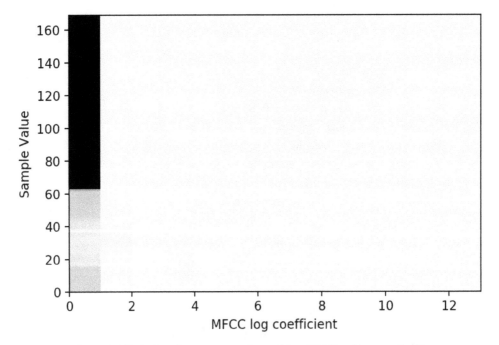

Figure 2.17: A visual representation of the MFCCs of an audio file

The preceding plot shows the MFCC values on the *x* axis and various points of the audio sample on the *y* axis. MFCCs are a different representation of the raw audio signal displayed in *Step 5* that has been proven to be useful in applications related to speech recognition.

8. Load **AUTOTUNE** so that you can use all the available threads of the CPU. Create a function that will take a dataset object, shuffle it, load the audio using the function you created in *Step 3*, generate the MFCCs using the function you created in *Step 6*, repeat the dataset object, batch it, and prefetch it. Use **AUTOTUNE** to prefetch with a buffer size based on your available CPU:

```
AUTOTUNE = tf.data.experimental.AUTOTUNE

def prep_ds(ds, shuffle_buffer_size=1024, \
            batch_size=64):
    # Randomly shuffle (file_path, label) dataset
    ds = ds.shuffle(buffer_size=shuffle_buffer_size)
    # Load and decode audio from file paths
    ds = ds.map(load_audio, num_parallel_calls=AUTOTUNE)
    # generate MFCCs from the audio data
    ds = ds.map(apply_mfccs)
    # Repeat dataset forever
    ds = ds.repeat()
    # Prepare batches
    ds = ds.batch(batch_size)
    # Prefetch
    ds = ds.prefetch(buffer_size=AUTOTUNE)

    return ds
```

9. Generate the training dataset using the function you created in *Step 8*. To do this, create a dataset object using TensorFlow's **from_tensor_slices** function and pass in the paths to the audio files. After that, you can use the function you created in *Step 8*:

```
ds = tf.data.Dataset.from_tensor_slices(paths)
train_ds = prep_ds(ds)
```

10. Take the first batch of the dataset and print it out:

```
for x in train_ds.take(1):\
    print(x)
```

The output will be as follows:

```
tf.Tensor(
[[[[-4.49402695e+01  6.66246712e-01  3.23954558e+00 ... -6.80628955e-01
     7.53509879e-01  1.03011954e+00]
   [-5.76376343e+01 -5.59932113e-01  1.81934524e+00 ...  1.11307764e+00
    -6.44906331e-03 -1.05823982e+00]
   [-3.97187042e+01 -3.85361761e-01  2.18436956e+00 ...  1.52272964e+00
     5.28275631e-02  6.03379756e-02]

   ...

   [-1.74754272e+02 -7.82595671e-05  8.29657256e-06 ...  8.80490370e-06
     6.48393325e-06 -2.29595958e-06]
   [-1.74754272e+02 -7.82595671e-05  8.29657256e-06 ...  8.80490370e-06
     6.48393325e-06 -2.29595958e-06]
   [-1.74754272e+02 -7.82595671e-05  8.29657256e-06 ...  8.80490370e-06
     6.48393325e-06 -2.29595958e-06]]]


  [[[-6.25524759e+00 -8.96556020e-01  1.73362285e-01 ... -1.67588305e+00
    -9.09781218e-01  7.50396788e-01]
   [-6.30506134e+00 -1.13798833e+00  2.74822772e-01 ... -6.23754025e-01
    -5.79600930e-01  5.42514861e-01]
   [-8.25854206e+00 -2.62160587e+00  1.48269266e-01 ... -7.27678180e-01
    -1.36761755e-01  1.12893879e+00]

   ...

   [-1.74754272e+02 -7.82595671e-05  8.29657256e-06 ...  8.80490370e-06
     6.48393325e-06 -2.29595958e-06]
   [-1.74754272e+02 -7.82595671e-05  8.29657256e-06 ...  8.80490370e-06
     6.48393325e-06 -2.29595958e-06]
   [-1.74754272e+02 -7.82595671e-05  8.29657256e-06 ...  8.80490370e-06
     6.48393325e-06 -2.29595958e-06]]]
```

Figure 2.18: A batch of the audio data after the MFCCs have been generated

The output shows the first batch of MFCC spectrum values in tensor form.

In this exercise, you imported audio data. You processed the dataset and batched the dataset so that it is appropriate for large-scale training. This method was a comprehensive approach in which the data was loaded and converted into the frequency domain, spectrograms were generated, and then finally the MFCCs were generated.

In the next activity, you will load in audio data and take the absolute value of the input, followed by scaling the values logarithmically. This will ensure that there are no negative values in the dataset. You will use the same audio dataset that you used in *Exercise 2.05, Loading Audio Data for TensorFlow Models*, that is, Google's speech commands dataset. This dataset consists of 1-second-long utterances of words.

ACTIVITY 2.03: LOADING AUDIO DATA FOR BATCH PROCESSING

In this activity, you will load audio data for batch processing. The audio preprocessing techniques that will be performed include taking the absolute value and using the logarithm of 1 plus the value. This will ensure the resulting values are non-negative and logarithmically scaled. The result will be a preprocessed dataset object that can be input into a TensorFlow model for training.

The steps for this activity are as follows:

1. Open a new Jupyter notebook to implement this activity.

2. Import the TensorFlow and **os** libraries.

3. Create a function that will load and then decode an audio file using TensorFlow's **read_file** function followed by the **decode_wav** function, respectively. Return the transpose of the resultant tensor from the function.

4. Load the file paths into the audio data as a list using **os.list_dir**.

5. Create a function that takes a dataset object, shuffles it, loads the audio using the function you created in *step 2*, and applies the absolute value and the **log1p** function to the dataset. This function adds **1** to each value in the dataset and then applies the logarithm to the result. Next, repeat the dataset object, batch it, and prefetch it with a buffer size equal to the batch size.

6. Create a dataset object using TensorFlow's **from_tensor_slices** function and pass in the paths to the audio files. Then, apply the function you created in *Step 4* to the dataset created in *Step 5*.

7. Take the first batch of the dataset and print it out.

8. Plot the first audio file from the batch.

The output will look as follows:

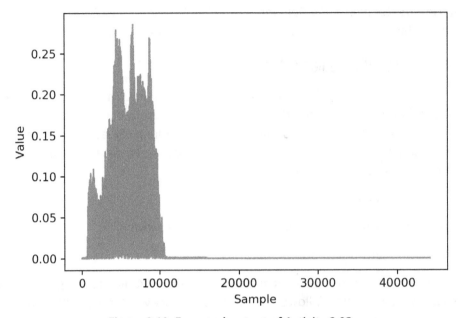

Figure 2.19: Expected output of Activity 2.03

> **NOTE**
>
> The solution to this activity can be found on page 464.

In this activity, you learned how to load and preprocess audio data in batches. You used most of the functions that you used in *Exercise 2.05, Loading Audio Data for TensorFlow Models*, to load in the data and decode the raw data. The difference between *Exercise 2.05, Loading Audio Data for TensorFlow Models*, and *Activity 2.03, Loading Audio Data for Batch Processing*, is the preprocessing steps; *Exercise 2.05, Loading Audio Data for TensorFlow Models*, involved generating MFCCs for the audio data, whereas *Activity 2.03, Loading Audio Data for Batch Processing*, involved scaling the data logarithmically. Both demonstrate common preprocessing techniques that can be used for all applications involving modeling on audio data.

In this section, you have explored how audio data can be loaded in batches for TensorFlow modeling. The comprehensive approach demonstrated many advanced signal processing techniques that should provide practitioners who wish to use audio data for their own applications with a good starting point.

SUMMARY

In this chapter, you learned how to load different forms of data and perform some preprocessing steps for a variety of data types. You began with tabular data in the form of a CSV file. Since the dataset consisted of a single CSV file, you utilized the pandas library to load the file into memory.

You then proceeded to preprocess the data by scaling the fields and converting all the fields into numerical data types. This is important since TensorFlow models can only be trained on numerical data, and the training process is improved in terms of speed and accuracy if all the fields are of the same scale.

Next, you explored how to load the image data. You batched the data so that you did not have to load in the entire dataset at once, which allowed you to augment the images. Image augmentation is useful as it increases the effective number of training examples and can help make a model more robust.

You then learned how to load in text data and took advantage of pretrained models. This helped you embed text into vectors that retain contextual information about the text. This allowed text data to be input into TensorFlow models since they require numerical tensors as inputs.

Finally, the final section covered how to load and process audio data and demonstrated some advanced signal processing techniques, including generating MFCCs, which can be used to generate informationally dense numerical tensors that can be input into TensorFlow models.

Loading and preprocessing data so that it can be input into machine learning models is an important and necessary first step to training any machine learning model. In the next chapter, you will explore many resources that TensorFlow provides to aid in the development of model building.

3

TENSORFLOW DEVELOPMENT

OVERVIEW

TensorFlow provides many resources for creating efficient workflows when developing data science and machine learning applications. In this chapter, you will learn how to use TensorBoard to visualize TensorFlow graphs and operations, TensorFlow Hub to access a community of users (a great source of pre-trained models), and Google Colab, which is a collaborative environment for developing code with others. You will use these tools to accelerate development by maximizing computational resources, transferring knowledge from pre-trained models, and visualizing all aspects of the model-building process.

INTRODUCTION

In the previous chapter, you learned how to load and process a variety of data types so that they can be used in TensorFlow modeling. This included tabular data from CSV files, image data, text data, and audio files. By the end of the chapter, you were able to process all these data types and produce numerical tensors from them that can be input for model training.

In this chapter, you will learn about TensorFlow resources that will aid you in your model building and help you create performant machine learning algorithms. You will explore the practical resources that practitioners can utilize to aid their development workflow, including TensorBoard, TensorFlow Hub, and Google Colab. TensorBoard is an interactive platform that offers a visual representation of the computational graphs and data produced during the TensorFlow development process. The platform solves the problem of visualizing various data types that is common in machine learning. The visualization toolkit can plot model evaluation metrics during the model-building process, display images, play audio data, and perform many more tasks that would otherwise require writing custom functions. TensorBoard provides simple functions for writing logs, which are subsequently visualized in a browser window.

TensorFlow Hub is an open source library of pre-trained machine learning models with a code base that's available for all to use and modify for their own applications. Models can be imported directly into code through dedicated libraries and can be viewed at https://tfhub.dev/. TensorFlow Hub allows users to use state-of-the-art models created by experts in the field and can result in massively reduced training times for models that incorporate pre-trained models as part of a user's model.

For example, the platform contains the ResNet-50 model, a 50-layer **Artificial Neural Network** (**ANN**) that achieved first place on the ILSVRC 2015 classification task, a competition to classify images into 1,000 distinct classes. The network has over 23 million trainable parameters and was trained on more than 14 million images. Training this model from scratch on an off-the-shelf laptop to achieve something close to the accuracy of the pre-trained model on TensorFlow Hub would take days. It is for this reason that the ability to utilize TensorFlow Hub models can accelerate development.

The final resource you will learn about in this chapter is Google Colab, which is an online development environment for executing Python code and creating machine learning algorithms on Google servers. The environment even has access to hardware that contains **Graphics Processing Units** (**GPUs**) and **Tensor Processing Units** (**TPUs**) that can speed up model training free of charge. Google Colab is available at https://colab.research.google.com/.

Google Colab resolves the issue of setting up a development environment for creating machine learning models that can be shared with others. For example, multiple machine learning practitioners can develop the same model and train the model on one hardware instance, as opposed to having to run the instance with their own resources. As the name suggests, the platform fosters collaboration among machine learning practitioners.

Now, let's explore TensorBoard, a resource that helps practitioners understand and debug their machine learning workflow.

TENSORBOARD

TensorBoard is a visualization toolkit used to aid in machine learning experimentation. The platform has dashboard functionality for visualizing many of the common data types that a data science or machine learning practitioner may need at once, such as scalar values, image batches, and audio files. While such visualizations can be created with other plotting libraries, such as **matplotlib** or **ggplot**, TensorBoard combines many visualizations in an easy-to-use environment. Moreover, all that is required to create the visualizations is to log the trace during the building, fitting, and evaluating steps. TensorBoard helps in the following tasks:

* Visualizing the model graph to view and understand the model's architecture:

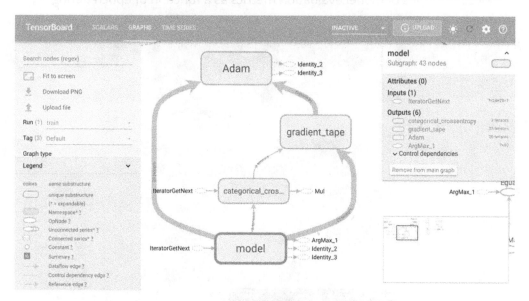

Figure 3.1: A visual representation of model graphs and functions in TensorBoard

* Viewing histograms and distributions of variables and tracking how they change over time.

- Displaying images, text, and audio data. For example, the following figure displays images from the Fashion MNIST dataset (https://www.tensorflow.org/datasets/catalog/fashion_mnist):

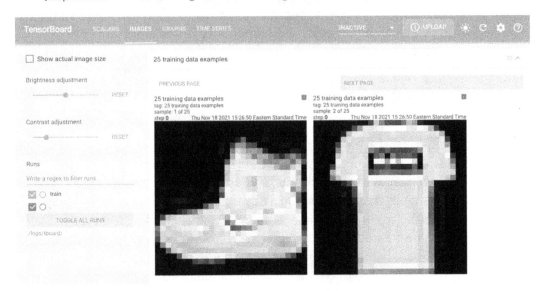

Figure 3.2: Viewing images in TensorBoard

- Plotting graphs of model evaluation metrics as a function of epoch during model training:

Figure 3.3: Plotting model evaluation metrics in TensorBoard

- Dimensionality reduction for visualizing embedding vectors:

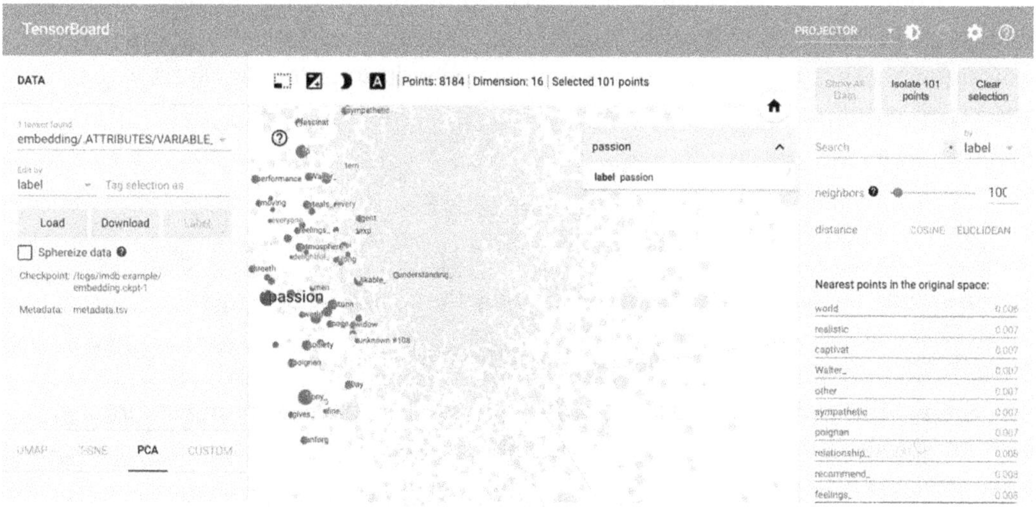

Figure 3.4: Visualizing embedding vectors in TensorBoard

TensorBoard creates visualizations from logs that are written during the development process. In order to create the logs to visualize the graph, a file writer object needs to be initialized within your development code, providing the location for the logs as an argument. The file writer is typically created at the beginning of a Jupyter notebook or equivalent development environment before any logs are written. This can be done as follows:

```
logdir = 'logs/'
writer = tf.summary.create_file_writer(logdir)
```

In the preceding code, the directory for writing the logs is set, and if this directory does not already exist a new one will be created automatically in the working directory after you run the preceding code. The file writer object writes a file to the log directory when the logs are exported. To begin tracing, the following code must be executed:

```
tf.summary.trace_on(graph=True, profiler=True)
```

The preceding command turns on the trace that records the computation graph that occurs from the time the command is executed. Without turning on the trace, nothing is logged, and so, nothing can be visualized in TensorBoard. Once the tracing of the computational graph is complete, the logs can be written to the log directory using the file writer object, as follows:

```
with writer.as_default():
    tf.summary.trace_export(name="my_func_trace",\
                            step=0, profiler_outdir=logdir)
```

When writing the logs, you will need to employ the following parameters:

- **name**: This parameter describes the name of the summary.

- **step**: This parameter describes the monotonic step value for the summary and can be set to **0** if the object does not change over time.

- **profiler_outdir**: This parameter describes the location to write the logs and is required if not provided when the file writer object is defined.

After logs have been written to a directory, TensorBoard can be launched through the command line using the following command, thereby passing in the directory for the logs as the **logdir** parameter:

```
tensorboard --logdir=./logs
```

Some versions of Jupyter Notebooks allow TensorBoard to be run directly within the notebook. However, library dependencies and conflicts can often prevent TensorBoard from running in notebook environments, in which case you can launch TensorBoard in a separate process from the command line. In this book, you will be using TensorFlow version 2.6 and TensorBoard version 2.1, and you will always use the command line to launch TensorBoard.

In the first exercise, you will learn how to use TensorBoard to visualize a graph process. You will create a function to perform tensor multiplication and then visualize the computational graph in TensorBoard.

EXERCISE 3.01: USING TENSORBOARD TO VISUALIZE MATRIX MULTIPLICATION

In this exercise, you will perform matrix multiplication of **7x7** matrices with random values and trace the computation graph and profiling information. Following that, you will view the computation graph using TensorBoard. This exercise will be performed in a Jupyter notebook. Launching TensorBoard will require running a command on the command line, as shown in the final step.

Follow these steps:

1. Open a new Jupyter notebook and import the TensorFlow library, and then set a seed for reproducibility. Since you are generating random values, setting a seed will ensure that the values generated are the same if the seed set is the same each time the code is run:

```
import tensorflow as tf
tf.random.set_seed(42)
```

2. Create a **file_writer** object and set the directory for which the logs will be stored:

```
logdir = 'logs/'
writer = tf.summary.create_file_writer(logdir)
```

3. Create a TensorFlow function to multiply two matrices together:

```
@tf.function
def my_matmult_func(x, y):
    result = tf.matmul(x, y)
    return result
```

4. Create sample data in the form of two tensors with the shape **7x7** with random variables:

```
x = tf.random.uniform((7, 7))
y = tf.random.uniform((7, 7))
```

5. Turn on graph tracing using TensorFlow's **summary** class:

```
tf.summary.trace_on(graph=True, profiler=True)
```

6. Apply the function that was created in *step 3* to the sample tensors that were created in *step 4*. Next, export the trace to the **log** directory, set the **name** argument for the graph for reference, and the **log** directory for the **profiler_outdir** argument. The **step** argument indicates the monotonic step value for the summary; the value should be nonzero if the values being traced vary, in which case they can be visualized with a step size dictated by this argument. For static objects, such as your graph trace here, it should be set to zero:

```
z = my_matmult_func(x, y)
with writer.as_default():
    tf.summary.trace_export(name="my_func_trace",\
                            step=0,\
                            profiler_outdir=logdir)
```

7. Finally, launch TensorBoard in the current working directory using the command line to view a visual representation of the graph. TensorBoard can be viewed in a web browser by visiting the URL that is provided after launching TensorBoard:

```
tensorboard --logdir=./logs
```

For those running Windows, in the Anaconda prompt, run the following:

```
tensorboard --logdir=logs
```

By running the preceding code, you will be able to visualize the following model graph:

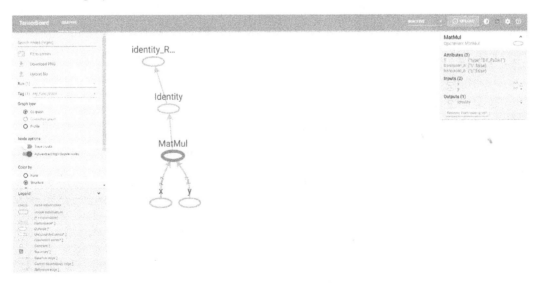

Figure 3.5: A visual representation of matrix multiplication in TensorBoard

In TensorBoard, you can view the process of a tensor multiplying the two matrices to produce another matrix. By selecting the various elements, you can view information about each individual object in the computational graph, depending on the type of object. Here, you have created two tensors, named **x** and **y**, represented by the nodes at the bottom. By selecting one of the nodes, you can view attributes about the tensor, including its data type (`float`), its user-specified name (**x** or **y**), and the name of the output node (`MatMul`). These nodes representing the input tensors are then input into another node representing the tensor multiplication process labeled `MatMul` after the TensorFlow function. Selecting this node reveals attributes of the function, including the input arguments, the input nodes (**x** and **y**), and the output node (`Identity`). The final two nodes, labeled `Identity` and `identity_RetVal`, represent the creation of the output tensor.

In this exercise, you used TensorBoard to visualize a computational graph. You created a simple function to multiply two tensors together and you recorded the process by tracing the graph and logging the results. After logging the graph, you were able to visualize it by launching TensorBoard and directing the tool to the location of the logs.

In the first activity, you will practice using TensorBoard to visualize a more complicated tensor transformation. In fact, any tensor process and transformation can be visualized in TensorBoard and the process demonstrated in the previous exercise is a good guide for creating and writing logs.

ACTIVITY 3.01: USING TENSORBOARD TO VISUALIZE TENSOR TRANSFORMATIONS

You are given two tensors of shape **5x5x5**. You are required to create TensorFlow functions to perform a tensor transformation and view a visual representation of the transformation.

The steps you will take are as follows:

1. Import the TensorFlow library and set the seed to **42**.

2. Set a log directory and initialize a file writer object to write the trace.

3. Create a TensorFlow function to multiply two tensors, add a value of **1** to all elements in the resulting tensor using the **ones_like** function to create a tensor of the same shape as the result of the matrix multiplication. Then, apply a **sigmoid** function to each value of the tensor.

4. Create two tensors with the shape **5x5x5**.

5. Turn on graph tracing.

6. Apply the function to the two tensors and export the trace to the log directory.

7. Launch TensorBoard in the command line and view the graph in a web browser:

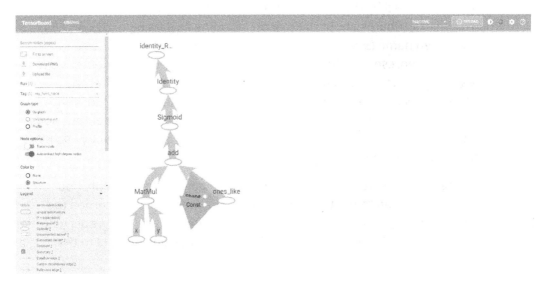

Figure 3.6: A visual representation of tensor transformation in TensorBoard

NOTE

The solution to this activity can be found on page 468.

However, TensorBoard is not only for visualizing computational graphs. Images, scalar variables, histograms, and distributions can all be viewed in TensorBoard by writing them to the log directory using the appropriate TensorFlow **summary** method. For example, images can be written to the logs as follows:

```
with file_writer.as_default():
    tf.summary.image("Training data", training_images, step=0)
```

The output of this will be a file added to the log directory named **Training data** that contains the images written by the file writer. Images can be viewed in TensorBoard by selecting the tab labeled **IMAGES**.

In the same manner, scalar variables can be written to the logs for viewing in TensorBoard as follows:

```
with file_writer.as_default():
    tf.summary.scalar('scalar variable', variable, step=0)
```

Audio files can be written to the logs for playback in TensorBoard in the following way:

```
with file_writer.as_default():
    tf.summary.audio('audio file', data, sample_rate=44100, \
                     step=0)
```

A histogram can be logged by passing in data as follows:

```
with file_writer.as_default():
    tf.summary.histogram('histogram', data, step=0)
```

In each of these examples of writing data to the logs, the **step** argument is set to zero since this is a required argument and must not be null. Setting the argument to zero indicates that the value is static and does not change with time. Each data type will be visible in a different tab in TensorBoard.

In the next exercise, you will write images to TensorBoard so that they can be viewed directly within the platform. With TensorBoard, this becomes a facile process that otherwise would require writing custom code to view images. You may want to visualize images of batches to verify the labels, check the augmentation process, or validate the images in general.

EXERCISE 3.02: USING TENSORBOARD TO VISUALIZE IMAGE BATCHES

In this exercise, you will use TensorBoard to view image batches. You will create a file writer and a data generator for the images, and then write one batch of images to the log files. Finally, you will view the images in TensorBoard.

> **NOTE**
>
> You can find the images in the **image_data** folder here:
> https://packt.link/1ue46.

Follow these steps:

1. Import the TensorFlow library and the **ImageDataGenerator** class:

```
import tensorflow as tf
from tensorflow.keras.preprocessing.image import \
    ImageDataGenerator
```

2. Create a **file_writer** object and set the directory to which the logs will be stored:

```
logdir = 'logs/'
writer = tf.summary.create_file_writer(logdir)
```

3. Initialize an **ImageDataGenerator** object:

```
train_datagen = ImageDataGenerator(rescale = 1./255)
```

4. Use the data generator's **flow_from_directory** method to create a batch image loader:

```
batch_size = 25
training_set = train_datagen.flow_from_directory\
                 ('image_data',\
                  target_size = (224, 224),\
                  batch_size = batch_size,\
                  class_mode = 'binary')
```

> **NOTE**
>
> Make sure you change the path (highlighted) to the location of the directory on your system. If you're running the Jupyter notebook from the same directory where the dataset is stored, you can run the preceding code without any modification.

5. Take the images from the first batch and write them to the logs using the file writer:

```
with file_writer.as_default():
    tf.summary.image("Training data", \
                     next(training_set)[0], \
                     max_outputs=batch_size, \
                     step=0)
```

6. Launch TensorBoard in the command line to view a visual representation of the graph. TensorBoard can be viewed in a web browser by visiting the URL that is provided after launching TensorBoard. The default URL provided is **http://localhost:6006/**:

```
tensorboard --logdir=./logs
```

For those running Windows, in the Anaconda prompt, run the following:

```
tensorboard --logdir=logs
```

Images in the directory will be displayed in TensorBoard as follows:

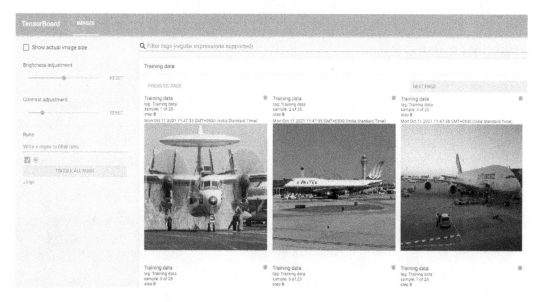

Figure 3.7: Viewing a batch of images in TensorBoard

NOTE

Images on your system may vary.

The result in TensorBoard is the images from the first batch. You can see that they are images of boats and planes. TensorBoard also provides you with the ability to adjust the brightness and contrast of the images; however, that affects only the images in TensorBoard and not the underlying image data.

In this exercise, you viewed a batch of images from an image data generator using TensorBoard. This is an excellent way to verify the quality of your training data. It may not be necessary to verify every image for quality, but sample batches can be viewed easily using TensorBoard.

This section has introduced one resource that TensorFlow offers to help data science and machine learning practitioners understand and visualize their data and algorithms: TensorBoard. You have used the resource to visualize computational graphs and image batches. In the next section, you will explore TensorFlow Hub, which is a repository for machine learning modules that can be accessed and incorporated into custom applications easily. The models are created by experts in the field, and you will learn how to access them for your own applications.

TENSORFLOW HUB

TensorFlow Hub is an online repository of machine learning modules. The modules contain assets with the associated weights that are needed to use any model (for instance, for predictions or transfer learning) where the knowledge gained in training one model is used to solve a different but related problem. These modules can be used directly to create applications that they were trained for, or they can be used as a starting point to build new applications. The platform can be visited at the following URL: https://tfhub.dev/. When you visit the website, you will be greeted by the following page:

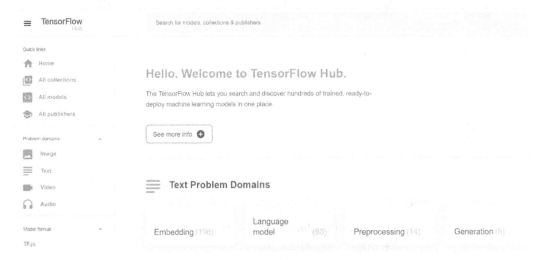

Figure 3.8: TensorFlow Hub home page

Once here, you can browse through models of various domains. The most popular domains include image, text, and video; many models exist for these domains:

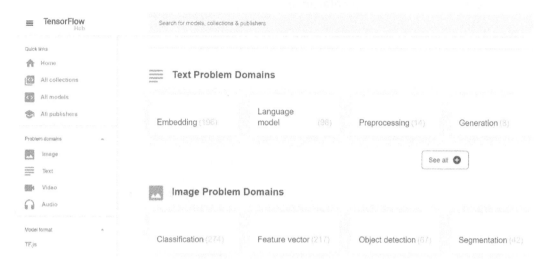

Figure 3.9: The model domains available on TensorFlow Hub

There are many models available on TensorFlow Hub that take images as their input data. These models are generally created for tasks including image classification, segmentation, embedding, generation, augmentation, and style transfer. Models created for text data are generally used for text embedding, and models used on video data are used for video classification. There are also audio data models for tasks including command detection and pitch extraction. TensorFlow Hub is consistently updated with new state-of-the-art models that can be used for all sorts of applications.

Selecting a model will land you on the following page, which will tell you information about the model, such as the size of the model, its architecture, the dataset on which it was trained, and the URL for reference:

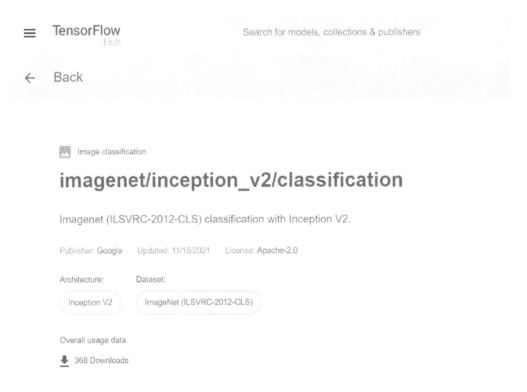

Figure 3.10: The page of a TensorFlow Hub model

When referencing models for your own applications, you will need the URL of the model's page to load it in.

Models can be accessed in notebook environments from TensorFlow Hub by utilizing the **tensorflow_hub** library. The library can be imported as follows:

```
import tensorflow_hub as hub
```

Models can be loaded by utilizing the library's **load** function and passing in the reference URL of the model:

```
module = hub.load("https://tfhub.dev/google/imagenet"\
                  "/inception_resnet_v2/classification/4")
```

Assets of the model's module, such as its architecture, can be viewed by accessing the **signatures** attribute. Each model may have different keys within the **signatures** attribute; however, much of the pertinent information will be contained within the **default** key:

```
model = module.signatures['default']
```

The model can also be used directly in training by treating the whole model like a single Keras layer using the **KerasLayer** method:

```
layer = hub.KerasLayer("https://tfhub.dev/google/imagenet"\
                    "/inception_resnet_v2/classification/4")
```

The process of using the model as layers for your own application is known as **transfer learning**, which will be explored in more depth in later chapters.

Viewing a model in TensorFlow Hub can be done by writing the model graph to the logs using a file writer as follows:

```
from tensorflow.python.client import session
from tensorflow.python.summary import summary
from tensorflow.python.framework import ops

with session.Session(graph=ops.Graph()) as sess:
    file_writer = summary.FileWriter(logdir)
    file_writer.add_graph(model.graph)
```

In the following exercise, you will download a model from TensorFlow Hub. After loading in the model, you will view the model's architecture using TensorBoard.

EXERCISE 3.03: DOWNLOADING A MODEL FROM TENSORFLOW HUB

In this exercise, you will download a model from TensorFlow Hub and then view the architecture of the model in TensorBoard. The model that will be downloaded is the **InceptionV3** model. This model was created in TensorFlow 1 and so requires some additional steps for displaying the model details as we're using TensorFlow 2. This model contains two parts: a part that includes convolutional layers to extract features from the images, and a classification part with fully connected layers.

The distinct layers will be visible in TensorBoard as they have been named appropriately by the original author.

> **NOTE**
>
> You can get the **InceptionV3** model here:
> https://tfhub.dev/google/imagenet/inception_v3/classification/5.

Follow these steps to complete this exercise:

1. Import the following libraries from TensorFlow:

    ```
    import tensorflow as tf
    import tensorflow_hub as hub
    from tensorflow.python.client import session
    from tensorflow.python.summary import summary
    from tensorflow.python.framework import ops
    ```

    The TensorFlow and TensorFlow Hub libraries are required to import and build the model, and the other classes from the TensorFlow library are required to visualize models that are created in TensorFlow 1 using TensorFlow 2, which is what you are using in this book.

2. Create a variable for the logs to be stored:

    ```
    logdir = 'logs/'
    ```

3. Load in a model module by using the **load** method from the **tensorflow_hub** library and pass in the URL for the model:

    ```
    module = hub.load('https://tfhub.dev/google/imagenet'\
                      '/inception_v3/classification/5')
    ```

4. Load the model from the **signatures** attribute of the module:

    ```
    model = module.signatures['default']
    ```

5. Write the model graph to TensorBoard using a file writer:

    ```
    with session.Session(graph=ops.Graph()) as sess:
        file_writer = summary.FileWriter(logdir)
        file_writer.add_graph(model.graph)
    ```

6. Launch TensorBoard in the command line to view a visual representation of the graph. TensorBoard can be viewed in a web browser by visiting the URL that is provided after launching TensorBoard:

```
tensorboard --logdir=./logs
```

For those running Windows, in the Anaconda prompt, run the following:

```
tensorboard --logdir=logs
```

You should get something like the following image:

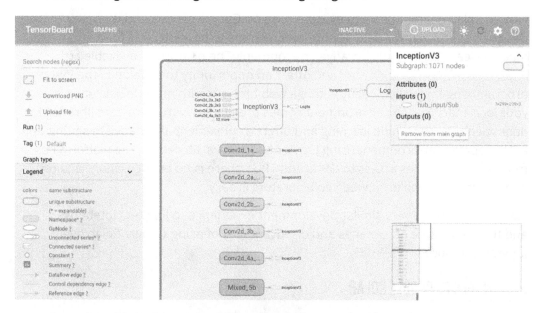

Figure 3.11: The architecture of the InceptionV3 model as viewed in TensorBoard

The result in TensorBoard is the architecture of the **InceptionV3** model. Here, you can view all the details about each layer of the model, including the input, output, and activation functions.

In this exercise, you successfully downloaded a model into a Jupyter notebook environment using the TensorFlow Hub library. Once the model was loaded into the environment, you visualized the architecture of the model using TensorBoard. This can be a helpful way to visualize your model's architecture for debugging purposes.

In this section, you have explored how to use TensorFlow Hub as a way to utilize the many brilliant models that have been created by experts in the machine learning field. As you will discover in later chapters, these models can be used to solve slightly different applications than those for which they were developed; this is known as transfer learning. In the next section, you will learn how to use Google Colab, an environment similar to Jupyter Notebooks that can be used to collaboratively develop applications in Python online, on Google servers.

GOOGLE COLAB

Google Colab enables users to execute code on Google servers and is designed specifically for data science practitioners to develop code for machine learning in a collaborative environment. The platform is available at https://colab.research.google.com/ and offers an opportunity to develop in the Python programming language directly within a web browser with no code executing on your local machine. The environment comes pre-loaded with up-to-date libraries for data science and machine learning and offers a convenient alternative to setting up a development environment using Jupyter Notebooks. Moreover, the platform has a free tier that includes access to GPUs and TPUs, there is no configuration required, and sharing notebooks between collaborators is easy.

Google Colab has a very similar development experience to Jupyter Notebooks, and there are some advantages and disadvantages of using Google Colab over Jupyter Notebooks.

ADVANTAGES OF GOOGLE COLAB

The following are a few of the main advantages of using Google Colab:

- **Collaborative**: Many users can access the same notebook and work collaboratively together.

- **Managed environment**: Google Colab runs on Google servers, which can be helpful if local computational resources are limited. There is no need to set up a development environment since many packages come pre-installed.

- **Easy accessibility**: Google Colab saves directly to Google Drive, offering seamless integration. Since the notebooks are saved in the cloud, they are available wherever Google Drive can be accessed.

- **Accelerated training times**: GPU and TPU servers are available, which can offer accelerated training times for training machine learning models, especially ANNs with many hidden layers.

- **Interactive widgets**: Widgets can be added to a notebook that can offer a way to easily vary input parameters and variables in an interactive manner.

DISADVANTAGES OF GOOGLE COLAB

The following are a few of the disadvantages of using Google Colab:

- **Restrained runtime**: Only two versions of TensorFlow are available on Google Colab, 1.X and 2.X, and they are updated, so specific functions may change over time, resulting in broken code. Additionally, the versions of TensorFlow may not interact well with other packages.

- **Internet dependence**: Since the Python code is executed on Google servers, Google Colab can only be accessed with an internet connection.

- **No automatic save**: Notebooks must be saved consistently, which is different from the automatic saving of Jupyter Notebooks.

- **Session timeout**: Notebooks running on the virtual machines have a maximum lifetime of 12 hours and environments that are left idle for too long will be disconnected.

- **Interactive library**: Libraries that contain interactive elements such as OpenCV or `geoplotlib` may not be capable of displaying interactive elements due to incompatibilities with the pre-loaded libraries.

DEVELOPMENT ON GOOGLE COLAB

Since Google Colab uses notebooks, the development environment is very similar to Jupyter Notebooks. In fact, IPython notebooks can be loaded into Google Colab. They can be loaded in via direct upload, Google Drive, or a GitHub repository. Alternatively, the platform provides example notebooks to get started. When you navigate to the platform, https://colab.research.google.com/, you will be greeted by the following screen, which provides notebooks to open or the option to select a new notebook to begin developing:

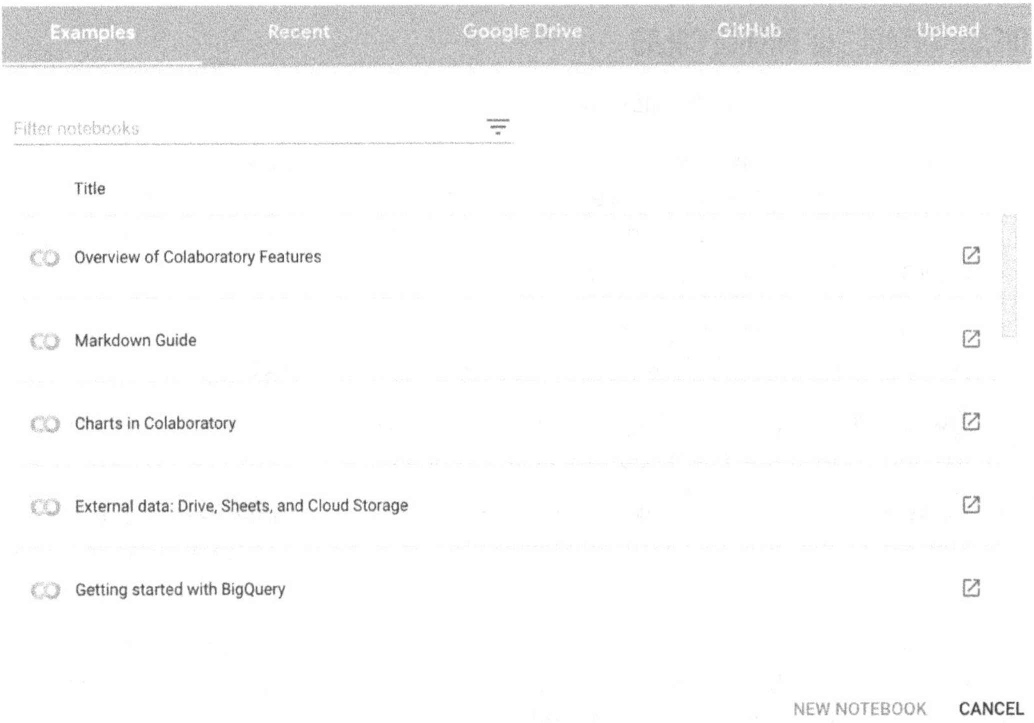

Figure 3.12: The home page of Google Colab

If a new notebook is selected, you are greeted by the following screen, which may be very reminiscent of developing in Jupyter Notebooks and has many of the same features. You can create code or text snippets in the exact same way and many practitioners find the transition from Jupyter seamless:

Figure 3.13: A blank notebook in Google Colab

In the next exercise, you will use Google Colab to import and manipulate data. One of the main differences between working in Google Colab compared to Jupyter Notebooks is that by working in Google Colab, you are developing on a remote server. This means that any data for analysis or training models must either be loaded on Google Drive or available directly online. In the following exercise, you will import CSV data directly from a GitHub repository for this book.

EXERCISE 3.04: USING GOOGLE COLAB TO VISUALIZE DATA

In this exercise, you will load a dataset from a GitHub repository that has bias correction data for next-day maximum and minimum air temperature forecasts for Seoul, South Korea.

> **NOTE**
>
> You can find the `Bias_correction_ucl.csv` file here: https://packt.link/8kP3j.

To perform the exercise, you will have to navigate to https://colab.research.google.com/ and create a new notebook to work in. You will need to connect to a GPU-enabled environment to speed up TensorFlow operations such as tensor multiplication. Once the data has been loaded into the development environment, you will view the first five rows. Next, you'll drop the `Date` field since matrix multiplication requires numerical fields. Then, you will perform tensor multiplication of the dataset with a tensor or uniformly random variables.

Follow these steps to complete this exercise:

1. Import TensorFlow and check the version of the library:

```
import tensorflow as tf
print('TF version:', tf.__version__)
```

You should get the version of the TensorFlow library:

$$TF \ version: \ 2.6.0$$

Figure 3.14: The output of the version of TensorFlow available in Google Colab

2. Navigate to the **Edit** tab, go to **Notebook Settings**, and then select **GPU** from the **Hardware Acceleration** dropdown. Verify that the GPU is enabled by displaying the GPU device name:

```
tf.test.gpu_device_name()
```

You should get the name of the GPU device:

$$'/device:GPU:0'$$

Figure 3.15: The GPU device name

3. Import the **pandas** library and load in the dataset directly from the GitHub repository:

```
import pandas as pd
df = pd.read_csv('https://raw.githubusercontent.com'\
                 '/PacktWorkshops/The-TensorFlow-Workshop'\
                 '/master/Chapter03/Datasets'\
                 '/Bias_correction_ucl.csv')
```

4. View the first five rows of the dataset using the **head** method:

```
df.head()
```

You should get the following output:

	station	Date	Present_Tmax	Present_Tmin	LDAPS_RHmin	LDAPS_RHmax	LDAPS_Tmax_lapse	LDAPS_Tmin_lapse
0	1.0	2013-06-30	28.7	21.4	58.255688	91.116364	28.074101	23.006936
1	2.0	2013-06-30	31.9	21.6	52.263397	90.604721	29.850689	24.035009
2	3.0	2013-06-30	31.6	23.3	48.690479	83.973587	30.091292	24.565633
3	4.0	2013-06-30	32.0	23.4	58.239788	96.483688	29.704629	23.326177
4	5.0	2013-06-30	31.4	21.9	56.174095	90.155128	29.113934	23.486480

5 rows × 25 columns

Figure 3.16: The output of the first five rows of the DataFrame

5. Drop the **Date** field since you'll be performing matrix multiplication, which requires numerical fields:

```
df.drop('Date', axis=1, inplace=True)
```

6. Import NumPy, convert the DataFrame to a NumPy array, and then create a TensorFlow tensor of uniform random variables. The value of the first axis of the tensor will be equal to the number of fields of the dataset, and the second axis will be equal to **1**:

```
import numpy as np
df = np.asarray(df).astype(np.float32)
random_tensor = tf.random.normal((df.shape[1],1))
```

7. Perform tensor multiplication on the dataset and the random tensor using TensorFlow's **matmul** function and print the result:

```
tf.matmul(df, random_tensor)
```

You should get output like the following:

```
<tf.Tensor: shape=(7752, 1), dtype=float32, numpy=
array([[2517.5198],
       [2308.5957],
       [2317.9607],
       ...,
       [1790.7307],
       [     nan],
       [     nan]], dtype=float32)>
```

Figure 3.17: The output of the tensor multiplication

The result from executing the multiplication is a new tensor with the shape **7752x1**.

In this exercise, you learned how to use Google Colab. You observed that Google Colab provides a convenient environment to build machine learning models and comes pre-loaded with many of the libraries that may be needed for any machine learning application. You can also see that the latest versions of the libraries are used. Unfortunately, the versions of TensorFlow cannot be modified, so using Google Colab in production environments may not be the most appropriate application. However, it is great for development environments.

In the following activity, you will practice further how to use Google Colab in a development environment. You will use TensorFlow Hub in the same way that was achieved in Jupyter Notebooks. This activity will be similar to what was achieved in *Exercise 2.04, Loading Text Data for TensorFlow Models*, in which text data was processed by using a pre-trained word embedding model. Utilizing pre-trained models will be covered in future chapters, but this activity will show how easy it is to utilize a pre-trained model from TensorFlow Hub.

ACTIVITY 3.02: PERFORMING WORD EMBEDDING FROM A PRE-TRAINED MODEL FROM TENSORFLOW HUB

In this activity, you will practice working in the Google Colab environment. You will download a universal sentence encoder from TensorFlow Hub from the following URL: https://tfhub.dev/google/universal-sentence-encoder/4. Once the model has been loaded into memory, you will use it to encode some sample text.

Follow these steps:

1. Import TensorFlow and TensorFlow Hub and print the version of the library.

2. Set the handle for the module as the URL for the universal sentence encoder.

3. Use the TensorFlow Hub **KerasLayer** class to create a hub layer, passing in the following arguments: **module_handle**, **input_shape**, and **dtype**.

4. Create a list containing a string, **The TensorFlow Workshop**, to encode with the encoder.

5. Apply **hub_layer** to the text to embed the sentence as a vector.

   Your final output should be like the following:

```
<tf.Tensor: shape=(1, 512), dtype=float32, numpy=
array([[-0.01592658, -0.01910833, -0.00460122, -0.04786165, -0.0090545 ,
        -0.05658781, -0.04260132,  0.06827556,  0.03513585,  0.01448399,
        -0.00549438,  0.04602941,  0.02016041,  0.0008662 , -0.01191864,
         0.07414375, -0.03241738, -0.04448074,  0.00137551, -0.06724778,
        -0.02604278,  0.01092253, -0.01246114,  0.03847544,  0.00819034,
         0.06088841, -0.02359939, -0.05117927, -0.01725158, -0.02764523,
         0.04102336, -0.03135261,  0.06100909, -0.02693282, -0.07294274,
        -0.00857984, -0.04100463, -0.01803453,  0.04117068, -0.01969654,
        -0.04563987,  0.0257121 , -0.03328102, -0.05113809, -0.03377022,
         0.07439086, -0.02235463, -0.00438892, -0.00755636,  0.07249703,
        -0.07135288, -0.05469208,  0.01436193,  0.0396053 , -0.01475235,
        -0.03984744,  0.05067959,  0.07571234,  0.03281045, -0.00155282,
        -0.07548428,  0.01494772, -0.04175217,  0.03947704, -0.0147364 ,
        -0.01756434, -0.00077199,  0.00788859, -0.07518636,  0.04074219,
        -0.02049077, -0.03601787, -0.01753781,  0.03299529,  0.05840027,
        -0.03444539,  0.0186691 ,  0.03436609,  0.05346094,  0.02573053,
        -0.05013486, -0.05430874, -0.04835197,  0.03301562, -0.03129521,
         0.04714367, -0.07143752, -0.02783648, -0.02234376, -0.0619083 ,
        -0.05527468,  0.02779463,  0.04658304, -0.02259884, -0.05570157,
         0.06667245, -0.02903359, -0.05355389,  0.06542732, -0.05243086,
        -0.03966407,  0.01379365, -0.03453102,  0.07174195, -0.00385802,
        -0.03642376, -0.01343285,  0.00164682, -0.05571308,  0.01775301,
         0.03831774,  0.00128905, -0.0665922 , -0.01266254,  0.00407203,
        -0.07047658,  0.04188056, -0.01210087, -0.04976766,  0.03678571,
```

Figure 3.18: Expected output of Activity 3.02

NOTE

The solution to this activity can be found on page 470.

This section introduced Google Colab, an online development environment used to run Python code on Google servers. This can allow any practitioner with an internet connection to begin building machine learning models. Moreover, you can browse the selection of pre-trained models to begin creating models for your own applications using another resource you learned about in this chapter, TensorFlow Hub. Google Colab provides practitioners with a zero-configuration, up-to-date environment, and even access to GPUs and TPUs for faster model training times.

SUMMARY

In this chapter, you used a variety of TensorFlow resources, including TensorBoard, TensorFlow Hub, and Google Colab. TensorBoard offers users a method to visualize computational model graphs, metrics, and any experimentation results. TensorFlow Hub allows users to accelerate their machine learning development using pre-trained models built by experts in the field. Google Colab provides a collaborative environment to develop machine learning models on Google servers. Developing performant machine learning models is an iterative process of trial and error, and the ability to visualize every step of the process can help practitioners debug and improve their models. Moreover, understanding how experts in the field have built their models and being able to utilize the pre-learned weights in the networks can drastically reduce training time. All of these resources are used to provide an environment to develop and debug machine learning algorithms in an efficient workflow.

In the next chapter, you will begin creating your own machine learning models in TensorFlow, beginning with regression models. Regression models aim to predict continuous variables from input data. You will make your regression models by utilizing Keras layers, which are useful for building ANNs.

4

REGRESSION AND CLASSIFICATION MODELS

OVERVIEW

In this chapter, you will learn how to build regression and classification models using TensorFlow. You will build models with TensorFlow utilizing Keras layers, which are a simple approach to model building that offer a high-level API for building and training models. You will create models to solve regression and classification tasks, including the classification of the binding properties of various molecules. You will also use TensorBoard to visualize the architecture of TensorFlow models and view the training process.

INTRODUCTION

In the previous chapter, you learned how to use some TensorFlow resources to aid in development. These included TensorBoard (for visualizing computational graphs), TensorFlow Hub (an online repository for machine learning modules), and Google Colab (an online Python development environment for running code on Google servers). All these resources help machine learning practitioners develop models efficiently.

In this chapter, you will explore how to create ANNs using TensorFlow. You will build ANNs with different architectures to solve regression and classification tasks. Regression tasks aim to predict continuous variables from the input training data, while classification tasks aim to classify the input data into two or more classes. For example, a model to predict whether or not it will rain on a given day is a classification task since the result of the model will be of two classes—rain or no rain. However, a model to predict the amount of rain on a given day would be an example of a regression task since the output of the model would be a continuous variable—the amount of rain.

Models that are used to tackle these tasks represent a large class of machine learning models, and a huge amount of machine learning problems fall into these two categories. This chapter will demonstrate how regression and classification models can be created, trained, and evaluated in TensorFlow. You will use much of the learning covered in the previous chapters (including using TensorBoard to monitor the model training process) to understand how to build performant models.

This chapter introduces the various parameters used to build ANNs (known as **hyperparameters**), which include activation functions, loss functions, and optimizers. Other hyperparameters to select in the model-fitting process include the number of epochs and batch size, which vary the number of times the entire dataset is used to update the weights and the number of data points for each update, respectively. You will also learn how to log variables during the model-fitting process so that they can be visualized in TensorBoard. This allows you to determine whether the model is under- or overfitting the training data. Finally, after building your model, you will learn how to evaluate it on the dataset to see how well it performs.

SEQUENTIAL MODELS

A sequential model is used to build regression and classification models. In sequential models, information propagates through the network from the input layer at the beginning to the output layer at the end. Layers are stacked in the model sequentially, with each layer having an input and an output.

Other types of ANN models exist, such as recurrent neural networks (in which the output feeds back into the input), which will be covered in later chapters. The difference between sequential and recurrent neural networks is shown in *Figure 4.01*. In both the models, the information flows from the input layer through the hidden layers to the output layer, as indicated by the direction of the arrows. However, in recurrent architectures, the output of the hidden layers feeds back into the input of the hidden layers:

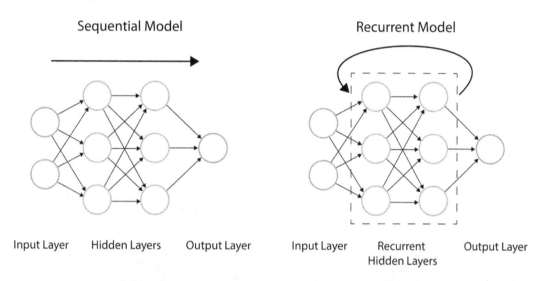

Figure 4.1: The architectures of sequential and recurrent ANNs

In the following section, you will learn how to create sequential models in TensorFlow that form the basis of regression and classification models. You will utilize the Keras API, which is now included as part of the TensorFlow library for sequential models, since the high-level API provides a simple interface for creating these models. Using the API, you will find that adding more layers to a model is incredibly easy and is great for new practitioners learning the field.

A sequential model can be initialized as follows:

```
model = tf.keras.Sequential()
```

Once the model has been initialized, layers can be added to the model. In this section, you will also explore how to add Keras layers to the model.

KERAS LAYERS

Keras layers are included in the TensorFlow package. Keras layers are a collection of commonly used layers that can be added easily to your sequential models.

> **NOTE**
>
> You can check out all the possible options for Keras layers here:
> https://www.tensorflow.org/api_docs/python/tf/keras/layers.

To add layers to a model of the **Sequential** class, you can use the model's **add** method. One optional layer that can be added to the beginning of a sequential model is an **input layer** as an entry point to the network. Input layers can take the following common input arguments:

- **input_shape** (required): The shape of the input tensor, not including the batch axis

- **batch_size**: An optional argument indicating the input batch size

- **name**: Optional name of the input layer

Input layers can be added to a model as follows. The following code snippet is used to add a layer, expecting inputs to have eight features:

```
model.add(tf.keras.layers.InputLayer(input_shape=(8,), \
                                     name='Input_layer'))
```

By providing a **name** argument, you can label the layers, which will be useful when visualizing the model in TensorBoard. Another type of layer that is commonly used when building regression and classification models is the **dense layer**. The dense layer is a fully connected layer, which means that all the nodes in the layer receive inputs from every node in the layer prior and then connect to every node of the next layer. A dense layer can be used as the first layer of the model with **input_shape** provided as an argument. The following are the common input arguments for layers of the **Dense** class:

- **units** (required): This is a positive integer denoting the number of units in the layer.

- **input_shape**: This is the shape of the input tensor but is not required unless it is the first layer of the model.

- **`activation`**: This is an optional argument indicating which activation function to apply to the output of the layer.

- **`use_bias`**: This is a Boolean argument indicating whether to use bias in the layer. The default is set to **True**.

- **`name`**: This refers to the name of the layer. One will be generated if this argument is not provided.

- **`kernel_initializer`**: This is the initializer for the kernel weights. The **Glorot uniform initializer**, which has a normal distribution centered on zero and a standard deviation that is dependent on the number of units in the layer, is used by default.

- **`bias_initializer`**: This is the initializer for the bias. The default of this parameter is used to set the bias values to zero.

- **`kernel_regularizer`**: This is the regularizer to use on the kernel weights. There are none applied by default.

- **`bias_regularizer`**: This is the regularizer to use on the bias. There are none applied by default.

The following is an example of adding a dense layer to a model with **12** units, adding a **sigmoid** activation function at the output of the layer, and naming the layer **Dense_layer_1**:

```
model.add(tf.keras.layers.Dense(units=12, name='Dense_layer_1', \
                                activation='sigmoid'))
```

Now that you understand how to initialize sequential models and add layers to them, you will create a Keras sequential model using TensorFlow in the first exercise. You will initialize a model, add layers to the model, add activation functions to the output of the model, and pass data through the model to simulate creating a prediction.

EXERCISE 4.01: CREATING AN ANN WITH TENSORFLOW

In this exercise, you will create your first sequential ANN in TensorFlow. You will have an input layer, a hidden layer with four units and a ReLU activation function, and an output layer with one unit. Then, you will create some simulation data by generating random numbers and passing it through the model, using the model's **predict** method to simulate a prediction for each data example.

Perform the following steps to complete the exercise:

1. Open a Jupyter notebook and import the TensorFlow library:

```
import tensorflow as tf
```

2. Initialize a Keras model of the sequential class:

```
model = tf.keras.Sequential()
```

3. Add an input layer to the model using the model's **add** method, and add the **input_shape** argument with size **(8,)** to represent input data with eight features:

```
model.add(tf.keras.layers.InputLayer(input_shape=(8,), \
                                    name='Input_layer'))
```

4. Add two layers of the **Dense** class to the model. The first will represent your hidden layer with four units and a ReLU activation function, and the second will represent your output layer with one unit:

```
model.add(tf.keras.layers.Dense(4, activation='relu', \
                                name='First_hidden_layer'))
model.add(tf.keras.layers.Dense(1, name='Output_layer'))
```

5. View the weights by calling the **variables** attribute of the model:

```
model.variables
```

You should get the following output:

```
Out[7]: [<tf.Variable 'Hidden_layer/kernel:0' shape=(8, 4) dtype=float32, numpy=
        array([[-0.57164866,  0.3538167 , -0.17721128,  0.21663547],
               [ 0.2400865 , -0.04494339, -0.25348833,  0.254395  ],
               [-0.13441253,  0.04241037,  0.16327792, -0.05576098],
               [ 0.49396342,  0.00978398,  0.55000216,  0.2663949 ],
               [-0.17641711, -0.3961743 , -0.5454148 , -0.23563156],
               [ 0.03186369,  0.00652903,  0.11141032, -0.20005405],
               [ 0.43056852,  0.64045316,  0.06564438, -0.56348765],
               [ 0.52928454, -0.41281876, -0.30966654,  0.3005106 ]],
        dtype=float32)>,
        <tf.Variable 'Hidden_layer/bias:0' shape=(4,) dtype=float32, numpy=array([0., 0., 0., 0.], dtype=float32)>,
        <tf.Variable 'Output_layer/kernel:0' shape=(4, 1) dtype=float32, numpy=
        array([[ 0.08683264],
               [-0.6753886 ],
               [ 0.72353196],
               [-0.0379988 ]], dtype=float32)>,
        <tf.Variable 'Output_layer/bias:0' shape=(1,) dtype=float32, numpy=array([0.], dtype=float32)>]
```

Figure 4.2: The variables of the ANN

This output shows all the variables that compose the model; they include the values for all weights and biases in each layer.

6. Create a tensor of size **32x8**, which represents a tensor with 32 records and 8 features:

```
data = tf.random.normal((32,8))
```

7. Call the **predict** method of the model and pass in the sample data:

```
model.predict(data)
prediction
```

You should get the following result:

```
Out[5]:  array([[ 0.07712938],
                [ 0.16851059],
                [-0.46971387],
                [-0.3231515 ],
                [-0.2860464 ],
                [ 0.01692858],
                [-0.2509554 ],
                [ 0.5132399 ],
                [ 0.0609077 ],
                [-0.27511594],
                [ 0.62906533],
                [-0.01987116],
                [-0.01488206],
                [-0.2381044 ],
                [ 0.05233095],
                [ 0.        ],
                [ 0.02039919],
                [-0.07099625],
                [ 0.00759757],
                [-0.00441622],
                [ 0.11206146],
                [ 0.50655746],
                [ 0.4938014 ],
                [ 0.        ],
                [-0.27319306],
                [-0.20734225],
                [ 0.6667459 ],
                [ 0.        ],
                [-0.45259422],
                [-0.01765753],
                [ 1.2930266 ],
                [ 0.02242063]], dtype=float32)
```

Figure 4.3: The output of the ANN after random inputs have been applied

Calling the **`predict()`** method on the sample data will propagate the data through the network. In each layer, there will be a matrix multiplication of the data with the weights, and the bias will be added before the data is passed as input data to the next layer. This process continues until the final output layer.

In this exercise, you created a sequential model with multiple layers. You initialized a model, added an input layer to accept data with eight features, added a hidden layer with four units, and added an output layer with one unit. Before fitting a model to training data, you must first compile the model with an optimizer and choose a loss function to minimize the value it computes by updating weights in the training process.

In the next section, you will explore how to compile models, then fit them to training data.

MODEL FITTING

Once a model has been initialized and layers have been added to the ANN, the model must be configured with an optimizer, losses, and any evaluation metrics through the compilation process. A model can be compiled using the model's **`compile`** method, as follows:

```
model.compile(optimizer='adam', loss='binary_crossentropy', \
              metrics=['accuracy'])
```

Optimizers can be chosen by simply naming the optimizer as the argument. The following optimizers are available as default for Keras models:

- **Stochastic gradient descent (SGD)**: This updates the weights for each example in the dataset. You can find more information about SGD here: https://keras.io/api/optimizers/sgd/.

- **RMSprop**: This is an adaptive optimizer that varies the weights during training by using a decaying average of the gradients at each update. You can find more information about RMSprop here: https://keras.io/api/optimizers/rmsprop/.

- **Adam**: This is also an adaptive optimizer that implements the Adam algorithm, updating the learning rates based on the first- and second-order gradients. You can find more information about Adam here: https://keras.io/api/optimizers/adam/.

- **Adagrad**: This adaptive gradient optimizer adapts the learning rate at each weight update. The learning rate is adapted for each feature using the prior gradients and observations. You can find more information about Adagrad here: https://keras.io/api/optimizers/adagrad/.

- **Adadelta**: This is a more robust version of Adagrad that uses a sliding window of gradient updates to adapt the learning rate. You can find more information about Adadelta here: https://keras.io/api/optimizers/adadelta/.

- **Adamax**: This is an adaptive optimizer that is a variant of the Adam optimizer. You can find more information about Adamax here: https://keras.io/api/optimizers/adamax/.

- **Nadam**: This is another adaptive optimizer that is a variant of the Adam optimizer with Nesterov momentum. You can find more information about Nadam here: https://keras.io/api/optimizers/Nadam/.

- **Ftrl**: This is an optimizer that implements the FTRL algorithm. You can find more information about Ftrl here: https://keras.io/api/optimizers/ftrl/.

Custom optimizers can also be added to Keras models if the provided ones are not relevant. Selecting the most appropriate optimizer is often a matter of trying each and identifying which optimizer produces the lowest error. This process is known as **hyperparameter tuning** and will be covered in a later chapter. In the next section, you will uncover another option when compiling models: the loss function. The goal of training a model is to minimize the value computed by the loss function.

THE LOSS FUNCTION

The loss function is the measure of error between the predicted results and the true results. You use the loss function during the training process to determine whether varying any of the weights and biases will create a better model by minimizing the loss function's value through the optimization process.

There are many different types of loss functions that can be used, and the specific one will depend on the problem and goal. In general, regression and classification tasks will have different loss functions. Since regression models predict continuous variables, loss functions for regression models typically aim to summarize how far, on average, the predictions are from the true values. For classification models, loss functions aim to determine how the quantity of true positive, true negative, false positive, and false negative classifications of the predicted classes vary compared to the true classes.

True positives are defined as correct predictions labeled positive by the classifier; similarly, **true negatives** are correct predictions labeled negative. **False positives** are predictions labeled positive where the true value is negative, and **false negatives** are predictions labeled negative that are actually positive. Loss functions that are directly available to use in Keras sequential models for regression include the following:

- **Mean squared error**: This is a loss function that calculates the squared difference between the true and predicted value for each data point, `(true value - predicted value)^2`, and returns the average across the entire dataset. This loss function is primarily used for regression problems, and the squaring of the difference between the two values ensures the loss function results in a positive number.

- **Mean absolute error**: This is another loss function primarily used for regression problems that calculates the absolute value of the difference between the true and predicted value for each data point, `|true value - predicted value|`, and returns the average across the dataset. This method also ensures that the result is a positive value.

- **Mean absolute percentage error**: This is another loss function used for regression problems that calculates the absolute value of the percentage error for each data point, `|(true value- predicted value) / true value|`, and returns the average across the dataset as a percentage.

For classification, loss functions that are available include the following:

- **Binary cross-entropy**: This is a loss function used for binary classification problems that outputs a value between **0** and **1**, with values closer to **1** representing a greater number of true positive classifications.

- **Categorical cross-entropy**: This is a loss function similar to binary cross-entropy; however, it is suitable for multi-class classification problems and also outputs values between **0** and **1**.

When compiling a model, other metrics can also be passed in as an argument to the method. They will be calculated after each epoch and saved during the training process. The metrics that are available to be calculated for Keras models include the following:

- **Accuracy**: This is the proportion of correct results out of the total results.

- **Precision**: This is the proportion of true positives out of the total positives predicted.

- **Recall**: This is the proportion of true positives out of the actual positives.

- **AUC**: This metric represents the area under the ROC curve.

These metrics can be incredibly valuable in understanding the performance of the model during the training process. All the metrics have values between **0** and **1**, with higher values representing better performance. Once the model has been compiled, it can be fit to the training data. This can be accomplished by calling the **fit** method and passing in the following arguments:

- **x**: This is the feature data as a TensorFlow tensor or NumPy array.

- **y**: This is the target data as a TensorFlow tensor or NumPy array.

- **epochs**: This refers to the number of epochs to run the model for. An epoch is an iteration over the entire training dataset.

- **batch_size**: This is the number of training data samples to use per gradient update.

- **validation_split**: This is the proportion of the training data to be used for validation that is evaluated after each epoch. This proportion of data is not used in the weight update process.

- **shuffle**: This indicates whether to shuffle the training data before each epoch.

To fit the model to the training data, the **fit** method can be applied to a model in the following way:

```
model.fit(x=features, y=target, epochs=10, batch_size=32, \
          validation_split=0.2, shuffle=False)
```

Once the **fit** method has been called, the model will begin fitting to the training data. After each epoch, the loss is returned for the training. If a validation split is defined, then the loss is also evaluated on the validation split.

MODEL EVALUATION

Once models are trained, they can be evaluated by utilizing the model's **evaluate** method. The **evaluate** method assesses the performance of the model according to the loss function used to train the model and any metrics that were passed to the model. The method is best used when determining how the model will perform on new, unseen data by passing in a feature and target dataset that has not been used in the training process or out-of-sample dataset. The method can be called as follows:

```
eval_metrics = model.evaluate(features, target)
```

The result of the method is first the loss calculated on the input data, and then, if any metrics were passed in the model compilation process, they will also be calculated when the **evaluate** method is executed. Model evaluation is an important step in determining how well your model is performing. Since there is an enormous number of hyperparameters (such as the number of hidden layers, the number of units in each layer, and the choice of activation functions, to name a few), model evaluation is necessary to determine which combination of hyperparameters is optimal. Effective model evaluation can help provide an unbiased view on which model architecture will perform best overall.

In the following exercise, you will undertake the process of creating an ANN, compiling the model, fitting the model to training data, and finally, evaluating the model on the training data. You will recreate the linear regression algorithm with an ANN, which can be interpreted as an ANN with only one layer and one unit. Furthermore, you will view the architecture of the model and model training process in TensorBoard.

EXERCISE 4.02: CREATING A LINEAR REGRESSION MODEL AS AN ANN WITH TENSORFLOW

In this exercise, you will create a linear regression model as an ANN using TensorFlow. The dataset, **Bias_correction_ucl.csv**, describes the bias correction of air temperature forecasts of Seoul, South Korea. The fields represent temperature measurements of the given date, the weather station at which the metrics were measured, model forecasts of weather-related metrics such as humidity, and projections for the temperature the following day. You are required to predict the next maximum and minimum temperature given measurements of the prior timepoints and attributes of the weather station.

> **NOTE**
>
> The **Bias_correction_ucl.csv** file can be found here: https://packt.link/khfeF.

Perform the following steps to complete this exercise:

1. Open a new Jupyter notebook to implement this exercise.

2. In a new Jupyter Notebook cell, import the TensorFlow and pandas libraries:

```
import tensorflow as tf
import pandas as pd
```

3. Load in the dataset using the pandas **read_csv** function:

```
df = pd.read_csv('Bias_correction_ucl.csv')
```

> **NOTE**
>
> Make sure you change the path (highlighted) to the CSV file based on its location on your system. If you're running the Jupyter notebook from the same directory where the CSV file is stored, you can run the preceding code without any modification.

4. Drop the **date** column and drop any rows that have null values since your model requires numerical values only:

```
df.drop('Date', inplace=True, axis=1)
df.dropna(inplace=True)
```

5. Create target and feature datasets. The target dataset will contain the columns named **Next_Tmax** and **Next_Tmin**, while the feature dataset will contain all columns except those named **Next_Tmax** and **Next_Tmin**:

```
target = df[['Next_Tmax', 'Next_Tmin']]
features = df.drop(['Next_Tmax', 'Next_Tmin'], axis=1)
```

6. Rescale the feature dataset:

```
from sklearn.preprocessing import MinMaxScaler
scaler = MinMaxScaler()
feature_array = scaler.fit_transform(features)
features = pd.DataFrame(feature_array, columns=features.columns)
```

7. Initialize a Keras model of the **Sequential** class:

```
model = tf.keras.Sequential()
```

8. Add an input layer to the model using the model's **add** method, and set **input_shape** to be the number of columns in the feature dataset:

```
model.add(tf.keras.layers.InputLayer\
         (input_shape=(features.shape[1],), \
                     name='Input_layer'))
```

9. Add the output layer of the **Dense** class to the model with a size of **2**, representing the two target variables:

```
model.add(tf.keras.layers.Dense(2, name='Output_layer'))
```

10. Compile the model with an RMSprop optimizer and a mean squared error loss:

```
model.compile(tf.optimizers.RMSprop(0.001), loss='mse')
```

11. Add a callback for TensorBoard:

```
tensorboard_callback = tf.keras.callbacks\
                        .TensorBoard(log_dir="./logs")
```

12. Fit the model to the training data:

```
model.fit(x=features.to_numpy(), y=target.to_numpy(),\
          epochs=50, callbacks=[tensorboard_callback])
```

You should get the following output:

```
Train on 7588 samples
Epoch 1/50
7588/7588 [==============================] - 1s 95us/sample - loss: 682.0368
Epoch 2/50
7588/7588 [==============================] - 0s 31us/sample - loss: 575.7078
Epoch 3/50
7588/7588 [==============================] - 0s 31us/sample - loss: 479.1454
Epoch 4/50
7588/7588 [==============================] - 0s 33us/sample - loss: 392.2625
Epoch 5/50
7588/7588 [==============================] - 0s 46us/sample - loss: 314.8800
Epoch 6/50
7588/7588 [==============================] - 0s 31us/sample - loss: 246.9707
Epoch 7/50
7588/7588 [==============================] - 0s 32us/sample - loss: 188.8617
Epoch 8/50
7588/7588 [==============================] - 0s 32us/sample - loss: 140.1037
Epoch 9/50
7588/7588 [==============================] - 0s 30us/sample - loss: 101.0663
Epoch 10/50
7588/7588 [==============================] - 0s 31us/sample - loss: 71.3585
Epoch 11/50
7588/7588 [==============================] - 0s 32us/sample - loss: 50.7607
Epoch 12/50
7588/7588 [==============================] - 0s 39us/sample - loss: 37.2300
```

Figure 4.4: The output of the fitting process showing the epoch, train time per sample, and loss after each epoch

13. Evaluate the model on the training data:

```
loss = model.evaluate(features.to_numpy(), target.to_numpy())
print('loss:', loss)
```

This results in the following output:

```
loss: 3.5468221449764012
```

14. View the model architecture and model-fitting process on TensorBoard by calling the following on the command line:

```
tensorboard --logdir=logs/
```

You can see its execution in a web browser by visiting the URL that is provided after launching TensorBoard. The default URL provided is **http://localhost:6006/**:

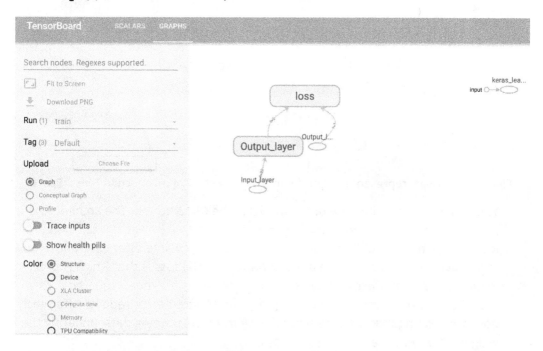

Figure 4.5: A visual representation of the model architecture in TensorBoard

The loss function can be visualized as shown in the following figure:

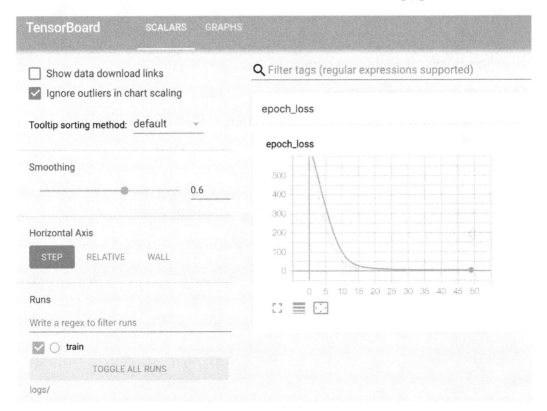

Figure 4.6: A visual representation of the loss as a function of an epoch in TensorBoard

You can see the architecture of the model in the **GRAPHS** tab. The architecture shows the input layer and output layer in the model, as well as the calculated loss. During the model-fitting process, the loss is calculated after each epoch and is displayed in TensorBoard in the **SCALARS** tab. The loss is that which is defined in the compilation process; so, in this case, the loss is the mean squared error. From TensorBoard, you can see that the mean squared error reduces after each epoch, indicating that the model is learning from the training data, updating the weights in order to reduce the total loss.

In this exercise, you have learned how to create, train, and evaluate an ANN with TensorFlow by using Keras layers. You recreated the linear regression algorithm by creating an ANN with an input layer and an output layer that has one unit for each output. Here, there were two outputs representing the maximum and minimum values of the temperature; thus, the output layer has two units.

In *Exercise 4.01*, *Creating an ANN with TensorFlow*, you created an ANN with only one layer containing weights and the output layer. This is an example of a **shallow neural network**. ANNs that have many hidden layers containing weights are called **deep neural networks**, and the process of training them is called **deep learning**. By increasing the number of layers and making the ANN deeper, the model becomes more flexible and will be able to model more complex functions. However, to gain this increase in flexibility, you need more training data and more computation power to train the model.

In the next exercise, you will create and train ANNs that have multiple hidden layers.

EXERCISE 4.03: CREATING A MULTI-LAYER ANN WITH TENSORFLOW

In this exercise, you will create a multi-layer ANN using TensorFlow. This model will have four hidden layers. You will add multiple layers to the model and activation functions to the output of the layers. The first hidden layer will have **16** units, the second will have **8** units, and the third will have **4** units. The output layer will have **2** units. You will utilize the same dataset as in *Exercise 4.02*, *Creating a Linear Regression Model as an ANN with TensorFlow*, which describes the bias correction of air temperature forecasts for Seoul, South Korea. The exercise aims to predict the next maximum and minimum temperature given measurements of the prior timepoints and attributes of the weather station.

Perform the following steps to complete this exercise:

1. Open a new Jupyter notebook to implement this exercise.

2. In a new Jupyter Notebook cell, import the TensorFlow and pandas libraries:

```
import tensorflow as tf
import pandas as pd
```

3. Load in the dataset using the pandas **read_csv** function:

```
df = pd.read_csv('Bias_correction_ucl.csv')
```

> **NOTE**
>
> Make sure you change the path (highlighted) to the CSV file based on its location on your system. If you're running the Jupyter notebook from the same directory where the CSV file is stored, you can run the preceding code without any modification.

4. Drop the **Date** column and drop any rows that have null values:

```
df.drop('Date', inplace=True, axis=1)
df.dropna(inplace=True)
```

5. Create target and feature datasets:

```
target = df[['Next_Tmax', 'Next_Tmin']]
features = df.drop(['Next_Tmax', 'Next_Tmin'], axis=1)
```

6. Rescale the feature dataset:

```
from sklearn.preprocessing import MinMaxScaler
scaler = MinMaxScaler()
feature_array = scaler.fit_transform(features)
features = pd.DataFrame(feature_array, columns=features.columns)
```

7. Initialize a Keras model of the **Sequential** class:

```
model = tf.keras.Sequential()
```

8. Add an input layer to the model using the model's **add** method, and set **input_shape** to the number of columns in the feature dataset:

```
model.add(tf.keras.layers.InputLayer\
                         (input_shape=(features.shape[1],), \
                          name='Input_layer'))
```

9. Add three hidden layers and an output layer of the **Dense** class to the model. The first hidden layer will have **16** units, the second will have **8** units, and the third will have **4** units. Label the layers appropriately. The output layer will have two units to match the target variable that has two columns:

```
model.add(tf.keras.layers.Dense(16, name='Dense_layer_1'))
model.add(tf.keras.layers.Dense(8, name='Dense_layer_2'))
model.add(tf.keras.layers.Dense(4, name='Dense_layer_3'))
model.add(tf.keras.layers.Dense(2, name='Output_layer'))
```

10. Compile the model with an RMSprop optimizer and mean squared error loss:

```
model.compile(tf.optimizers.RMSprop(0.001), loss='mse')
```

11. Add a callback for TensorBoard:

```
tensorboard_callback = tf.keras.callbacks\
                        .TensorBoard(log_dir="./logs")
```

12. Fit the model to the training data for **50** epochs and add a validation split equal to 20%:

```
model.fit(x=features.to_numpy(), y=target.to_numpy(),\
          epochs=50, callbacks=[tensorboard_callback] , \
          validation_split=0.2)
```

You should get the following output:

```
Train on 6070 samples, validate on 1518 samples
Epoch 1/50
6070/6070 [==============================] - 1s 175us/sample - loss: 640.6369 - val_loss: 597.5753
Epoch 2/50
6070/6070 [==============================] - 0s 44us/sample - loss: 579.3356 - val_loss: 551.4062
Epoch 3/50
6070/6070 [==============================] - 0s 42us/sample - loss: 534.9167 - val_loss: 508.4470
Epoch 4/50
6070/6070 [==============================] - 0s 44us/sample - loss: 492.8038 - val_loss: 467.4198
Epoch 5/50
6070/6070 [==============================] - 0s 43us/sample - loss: 452.4671 - val_loss: 428.1580
Epoch 6/50
6070/6070 [==============================] - 0s 44us/sample - loss: 413.9526 - val_loss: 390.6964
Epoch 7/50
6070/6070 [==============================] - 0s 44us/sample - loss: 377.2294 - val_loss: 355.0530
Epoch 8/50
6070/6070 [==============================] - 0s 45us/sample - loss: 342.3214 - val_loss: 321.2107
Epoch 9/50
6070/6070 [==============================] - 0s 43us/sample - loss: 309.2548 - val_loss: 289.1745
Epoch 10/50
6070/6070 [==============================] - 0s 43us/sample - loss: 277.8973 - val_loss: 258.9163
Epoch 11/50
6070/6070 [==============================] - 0s 43us/sample - loss: 248.4023 - val_loss: 230.4784
Epoch 12/50
6070/6070 [==============================] - 0s 47us/sample - loss: 220.7003 - val_loss: 203.8202
Epoch 13/50
6070/6070 [==============================] - 0s 42us/sample - loss: 194.8023 - val_loss: 178.9795
Epoch 14/50
6070/6070 [==============================] - 0s 42us/sample - loss: 170.6718 - val_loss: 155.9259
Epoch 15/50
6070/6070 [==============================] - 0s 47us/sample - loss: 148.3619 - val_loss: 134.6750
Epoch 16/50
6070/6070 [==============================] - 0s 62us/sample - loss: 127.8401 - val_loss: 115.1973
Epoch 17/50
6070/6070 [==============================] - 0s 49us/sample - loss: 109.1198 - val_loss: 97.5518
Epoch 18/50
6070/6070 [==============================] - 0s 46us/sample - loss: 92.2168 - val_loss: 81.6851
```

Figure 4.7: The output of the fitting process showing the epoch, training time per sample, and loss after each epoch

13. Evaluate the model on the training data:

```
loss = model.evaluate(features.to_numpy(), target.to_numpy())
print('loss:', loss)
```

This will display the following result:

```
loss: 1.664448248190068
```

14. View the model architecture and model-fitting process in TensorBoard:

```
tensorboard --logdir=logs/
```

You should get something like the following:

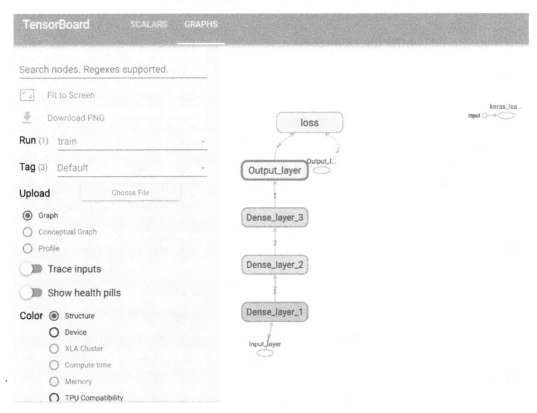

Figure 4.8: A visual representation of the model architecture in TensorBoard

You can visualize the loss function as shown in the following screenshot:

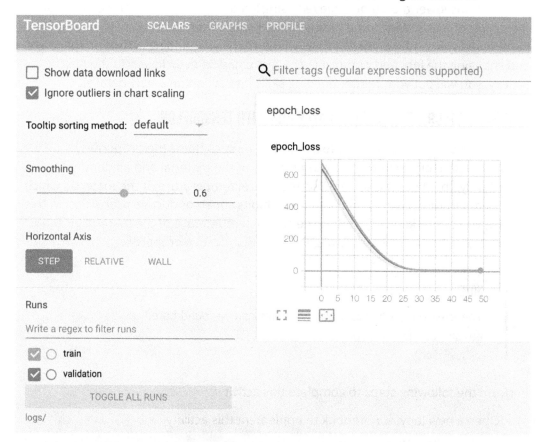

Figure 4.9: A visual representation of the loss as a function of an epoch in TensorBoard on the training and validation split

The network architecture shows the input layer and the four hidden layers of the model as well as the calculated loss at the end. During the model-fitting process, the loss is calculated after each epoch and is displayed in TensorBoard in the **SCALARS** tab. Here, the loss is the mean squared error. From TensorBoard, you can see that the mean squared error reduces on the training set (the orange line) and the validation set (the blue line), after each epoch, indicating that the model is learning effectively from the training data.

In this exercise, you have created an ANN with multiple hidden layers. The loss you obtained was lower than that achieved using linear regression, which demonstrates the power of ANNs. With some tuning to the hyperparameters (such as varying the number of layers, the number of units within each layer, adding activation functions, and changing the loss and optimizer), the loss could be even lower. In the next activity, you will put your model-building skills into action on a new dataset.

ACTIVITY 4.01: CREATING A MULTI-LAYER ANN WITH TENSORFLOW

The feature dataset, **superconductivity.csv**, contains the properties of superconductors including the atomic mass of the material and its density. Importantly, the dataset also contains the critical temperature of the material, which is the temperature at which the material exhibits superconductive properties. In this activity, you are tasked with finding the critical temperature of the material or the temperature at which the material gains superconductive properties.

> **NOTE**
>
> The **superconductivity.csv** file can be found here: https://packt.link/sOCPh.

Perform the following steps to complete this activity:

1. Open a new Jupyter notebook to implement this activity.

2. Import the TensorFlow and pandas libraries.

3. Load in the **superconductivity.csv** dataset.

4. Drop any rows that have null values.

5. Set the target as the **critical_temp** column and the feature dataset as the remaining columns.

6. Rescale the feature dataset using a standard scaler.

7. Initialize a model of the Keras **Sequential** class.

8. Add an input layer, four hidden layers of sizes **64**, **32**, **16**, and **8**, and an output layer of size **1** to the model. Add a ReLU activation function to the first hidden layer.

9. Compile the model with an RMSprop optimizer with a learning rate equal to **0.001** and the mean squared error for the loss.

10. Add a callback to write logs to TensorBoard.

11. Fit the model to the training data for **100** epochs, with a batch size equal to **32** and a validation split equal to 20%.

12. Evaluate the model on the training data.

13. View the model architecture in TensorBoard.

    You should get an output like the following:

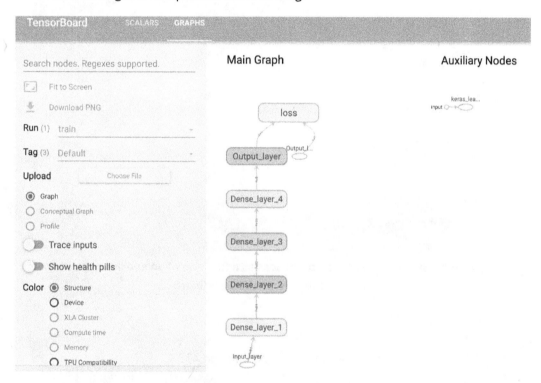

Figure 4.10: A visual representation of the model architecture in TensorBoard

14. Visualize the model-fitting process in TensorBoard. You should get the
following output:

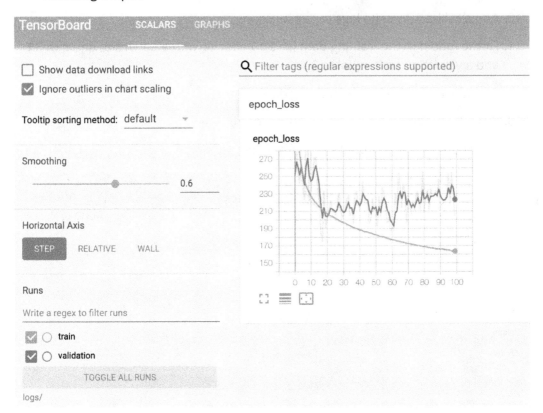

Figure 4.11: A visual representation of the loss as a function of an epoch on the training and validation split in TensorBoard

> **NOTE**
>
> The solution to this activity can be found on page 472.

In the next section, you will explore classification models, which attempt to classify data into distinct classes. You will begin with binary classification models that classify data into just two classes. This is the simplest form of a classification model. Once binary classifiers are mastered, more complicated models can be tackled, such as multi-label and multi-class classification.

CLASSIFICATION MODELS

The goal of classification models is to classify data into distinct classes. For example, a spam filter is a classification model that aims to classify emails into "spam" (referring to unsolicited and unwanted email) or "ham" (a legitimate email). Spam filters are an example of a binary classifier since there are two classes. The input to the filter may include the content of the email, the email address of the sender, and the subject line, among other features, and the output will be the predicted class, **spam** or **ham**. Classification models can classify data into more than two distinct classes (known as **multi-class classification**) or classify data with multiple positive labels (known as **multi-label classification**).

There are several different algorithms that can be used for classification tasks. Some popular ones include logistic regression, decision trees, and ANNs. ANNs are a great choice for classification models since they can learn complex relationships between the features and the target, and results can be achieved with the appropriate activation function on the output layer of the ANN.

A common activation function to use for classification models is the sigmoid function, which is the same function used in logistic regression. In fact, a logistic regression model can be created by building an ANN with a single layer with one unit and a sigmoid activation function. The sigmoid function is a transformation in which the input is any real value, and the output is a number strictly between **0** and **1**. A visual representation is shown in the following figure.

The output of the sigmoid transformation can be interpreted as a probability of a value being in the positive class; a value closer to a value of **1** indicates a higher probability of being in the positive class:

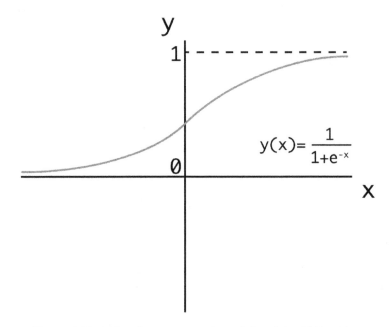

Figure 4.12: A visual representation of the sigmoid function

After the sigmoid function has been applied, a threshold is applied, above which the data is classified as the positive class and below as the negative class. The default threshold for a sigmoid function is **0.5**, meaning that any value at or above **0.5** is classified as positive.

In the next exercise, you will create a logistic regression model with TensorFlow. You will achieve this by creating a single-layer ANN, the process of which is similar to that of the linear regression model in *Exercise 4.02, Creating a Linear Regression Model as an ANN with TensorFlow*. The difference is that you will add a sigmoid activation function to the output of the ANN. Another difference that separates the two exercises is the loss function that you will use to calculate the loss.

EXERCISE 4.04: CREATING A LOGISTIC REGRESSION MODEL AS AN ANN WITH TENSORFLOW

In this exercise, you will create a logistic regression model as an ANN using TensorFlow. The dataset, **qsar_androgen_receptor.csv**, is used to develop classification models for the discrimination of binder/non-binder molecules given various attributes of the molecules. Here, the molecule attributes represent the features of your dataset, and their binding properties represent the target variable, in which a positive value represents a binding molecule, and a negative value represents a non-binding molecule. You will create a logistic regression model to predict the binding properties of the molecule given attributes of the molecule provided in the dataset.

> **NOTE**
>
> The **qsar_androgen_receptor.csv** file can be found here:
> https://packt.link/hWvjc.

Perform the following steps to complete this exercise:

1. Open a new Jupyter notebook to implement this exercise.

2. Import the TensorFlow and pandas libraries:

```
import tensorflow as tf
import pandas as pd
```

3. Load in the dataset using the pandas **read_csv** function:

```
df = pd.read_csv('qsar_androgen_receptor.csv', \
                 sep=';')
```

> **NOTE**
>
> Make sure you change the path (highlighted) to the CSV file based on its location on your system. If you're running the Jupyter notebook from the same directory where the CSV file is stored, you can run the preceding code without any modification.

4. Drop any rows that have null values:

```
df.dropna(inplace=True)
```

5. Create target and feature datasets:

```
target = df['positive'].apply(lambda x: 1 if x=='positive' else 0)
features = df.drop('positive', axis=1)
```

6. Initialize a Keras model of the **Sequential** class:

```
model = tf.keras.Sequential()
```

7. Add an input layer to the model using the model's **add** method and set **input_shape** to be the number of columns in the feature dataset:

```
model.add(tf.keras.layers.InputLayer\
          (input_shape=(features.shape[1],), \
                   name='Input_layer'))
```

8. Add the output layer of the **Dense** class to the model with a size of **1**, representing the target variable:

```
model.add(tf.keras.layers.Dense(1, name='Output_layer', \
                          activation='sigmoid'))
```

9. Compile the model with an RMSprop optimizer and binary cross-entropy for the loss, and compute the accuracy:

```
model.compile(tf.optimizers.RMSprop(0.0001), \
              loss='binary_crossentropy', metrics=['accuracy'])
```

10. Create a TensorBoard callback:

```
tensorboard_callback = tf.keras.callbacks.TensorBoard\
                       (log_dir="./logs")
```

11. Fit the model to the training data for **50** epochs, adding the TensorBoard callback with a validation split of 20%:

```
model.fit(x=features.to_numpy(), y=target.to_numpy(), \
          epochs=50, callbacks=[tensorboard_callback] , \
          validation_split=0.2)
```

Your output should be similar to the following figure:

```
Train on 7588 samples
Epoch 1/50
7588/7588 [==============================] - 1s 95us/sample - loss: 682.0368
Epoch 2/50
7588/7588 [==============================] - 0s 31us/sample - loss: 575.7078
Epoch 3/50
7588/7588 [==============================] - 0s 31us/sample - loss: 479.1454
Epoch 4/50
7588/7588 [==============================] - 0s 33us/sample - loss: 392.2625
Epoch 5/50
7588/7588 [==============================] - 0s 46us/sample - loss: 314.8800
Epoch 6/50
7588/7588 [==============================] - 0s 31us/sample - loss: 246.9707
Epoch 7/50
7588/7588 [==============================] - 0s 32us/sample - loss: 188.8617
Epoch 8/50
7588/7588 [==============================] - 0s 32us/sample - loss: 140.1037
Epoch 9/50
7588/7588 [==============================] - 0s 30us/sample - loss: 101.0663
Epoch 10/50
7588/7588 [==============================] - 0s 31us/sample - loss: 71.3585
Epoch 11/50
7588/7588 [==============================] - 0s 32us/sample - loss: 50.7607
Epoch 12/50
7588/7588 [==============================] - 0s 39us/sample - loss: 37.2300
```

Figure 4.13: The output of the fitting process showing the epoch, training time per sample, and loss after each epoch

12. Evaluate the model on the training data:

```
loss, accuracy = model.evaluate(features.to_numpy(), \
                                target.to_numpy())
print(f'loss: {loss}, accuracy: {accuracy}')
```

You should get output something like the following:

```
loss: 0.2781583094794838, accuracy: 0.9110320210456848
```

13. Visualize the model-fitting process in TensorBoard by calling the following command on the command line:

```
tensorboard --logdir=logs/
```

You should get a screen similar to the following in the browser:

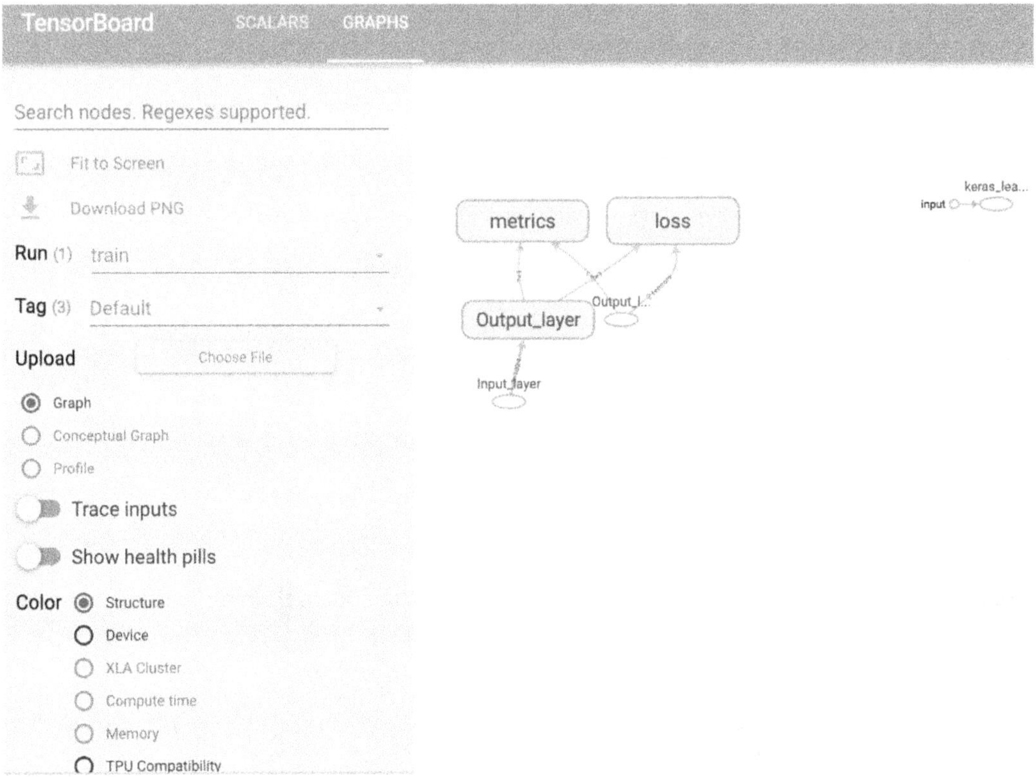

Figure 4.14: A visual representation of the model architecture in TensorBoard

The loss function can be represented as follows:

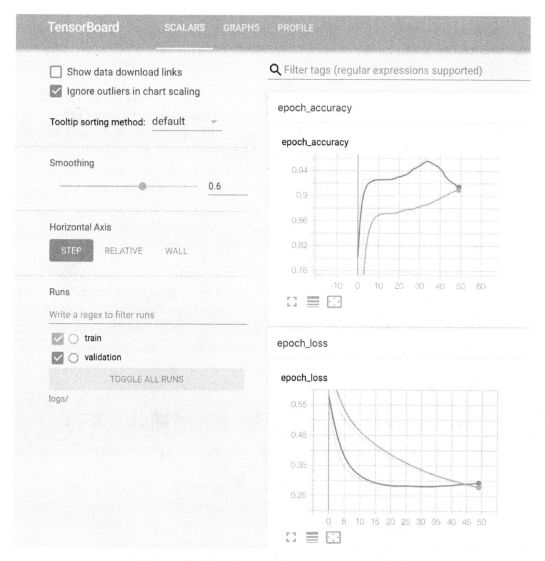

Figure 4.15: A visual representation of the loss and accuracy as a function of an epoch evaluated on the training and validation split in TensorBoard

You can see from TensorBoard that, with the addition of the **metrics** argument that was added in the model compilation process, there is an additional node in the architecture for the calculation of the accuracy metric. There is also an additional chart in the **SCALARS** tab showing the accuracy metric as a function of the epoch for the training and validation split.

You can see from the charts that, for the training set, the accuracy increases, and the loss decreases over time, which is a positive indication that the model is learning. However, on the validation split, the accuracy begins to decrease, and the loss begins to increase, which is a sign that the model may be overfitting to the training data.

In this exercise, you have learned how to build a classification model to discriminate between the binding properties of various molecules based on their other molecular attributes. The classification model was equivalent to a logistic regression model since it had only one layer and was preceded by a sigmoid activation function. With only one layer, there is a weight for each input feature and a single value for the bias. The sigmoid activation function transforms the output of the layer into a value between **0** and **1**, which is then rounded to represent your two classes. **0.5** and above represents one class, the molecule with binding properties, and below **0.5** represents the other class, molecules with non-binding properties.

The next activity will summarize your learning in this chapter by combining your knowledge of creating multi-layer ANNs as you accomplished in *Exercise 4.03, Creating a Multi-Layer ANN with TensorFlow*, and *Activity 4.01, Creating a Multi-Layer ANN with TensorFlow*, with your knowledge of creating classification models from *Exercise 4.04, Creating a Logistic Regression Model as an ANN with TensorFlow*. You will use the same dataset as in the preceding activity but change the target variable to make it more suitable for a classification task.

ACTIVITY 4.02: CREATING A MULTI-LAYER CLASSIFICATION ANN WITH TENSORFLOW

The feature dataset, `superconductivity.csv`, contains the properties of superconductors including the atomic mass of the material and its density. Importantly, the dataset also contains the critical temperature of the material, which is the temperature at which the material exhibits superconductive properties. You are required to determine which superconductors will express superconductive properties above the boiling point of nitrogen (77.36 K), thereby allowing superconductivity using liquid nitrogen, which is readily available. Your target variable will have a **true** value when the critical temperature is above 77.36 K and **false** below, indicating whether the material expresses superconductive properties above the boiling point of nitrogen.

> **NOTE**
>
> The `superconductivity.csv` file can be found here: http://packt.link/sOCPh.

Perform the following steps to complete this activity:

1. Open a Jupyter notebook to complete the activity.

2. Import the TensorFlow and pandas libraries.

3. Load in the **superconductivity.csv** dataset.

4. Drop any rows that have null values.

5. Set the target values to **true** when values of the **critical_temp** column are above **77.36** and **false** when below. The feature dataset is the remaining columns in the dataset.

6. Rescale the feature dataset using a standard scaler.

7. Initialize a model of the Keras **Sequential** class.

8. Add an input layer, three hidden layers of sizes **32**, **16**, and **8**, and an output layer with a **sigmoid** activation function of size **1** to the model.

9. Compile the model with an RMSprop optimizer with a learning rate equal to **0.0001** and binary cross-entropy for the loss and compute the accuracy metric.

10. Add a callback to write logs to TensorBoard.

11. Fit the model to the training data for **50** epochs and a validation split equal to 0%.

12. Evaluate the model on the training data.

13. View the model architecture and model-fitting process in TensorBoard.

> **NOTE**
>
> The solution to the activity can be found on page 477.

In this section, you have begun your foray into building, training, and evaluating classification models using TensorFlow. You have seen that they are built in much the same way as ANNs for regression tasks with the primary difference being the activation function on the output layer.

SUMMARY

In this chapter, you began your journey into creating ANNs in TensorFlow. You saw how simple it is to create regression and classification models by utilizing Keras layers. Keras layers are distinct classes that exist in a separate library that uses TensorFlow in the backend. Due to their popularity and ease of use, they are now included in TensorFlow and can be called in the same way as any other TensorFlow class.

You created ANNs with fully connected layers, varying layers, beginning with an ANN that resembles a linear regression algorithm, which is equivalent to a single-layer ANN. Then, you added layers to your ANN and added activation functions to the output of the layers. Activation functions can be used to determine whether a unit is fired or can be used to bind the value of the output from a given unit. Regression models aim to predict a continuous variable from the data provided. In the exercises and activities throughout this chapter, you attempted to predict the temperature in Seoul given data from weather stations, and the critical temperature of superconducting materials given various material properties.

Finally, you explored classification models, which aim to classify data into distinct classes. These models are similar to regression models in the way they are set up; however, an activation is used on the final output to bind the output values between two numbers that represent whether or not the data point is classified into the class. You began with binary classification models, which aim to classify the data into two classes, and demonstrated the concept of binary classification with an exercise in which you classified molecules into classes that represent their binding properties based on other attributes of the molecules' properties.

In the next chapter, you will explore classification models in more depth. You will learn some of the intricacies and capabilities of classification models, including how to classify data that has more than two distinct classes (known as multi-class classification), and whether data points can have more than one positive label (known as multi-label classification). You will address how to structure the architecture to create these models, the appropriate loss functions to use when training, and the relevant metrics to calculate to understand whether models are performing well.

5

CLASSIFICATION MODELS

OVERVIEW

In this chapter, you will explore different types of classification models. You will gain hands-on experience of using TensorFlow to build binary, multi-class, and multi-label classifiers. Finally, you will learn the concepts of model evaluation and how you can use different metrics to assess the performance of a model.

By the end of this chapter, you will have a good understanding of what classification models are and how programming with TensorFlow works.

INTRODUCTION

In the previous chapter, you learned about regression problems where the target variable is continuous. A continuous variable can take any value between a minimum and maximum value. You learned how to train such models with TensorFlow.

In this chapter, you will look at another type of supervised learning problem called classification, where the target variable is discrete — meaning it can only take a finite number of values. In industry, you will most likely encounter such projects where variables are aggregated into groups such as product tiers, or classes of users, customers, or salary ranges. The objective of a classifier is to learn the patterns from the data and predict the right class for observation.

For instance, in the case of a loan provider, a classification model will try to predict whether a customer is most likely to default in the coming year based on their profile and financial position. This outcome can only take two possible values (**yes** or **no**), which is a binary classification. Another classifier model could predict the ratings from 1 to 5 of a new movie for a user given their previous ratings and the information about this movie. When the outcome can be more than two possible values, you are dealing with a multi-class classification. Finally, there is a third type of classifier called multi-label where the model will predict more than a class. For example, a model will analyze an input image and predict whether there is a cat, a dog, or a mouse in the image. In such a case, the model will predict three different binary outputs (or labels).

You will go through each type of classifier in this chapter, detail their specificities, and explore how to measure the performance of these models.

BINARY CLASSIFICATION

As mentioned previously, binary classification refers to a type of supervised learning where the target variable can only take two possible values (or classes) such as true/false or yes/no. For instance, in the medical industry, you may want to predict whether a patient is more likely to have a disease based on their personal information such as age, height, weight, and/or medical measurements. Similarly, in marketing, advertisers might utilize similar information to optimize email campaigns.

Machine learning algorithms such as the random forest classifier, support vector classifier, or logistic regression work well for classification. Neural networks can also achieve good results for binary classification. It is extremely easy to turn a regression model such as those in the previous chapter into a binary classifier. There are only two key changes required: the activation function for the last layer and the loss function.

LOGISTIC REGRESSION

Logistic regression is one of the most popular algorithms for dealing with binary classification. As its name implies, it is an extension of the linear regression algorithm. A linear regression model predicts an output that can take an infinite number of values within a range. For logistic regression, you want your model to predict values between **0** and **1**. The value **0** usually corresponds to **false** (or **no**) while the value **1** refers to **true** (or **yes**).

In other terms, the output of logistic regression will be the probability of it being true. For example, if the output is **0.3**, you can say there is a probability of 30% that the result should be true (or yes). But as there are only two possible values, this will also mean there is a probability of 70% (100% – 30%) of having the outcome of false (or no):

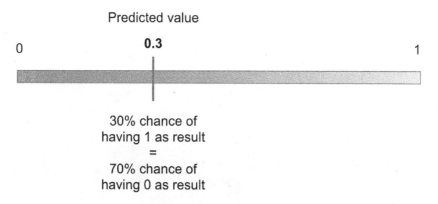

Figure 5.1: Output of logistic regression

Now that you know what the output of logistic regression is, you just need to find a function that can transform an input value that is continuous into a value between **0** and **1**. Luckily, such a mathematical function does exist, and it is called the **sigmoid function**. The formula for this function is as follows:

$$\sigma(x) = \frac{1}{1 + e^{-x}} = \frac{e^x}{e^x + 1}$$

Figure 5.2: Formula of the sigmoid function

e^x corresponds to the exponential function applied to **x**. The exponential function ranges from 0 to positive infinity. So, if **x** has a value close to positive infinite, the value of sigmoid will tend to **1**. On the other hand, if **x** is very close to negative infinite, then the value of sigmoid will tend to **0**:

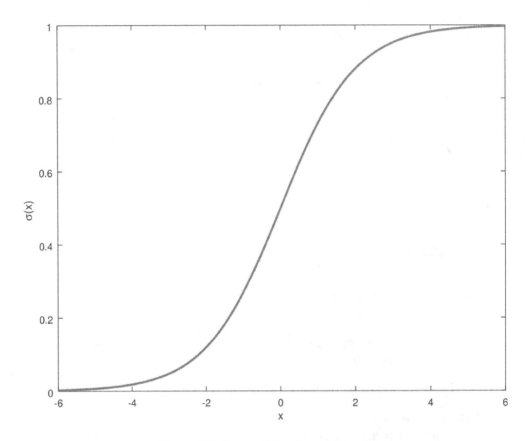

Figure 5.3: Curve of the sigmoid function

So, applying the sigmoid function on the output of a linear regression model turns it into logistic regression. The same logic holds for neural networks: if you apply the sigmoid function on a perceptron model (linear regression), you will get a binary classifier. To do so, you just need to specify sigmoid as the activation function of the last fully connected layer of a perceptron model. In TensorFlow, you specify the **activation** parameter as:

```
from tensorflow.keras.layers import Dense

Dense(1, activation='sigmoid')
```

The preceding code snippet shows how to define a fully connected layer with a single unit that can output any value and apply the sigmoid activation function to it. The result will then be within **0** and **1**. Now that you know how to modify a neural network's regression model to turn it into a binary classifier, you need to specify the relevant loss function.

BINARY CROSS-ENTROPY

In the previous section, you learned how to turn a linear regression model into a binary classifier. With neural networks, it is as simple as adding sigmoid as the activation function for the last fully connected layer. But there is another consideration that will impact the training of this model: the choice of the loss function.

For linear regression, the most frequently used loss functions are **mean squared error** and **mean absolute error** as seen in *Chapter 4, Regression and Classification Models*. These functions will calculate the difference between the predicted and the actual values, and the neural network model will update all its weights accordingly during backpropagation. For a binary classification, the typical loss function is **binary cross-entropy** (also called **log loss**). The formula for this function is as follows:

$$L(y, \hat{y}) = -\frac{1}{N} \sum_{i=0}^{N} (y_i * log(\hat{y}_i) + (1 - y_i) * log(1 - \hat{y}_i))$$

Figure 5.4: Formula of binary cross-entropy

y_i represents the actual value for the observation **i**.

$\hat{y}_i$ represents the predicted probability for the observation **i**.

N represents the total number of observations.

This formula looks quite complicated, but its logic is quite simple. Consider the following example of a single observation: the actual value is **1** and the predicted probability is **0.8**. If the preceding formula is applied, the result will be as follows:

$L(1, 0.8) = -1 * (1 * log(0.8) + (1 - 1) * log(1 - 0.8))$

Notice that the right-hand side of the equation is approximately zero:

$L(1, 0.8) = -(1 * log(0.8)) = 0.097$

So, the loss value will be very small as the predicted value is very close to the actual one.

Now consider another example where the actual value is **0** and the predicted probability is **0.99**. The result will be as follows:

$$L(0,0.99) = -1 * (0 * log(0.99) + (1 - 0) * log(1 - 0.99))$$

$$L(0,0.99) = -(log(1 - 0.99) = 2$$

The loss will be high in this case as the prediction is very different from the actual value.

The **binary cross-entropy function** is a good fit for assessing the difference between predicted and actual values for a binary classification. TensorFlow provides a class called **BinaryCrossentropy** that computes this loss:

```
from tensorflow.keras.losses import BinaryCrossentropy

bce = BinaryCrossentropy()
```

BINARY CLASSIFICATION ARCHITECTURE

The architecture for binary classifiers is extremely similar to that of linear regression as seen in *Chapter 4, Regression and Classification Models*. It is composed of an input layer that reads each observation of the input dataset, an output layer responsible for predicting the response variable, and some hidden layers that learn the patterns leading to the correct predictions. The following diagram shows an example of such an architecture:

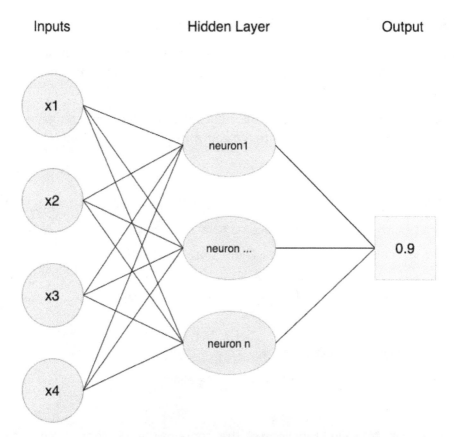

Figure 5.5: Architecture of the binary classifier

The only difference compared to linear regression is the output, which is a probability value between **0** and **1**. This probability indicates the likelihood of the occurrence for one of the two possible values. As seen previously, this is achieved using the sigmoid activation function and binary cross-entropy for backpropagation.

Now that you have seen all the elements to build a binary classifier, you can put this into practice with an exercise.

EXERCISE 5.01: BUILDING A LOGISTIC REGRESSION MODEL

In this exercise, you will build and train a logistic regression model in TensorFlow that will predict the winning team in a game of Dota 2 using some information about the game, such as the mode and type used.

You will be working on the Dota 2 dataset. Dota 2 is a popular computer game. The dataset contains information related to the game and the target variable indicates which team won.

> **NOTE**
>
> The training dataset can be accessed here: https://packt.link/Tdvdj.
>
> The test dataset can be accessed here: https://packt.link/4PsPN.
>
> The original dataset can be found here:
> https://archive.ics.uci.edu/ml/datasets/Dota2+Games+Results.

1. Open a new Jupyter notebook.

2. Import the pandas library and use **pd** as the alias:

```
import pandas as pd
```

3. Create a variable called **train_url** that contains the URL to the training set:

```
train_url = 'https://raw.githubusercontent.com/PacktWorkshops'\
            '/The-TensorFlow-Workshop/master/Chapter05'\
            '/dataset/dota2Train.csv'
```

4. Load the training dataset into a **DataFrame()** function called **X_train** using **read_csv()** method, provide the URL to the CSV file, and set **header=None** as the dataset doesn't provide column names. Print the first five rows of the DataFrame using **head()** method:

```
X_train = pd.read_csv(train_url, header=None)
X_train.head()
```

The expected output will be as follows:

	0	1	2	3	4	5	6	7	8	9	10	11	12	13	14	15	16	17	18	19	20	21	22	23	24	25	26
0	-1	223	2	2	0	0	0	0	0	0	0	0	0	0	1	0	0	0	1	0	0	0	-1	0	0	0	-1	0
1	1	152	2	2	0	0	0	1	0	-1	0	0	0	0	0	0	0	1	0	0	0	0	0	0	-1	0	0
2	1	131	2	2	0	0	0	1	0	-1	0	0	0	0	0	0	0	0	0	0	0	0	0	-1	0	1	0
3	1	154	2	2	0	0	0	0	0	0	-1	0	0	0	0	0	0	0	0	0	1	0	0	0	0	0	-1
4	-1	171	2	3	0	0	0	0	0	-1	0	0	-1	0	-1	0	0	0	0	1	0	0	0	0	0	0	0

5 rows × 117 columns

Figure 5.6: The first five rows of the Dota 2 training set

You can see that the dataset contains 117 columns, and they are all numeric. Note also that the target variable (column **0**) contains two different values: **-1** and **1**. As you will train a logistic regression model, the possible values should be **0** and **1**. You will need to replace the **-1** values with **0**.

5. Extract the target variable (column 0) using the **pop()** method and save it in a variable called **y_train**:

```
y_train = X_train.pop(0)
```

6. Replace all values with **-1** with **0** from the target variable using **replace()**, and print the first five rows using **head()** method:

```
y_train = y_train.replace(-1,0)
y_train.head()
```

The expected output will be as follows:

```
0    0
1    1
2    1
3    1
4    0
Name: 0, dtype: int64
```

Figure 5.7: The first five rows of the Dota 2 target variable from the training set

Now all the values from the target variable of the training set are either **0** or **1**.

7. Create a variable called **test_url** that contains the URL to the test set:

```
test_url = 'https://raw.githubusercontent.com/PacktWorkshops'\
           '/The-TensorFlow-Workshop/master/Chapter05/dataset'\
           '/dota2Test.csv'
```

8. Load the test dataset into a **DataFrame()** function called **X_test** using **read_csv()** method, provide the URL to the CSV file, and set **header=None** as the dataset doesn't provide column names. Print the first five rows using **head()** method:

```
X_test = pd.read_csv(test_url, header=None)
X_test.head()
```

The expected output will be as follows:

	0	1	2	3	4	5	6	7	8	9	10	11	12	13	14	15	16	17	18	19	20
0	-1	223	8	2	0	-1	0	0	0	0	0	0	1	0	0	0	0	0	0	0	0
1	1	227	8	2	0	0	0	0	0	0	0	0	0	1	0	0	0	0	1	0	0	0
2	-1	136	2	2	1	0	0	0	-1	0	0	0	1	1	0	0	0	0	0	0	0
3	1	227	2	2	-1	0	0	0	0	0	0	0	1	0	0	0	0	0	0	0	0
4	1	184	2	3	0	0	0	-1	0	0	0	-1	0	0	0	0	0	0	0	0	0

5 rows × 117 columns

Figure 5.8: The first five rows of the Dota 2 test set

The test set is very similar to the training one, and you will need to perform the same transformation on it.

9. Extract the target variable (column 0) using the **pop()** method and save it in a variable called **y_test**:

```
y_test = X_test.pop(0)
```

10. Replace all values with **-1** with **0** from the target variable using **replace()** method and print the first five rows using **head()** method:

```
y_test = y_test.replace(-1,0)
y_test.head()
```

The expected output will be as follows:

```
0    0
1    1
2    0
3    1
4    1
Name: 0, dtype: int64
```

Figure 5.9: The first five rows of the Dota 2 target variable from the test set

11. Import TensorFlow library and use **tf** as the alias:

```
import tensorflow as tf
```

12. Set the seed for TensorFlow as **8**, using **tf.random.set_seed()** to get reproducible results:

```
tf.random.set_seed(8)
```

13. Instantiate a sequential model using **tf.keras.Sequential()** and store it in a variable called **model**:

```
model = tf.keras.Sequential()
```

14. Import the **Dense()** class from **tensorflow.keras.layers**:

```
from tensorflow.keras.layers import Dense
```

15. Create a fully connected layer of **512** units with **Dense()** and specify ReLu as the activation function and the input shape as **(116,)**, which corresponds to the number of features from the dataset. Save it in a variable called **fc1**:

```
fc1 = Dense(512, input_shape=(116,), activation='relu')
```

16. Create a fully connected layer of **512** units with **Dense()** and specify ReLu as the activation function. Save it in a variable called **fc2**:

```
fc2 = Dense(512, activation='relu')
```

17. Create a fully connected layer of **128** units with **Dense()** and specify ReLu as the activation function. Save it in a variable called **fc3**:

```
fc3 = Dense(128, activation='relu')
```

18. Create a fully connected layer of **128** units with **Dense ()** and specify ReLu as the activation function. Save it in a variable called **fc4**:

```
fc4 = Dense(128, activation='relu')
```

19. Create a fully connected layer of **128** units with **Dense ()** and specify sigmoid as the activation function. Save it in a variable called **fc5**:

```
fc5 = Dense(1, activation='sigmoid')
```

20. Sequentially add all five fully connected layers to the model using **add ()** method:

```
model.add(fc1)
model.add(fc2)
model.add(fc3)
model.add(fc4)
model.add(fc5)
```

21. Print the summary of the model using **summary ()** method:

```
model.summary()
```

The expected output will be as follows:

```
Model: "sequential"
```

Layer (type)	Output Shape	Param #
dense (Dense)	(None, 512)	59904
dense_1 (Dense)	(None, 512)	262656
dense_2 (Dense)	(None, 128)	65664
dense_3 (Dense)	(None, 128)	16512
dense_4 (Dense)	(None, 1)	129

```
Total params: 404,865
Trainable params: 404,865
Non-trainable params: 0
```

Figure 5.10: Summary of the model architecture

The preceding output shows that there are five layers in your model (as expected) and displays the number of parameters at each layer. For example, the first layer contains 59,904 parameters, and the total number of parameters for this model is 404,855. All these parameters will be trained while fitting the model.

22. Instantiate a **BinaryCrossentropy()** function from **tf.keras.losses** and save it in a variable called **loss**:

```
loss = tf.keras.losses.BinaryCrossentropy()
```

23. Instantiate **Adam()** from **tf.keras.optimizers** with **0.001** as the learning rate and save it in a variable called **optimizer**:

```
optimizer = tf.keras.optimizers.Adam(0.001)
```

24. Compile the model using the **compile()** function and specify the optimizer and loss you just created in previous steps:

```
model.compile(optimizer=optimizer, loss=loss)
```

25. Start the model training process using **fit()** method on the training set for five epochs:

```
model.fit(X_train, y_train, epochs=5)
```

The expected output will be as follows:

```
Epoch 1/5
2896/2896 [==============================] - 22s 7ms/step - loss: 0.6923
Epoch 2/5
2896/2896 [==============================] - 21s 7ms/step - loss: 0.6695
Epoch 3/5
2896/2896 [==============================] - 22s 7ms/step - loss: 0.6663
Epoch 4/5
2896/2896 [==============================] - 22s 8ms/step - loss: 0.6658
Epoch 5/5
2896/2896 [==============================] - 22s 8ms/step - loss: 0.6650
<keras.callbacks.History at 0x7f0a611f4d10>
```

Figure 5.11: Logs of the training process

The preceding output shows the logs of each epoch during the training of the model. Note that it took around 15 seconds to process a single epoch and the loss value decreased from **0.6923** (first epoch) to **0.6650** (fifth epoch), so the model is slowly improving its performance by reducing the binary cross-entropy loss.

26. Predict the results of the test set using **predict()** method. Save it in a variable called **preds** and display its first five values:

```
preds = model.predict(X_test)
preds[:5]
```

The expected output will be as follows:

```
array([[0.45389563],
       [0.55355024],
       [0.6289436 ],
       [0.6477938 ],
       [0.34938103]], dtype=float32)
```

Figure 5.12: Predictions of the first five rows of the test set

The preceding output shows the probability of each prediction. Each value below **0.5** will be classified as **0** (first and last observation in this output) and all values greater than or equal to **0.5** will be **1** (second to fourth observations).

27. Display the first five true labels of the test set:

```
y_test[:5]
```

The expected output will be as follows:

```
0    0
1    1
2    0
3    1
4    1
Name: 0, dtype: int64
```

Figure 5.13: True labels of the first five rows of the test set

Comparing this output with the model predictions on the first five rows of the test set, there are some incorrect values: the third prediction (index **2**) should be a value of **0** and the last one should be **0**. So, out of these five observations, your binary classifiers made two mistakes.

In the section ahead, you will see how to properly evaluate the performance of a model with different metrics.

METRICS FOR CLASSIFIERS

In the previous section, you learned how to train a binary classifier to predict the right output: either **0** or **1**. In *Exercise 5.01, Building a Logistic Regression Model*, you looked at a few samples to assess the performance of the models that were built. Usually, you would evaluate a model not just on a small subset but on the whole dataset using a performance metric such as accuracy or F1 score.

ACCURACY AND NULL ACCURACY

One of the most widely used metrics for classification problems is accuracy. Its formula is quite simple:

$$accuracy = \frac{number\ of\ correct\ predictions}{total\ number\ of\ observations}$$

Figure 5.14: Formula of the accuracy metric

The maximum value for accuracy is **1**, which means the model correctly predicts 100% of the cases. Its minimum value is **0**, where the model can't predict any case correctly.

For a binary classifier, the number of correct predictions is the number of observations with a value of **0** or **1** as the correctly predicted value:

$$accuracy = \frac{number\ of\ correct\ predictions\ for\ value\ 0\ +\ number\ of\ correct\ predictions\ for\ value\ 1}{total\ number\ of\ observations}$$

Figure 5.15: Formula of the accuracy metric for a binary classifier

Say you are assessing the performance of two different binary classifiers predicting the outcome on 10,000 observations on the test set. The first model correctly predicted 5,000 instances of value **0** and 3,000 instances of value **1**. Its accuracy score will be as follows:

$$accuracy_{model1} = \frac{5000 + 3000}{10000} = \frac{8000}{10000} = 0.8$$

Figure 5.16: Formula for the accuracy of model1

The second model correctly predicted the value **0** for 500 cases and the value **1** for 1,500 observations. Its accuracy score will be as follows:

$$accuracy_{model2} = \frac{500 + 1500}{10000} = \frac{2000}{10000} = 0.2$$

Figure 5.17: Formula for the accuracy of model2

The first model predicts accurately 80% of the time, while the second model is only 20% accurate. In this case, you can say that model 1 is better than model 2.

Even though **0.8** is usually a relatively good score, this does not necessarily mean your model is performing well. For instance, say your dataset contains 9,000 cases of value **0** and 1,000 cases of value **1**. A very simple model that always predicts value **0** will achieve an accuracy score of 0.9. In this case, the first model is performing even less well than this extremely simple model. This characteristic of such a model that always predicts the most frequent value of a dataset is called the **null accuracy**. It is used as a baseline to compare with other trained models. In the preceding example, the null accuracy is **0.9** since the simple model predicts **0**, which is correct 90% of the time.

> **NOTE**
>
> The accuracy and null accuracy metrics are not specific to binary classification but can also be applied to other types of classification.

TensorFlow provides a class, **tf.keras.metrics.Accuracy**, that can calculate the accuracy score from tensors. This class has a method called **update_state()** that takes two tensors as input parameters and will compute the accuracy score between them. You can access this score by calling the **result()** method. The output result will be a tensor. You can use the **numpy()** method to convert it into a NumPy array. Here is an example of how to calculate the accuracy score:

```
from tensorflow.keras.metrics import Accuracy

preds = [1, 1, 1, 1, 0, 0]
target = [1, 0, 1, 0, 1, 0]

acc = Accuracy()
acc.update_state(preds, target)
acc.result().numpy()
```

This will result in the following accuracy score:

```
0.5
```

> **NOTE**
>
> TensorFlow doesn't provide a class for the null accuracy metric, but you can easily compute it using **Accuracy()** and provide a tensor with only **1** (or **0**) as the predictions.

PRECISION, RECALL, AND THE F1 SCORE

In the previous section, you learned how to use the accuracy metric to assess the performance of a model and compare it against a baseline called the null accuracy. The accuracy score is widely used as it is well known to non-technical audiences, but it does have some limitations. Consider the following example.

Predicted Value

		False	True
Actual Value	False	980	10
	True	9	1

Figure 5.18: Example of model predictions versus actual values

This model achieves an accuracy score of 0.981 $(\frac{980 + 1}{1000})$, which is quite high. But if this model is used to predict whether a person has a disease, it will only predict correctly in a single case. It incorrectly predicted in nine cases that these people are not sick while they actually have the given disease. At the same time, it incorrectly predicted sickness for 10 people who were actually healthy. This model's performance, then, is clearly unsatisfactory. Unfortunately, the accuracy score is simply an overall score, and it doesn't tell you where the model is performing badly.

Luckily, other metrics provide a better assessment of a model, such as precision, recall, or F1 score. All three of these metrics have the same range of values as the accuracy score: **1** is the perfect score, wherein all observations are predicted correctly, and **0** is the worst, wherein there is no correct prediction at all.

But before looking at them, you need to be familiar with the following definitions:

- **True Positive (TP)**: All the observations where the actual value and the corresponding prediction are both true

- **True Negative (TN)**: All the observations where the actual value and the corresponding prediction are both false

- **False Positive (FP)**: All the observations where the prediction is true, but the values are actually false

- **False Negative (FN)**: All the observations where the prediction is false, but the values are actually true

Taking the same example as *Figure 5.18*, you will get the following:

- TP = 1

- TN = 980

- FP = 10

- FN = 9

This is seen in the following table:

Predicted Value

		False	True
Actual Value	False	TN = 980	FP = 10
	True	FN = 9	TP = 1

Figure 5.19: Example of TP, TN, FP, and FN

The precision score is a metric that assesses whether a model has predicted a lot of FPs. Its formula is as follows:

$$precision = \frac{TP}{TP + FP}$$

Figure 5.20: Formula of precision

In the preceding example, the precision score will be $\frac{1}{1+10} = 0.09$. You can see this model is making a lot of mistakes and has predicted a lot of FPs compared to the actual TP.

Recall is used to assess the number of FNs compared to TPs. Its formula is as follows:

$$recall = \frac{TP}{TP + FN}$$

Figure 5.21: Formula of recall

In the preceding example, the recall score will be $\frac{1}{1+9} = 0.1$. With this metric, you can see that the model is not performing well and is predicting a lot of FNs.

Finally, the F1 score is a metric that combines both precision and recall (it is the harmonic mean of precision and recall). Its formula is as follows:

$$F1\ score\ = 2 * \frac{Precision * Recall}{Precision + Recall}$$

Figure 5.22: Formula for the F1 score

Taking the same example as the preceding, the F1 score will be

$2 \times \frac{0.09 \times 0.1}{0.09 + 0.1} = 2 \times \frac{0.009}{0.19} = 0.095$

The model has achieved an F1 score of **0.095**, which is very different from its accuracy score of **0.981**. So, the F1 score is a good performance metric when you want to emphasize the incorrect predictions—the score considers the number of FNs and FPs in the score, as well as the TPs and TNs.

> **NOTE**
>
> As with accuracy, precision, and recall performance metrics, the F1 score can also be applied to other types of classification.

You can easily calculate precision and recall with TensorFlow by using the respective classes of **Precision()** and **Recall()**:

```
from tensorflow.keras.metrics import Precision, Recall

preds = [1, 1, 1, 1, 0, 0]
target = [1, 0, 1, 0, 1, 0]

prec = Precision()
prec.update_state(preds, target)
print(f"Precision: {prec.result().numpy()}")

rec = Recall()
rec.update_state(preds, target)
print(f"Recall: {rec.result().numpy()}")
```

This results in the following output:

```
Precision: 0.6666666865348816
Recall: 0.5
```

Figure 5.23: Precision and recall scores of the provided example

> **NOTE**
>
> TensorFlow doesn't provide a class to calculate the F1 score, but this can easily be done by creating a custom metric. This will be covered in *Exercise 5.02, Classification Evaluation Metrics*.

CONFUSION MATRICES

A confusion matrix is not a performance metric *per se*, but more a graphical tool used to visualize the predictions of a model against the actual values. You have actually already seen an example of this in the previous section with *Figure 5.18*.

A confusion matrix will show all the possible values of the predictions on one axis (for example, the horizontal axis) and the actual values on the other axis (the vertical axis). At the intersection of each combination of predicted and actual values, you will record the number of observations that fall under this case.

For a binary classification, the confusion matrix will look like the following:

		Predicted Value	
		False	**True**
Actual Value	**False**	TN	FP
	True	FN	TP

Figure 5.24: Confusion matrix for a binary classification

The ideal situation will be that all the values sit on the diagonal of this matrix. This will mean your model is correctly predicting all possible values. All values outside of this diagonal are where your model made some mistakes.

> **NOTE**
>
> Confusion matrices can also be used for multi-class classification and are not specific to binary classification only.

Run the code below to see the confusion matrix:

```
from tensorflow.math import confusion_matrix

preds = [1, 1, 1, 1, 0, 0]
target = [1, 0, 1, 0, 1, 0]

print(confusion_matrix(target, preds))
```

This will display the following output:

```
tf.Tensor(
[[1 2]
 [1 2]], shape=(2, 2), dtype=int32)
```

Figure 5.25: TensorFlow confusion matrix

The preceding output shows the confusion matrix. From it, you can see that the model has predicted the following results: two TPs, one TN, two FPs, and one FN.

In the next exercise, you will apply these performance metrics to the same logistic regression model that you created in *Exercise 5.01, Building a Logistic Regression Model*.

EXERCISE 5.02: CLASSIFICATION EVALUATION METRICS

In this exercise, you will reuse the same logistic regression model as in *Exercise 5.01, Building a Logistic Regression Model*, and assess its performance by looking at different performance metrics: accuracy, precision, recall, and F1 score.

The original dataset was shared by Stephen Tridgell from the University of Sydney.

> **NOTE**
>
> The training dataset can be accessed here: https://packt.link/QJGpA.
>
> The test dataset can be accessed here: https://packt.link/ix5rW.
>
> The model from *Exercise 5.01, Building a Logistic Regression Model*, can be found here: https://packt.link/sSRQL.

Now, run the following instructions:

1. Open a new Jupyter notebook.

2. Import the pandas library and use **pd** as the alias:

```
import pandas as pd
```

3. Create a variable called **train_url** that contains the URL to the training set:

```
train_url = 'https://raw.githubusercontent.com/PacktWorkshops'\
            '/The-TensorFlow-Workshop/master/Chapter05/dataset'\
            '/dota2PreparedTrain.csv'
```

4. Load the training dataset into a **DataFrame()** function called **X_train** using **read_csv()** method, provide the URL to the CSV file, and set **header=None** as the dataset doesn't provide column names:

```
X_train = pd.read_csv(train_url, header=None)
```

5. Extract the target variable (column **0**) using the **pop()** method and save it in a variable called **y_train**:

```
y_train = X_train.pop(0)
```

6. Create a variable called **test_url** that contains the URL to the test set:

```
test_url = 'https://raw.githubusercontent.com/PacktWorkshops'\
           '/The-TensorFlow-Workshop/master/Chapter05/dataset'\
           '/dota2PreparedTest.csv'
```

7. Load the test dataset into a **DataFrame()** function called **X_test** using **read_csv()** method, provide the URL to the CSV file, and set **header=None** as the dataset doesn't provide column names:

```
X_test = pd.read_csv(test_url, header=None)
```

8. Extract the target variable (column **0**) using the **pop()** method and save it in a variable called **y_test**:

```
y_test = X_test.pop(0)
```

9. Import the **tensorflow** library using **tf** as the alias and import the **get_file()** method from **tensorflow.keras.utils**:

```
import tensorflow as tf
from tensorflow.keras.utils import get_file
```

10. Create a variable called **model_url** that contains the URL to the model:

```
model_url = 'https://github.com/PacktWorkshops'\
            '/The-TensorFlow-Workshop/blob/master/Chapter05'\
            'model/exercise5_01_model.h5?raw=true'
```

11. Download the model locally using the **get_file()** method by providing the name (**exercise5_01_model.h5**) of the file and its URL. Save the output to a variable called **model_path**:

```
model_path = get_file('exercise5_01_model.h5', model_url)
```

12. Load the model with **tf.keras.models.load_model()** and specify the local path to the model:

```
model = tf.keras.models.load_model(model_path)
```

13. Print the model summary using the **summary()** method:

```
model.summary()
```

The expected output will be as follows:

```
Model: "sequential_1"
```

Layer (type)	Output Shape	Param #
dense_5 (Dense)	(None, 512)	59904
dense_6 (Dense)	(None, 512)	262656
dense_7 (Dense)	(None, 128)	65664
dense_8 (Dense)	(None, 128)	16512
dense_9 (Dense)	(None, 1)	129

```
Total params: 404,865
Trainable params: 404,865
Non-trainable params: 0
```

Figure 5.26: Summary of the model

The preceding output shows the same architecture as the model from *Exercise 5.01, Building a Logistic Regression Model.*

14. Predict the results of the test set using **predict()** method. Save it in a variable called **preds_proba** and display its first five values:

```
preds_proba = model.predict(X_test)
preds_proba[:5]
```

The expected output will be as follows:

```
array([[0.        ],
       [0.43662974],
       [0.52438205],
       [0.61323625],
       [0.6282078 ]], dtype=float32)
```

Figure 5.27: Predicted probabilities of the test set

The outputs are the predicted probabilities of being **1** (or true) for each observation. You need to convert these probabilities into **0** and **1** only. To do so, you will need to consider all cases with a probability greater than or equal to **0.5** to be **1** (or true), and **0** (or false) for the records with a probability lower than **0.5**.

15. Convert the predicted probabilities into **1** when the probability is greater than or equal to **0.5**, and **0** when below **0.5**. Save the results in a variable called **preds** and print its first five rows:

```
preds = preds_proba >= 0.5
preds[:5]
```

The expected output will be as follows:

```
array([[False],
       [False],
       [ True],
       [ True],
       [ True]])
```

Figure 5.28: Predictions of the test set

Now the predictions have been converted to binary values: true (which equals **1**) and false (which equals **0**).

16. Import **Accuracy**, **Precision**, and **Recall** from **tensorflow.keras.metrics**:

```
from tensorflow.keras.metrics import Accuracy, Precision, Recall
```

17. Instantiate **Accuracy**, **Precision**, and **Recall** objects and save them in variables called **acc**, **pres**, and **rec**, respectively:

```
acc = Accuracy()
prec = Precision()
rec = Recall()
```

18. Calculate the accuracy score on the test set with the **update_state()**, **result()**, and **numpy()** methods. Save the results in a variable called **acc_results** and print its content:

```
acc.update_state(preds, y_test)
acc_results = acc.result().numpy()
acc_results
```

The expected output will be as follows:

```
0.59650314
```

This model achieved an accuracy score of **0.597**.

19. Calculate the precision score on the test set with the **update_state()**, **result()**, and **numpy()** methods. Save the results in a variable called **prec_results** and print its content:

```
prec.update_state(preds, y_test)
prec_results = prec.result().numpy()
prec_results
```

The expected output will be as follows:

```
0.59578335
```

This model achieved a precision score of **0.596**.

20. Calculate the recall score on the test set with the **update_state()**, **result()**, and **numpy()** methods. Save the results in a variable called **rec_results** and print its content:

```
rec.update_state(preds, y_test)
rec_results = rec.result().numpy()
rec_results
```

The expected output will be as follows:

```
0.6294163
```

This model achieved a recall score of **0.629**.

21. Calculate the F1 score by applying the formula shown in the previous section. Save the result in a variable called **f1** and print its content:

```
f1 = 2*(prec_results * rec_results) / (prec_results + rec_results)
f1
```

The expected output will be as follows:

```
0.6121381493171637
```

Overall, the model has achieved quite a low score close to **0.6** for all four different metrics: accuracy, precision, recall, and F1 score. So, this model is making almost as many correct predictions as bad ones. You may try on your own to build another model and see whether you can improve its performance.

In the section ahead, you will be looking at expanding classification to more than two possible values with multi-class classification.

MULTI-CLASS CLASSIFICATION

With binary classification, you were limited to dealing with target variables that can only take two possible values: **0** and **1** (false or true). Multi-class classification can be seen as an extension of this and allows the target variable to have more than two values (or you can say binary classification is just a subset of multi-class classification). For instance, a model that predicts different levels of disease severity for a patient or another one that classifies users into different groups based on their past shopping behaviors will be multi-class classifiers.

In the next section, you will dive into the softmax function, which is used for multi-class classification.

THE SOFTMAX FUNCTION

Binary classifiers require a specific activation function for the last fully connected layer of a neural network, which is sigmoid. The activation function specific to multi-class classifiers is different. It is softmax. Its formula is as follows:

$$softmax(y_i) = \frac{e^{y_i}}{\Sigma_j e^{y_j}}$$

Figure 5.29: Formula of softmax function

y_i corresponds to the predicted value for class **i**.

y_j corresponds to the predicted value for class **j**.

This formula will be applied to each possible value of the target variable. If you have 10 possible values, then this activation function will calculate 10 different softmax values.

Note that softmax exponentiates the predicted values on both the numerator and the denominator. The reason behind this is that the exponential function magnifies small changes between predicted values and makes probabilities lie closer to **0** or **1** for the purpose of interpreting the resulting output. For instance, **exp(2)** = **7.39** while **exp(2.2)** = **9.03**. So, if two classes have predicted values close to each other, the difference between their exponentiated values will be much bigger and therefore it will be easier to select the higher one.

The result of the softmax function is between **0** and **1** as the method divides the value for one class by the sum of all the classes. So, the actual output of a softmax function is the probability of the relevant class being the final prediction:

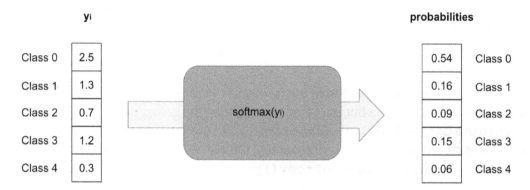

Figure 5.30: Example of softmax transformation

In the preceding example, the target variable has five different values, and the softmax function transforms them into probabilities. The first class (**0**) is the one with the highest probability, and this will be the final prediction.

CATEGORICAL CROSS-ENTROPY

Multi-class classification also requires a specific loss function that is different from the binary cross-entropy for binary classifiers. For multi-class classification, the loss function is categorical cross-entropy. Its formula is as follows:

$$L(y, \hat{y}) = -\frac{1}{N} \sum_{i=0}^{N} \sum_{j=0}^{M} (y_{ij} * log(\hat{y}_{ij}))$$

Figure 5.31: Formula of categorical cross-entropy

y_{ij} represents the probability of the actual value for the observation **i** to be of class **j**.

$\hat{y}_i$ represents the predicted probability for the observation **i** to be of class **j**.

TensorFlow provides two different classes for this loss function: `CategoricalCrossentropy()` and `SparseCategoricalCrossentropy()`:

```
from tensorflow.keras.losses import CategoricalCrossentropy,
                            SparseCategoricalCrossentropy


cce = CategoricalCrossentropy()
scce = SparseCategoricalCrossentropy()
```

The difference between them lies in the format of the target variable. If the actual values are stored as a one-hot encoding representing the actual class, then you will need to use `CategoricalCrossentropy()`. On the other hand, if the response variable is stored as integers for representing the actual classes, you will have to use `SparseCategoricalCrossentropy()`:

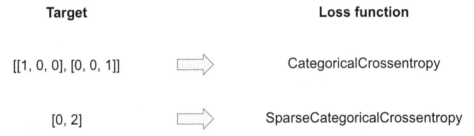

Target		Loss function
[[1, 0, 0], [0, 0, 1]]		CategoricalCrossentropy
[0, 2]		SparseCategoricalCrossentropy

Figure 5.32: Loss function to be used depending on the format of the target variable

The output of a multi-class model will be a vector containing probabilities for each class of the target variable, such as the following:

```
import numpy as np
preds_proba = np.array([0.54, 0.16, 0.09, 0.15, 0.06])
```

The first value (**0.54**) corresponds to the probability of having the class at index 0, **0.016** is the probability of the class at index 1, while **0.09** corresponds to the probability for the class of index 2, and so on.

In order to get the final prediction (that is, the class with the highest probability), you need to use the **argmax()** function, which will look at all the values from a vector, find the maximum one, and return the index associated with it:

```
preds_proba.argmax()
```

This will display the following output:

```
0
```

In the preceding example, the final prediction is **class** 0, which corresponds to the vector index with the highest probability (**0.54**).

MULTI-CLASS CLASSIFICATION ARCHITECTURE

The architecture for a multi-class classifier is very similar to logistic regression, except that the last layer will contain more units. Each of them corresponds to a class of the target variable. For instance, if you are building a model that takes as input a vector of size 6 and predicts a response with three different values with a single hidden layer, its architecture will look like the following:

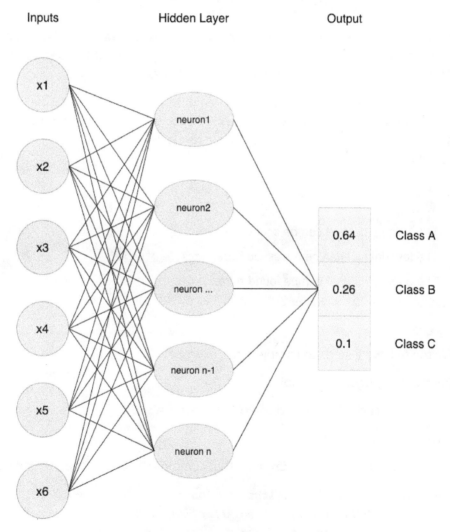

Figure 5.33: Architecture of a multi-class classifier

The softmax activation function at the last layer provides a probability of occurrence for each of the possible classes: **A**, **B**, and **C**. These probabilities are dependent on each other as there should be only one class predicted at the end. If class **A** is more likely to be the prediction (as in the preceding example), then the probabilities for the remaining classes (**B** and **C**) should be lower. Note that the sum of all the class probabilities equals **1**. So, they are indeed dependent on one another.

Now that you know all the building blocks, you can build a multi-class classifier in the following exercise.

EXERCISE 5.03: BUILDING A MULTI-CLASS MODEL

In this exercise, you will build and train a multi-class classifier in TensorFlow that will predict the radiator position of a space shuttle from eight different values using the nine different numerical features provided in this dataset.

The target variable (last column) contains seven different levels: **Rad.Flow**, **Fpv.Close**, **Fpv.Open**, **High**, **Bypass**, **Bpv.Close**, and **Bpv.Open**. Your goal is to accurately predict one of these seven levels using the nine features from the dataset.

> **NOTE**
>
> The training dataset can be accessed here: https://packt.link/46iMY.
>
> The test dataset can be accessed here: https://packt.link/dcNPt.
>
> The original dataset can be found here: http://archive.ics.uci.edu/ml/datasets/Statlog+%28Shuttle%29.

Perform the following steps to complete the exercise:

1. Open a new Jupyter notebook.

2. Import the pandas library and use **pd** as the alias:

```
import pandas as pd
```

3. Create a variable called **train_url** that contains the URL to the training set:

```
train_url = 'https://raw.githubusercontent.com/PacktWorkshops'\
            '/The-TensorFlow-Workshop/master/Chapter05'\
            '/dataset/shuttle.trn'
```

4. Load the training dataset into a DataFrame called **X_train** using the
    read_table() method, provide the URL to the CSV file, use **header=None**
    as the dataset doesn't provide column names, and use **sep=' '** as each
    column is separated by spaces in this dataset. Print the first five rows using
    head() method:

```
X_train = pd.read_table(train_url, header=None, sep=' ')
X_train.head()
```

The expected output will be as follows:

	0	1	2	3	4	5	6	7	8	9
0	50	21	77	0	28	0	27	48	22	2
1	55	0	92	0	0	26	36	92	56	4
2	53	0	82	0	52	-5	29	30	2	1
3	37	0	76	0	28	18	40	48	8	1
4	37	0	79	0	34	-26	43	46	2	1

Figure 5.34: The first five rows of the training set

You can see that the dataset contains 10 columns, and they are all numeric. Also,
note that the target variable (column **9**) contains different class values.

5. Extract the target variable (column **9**) using the **pop()** method and save it in a
    variable called **y_train**:

```
y_train = X_train.pop(9)
```

6. Create a variable called **test_url** that contains the URL to the test set:

```
test_url = 'https://raw.githubusercontent.com/PacktWorkshops'\
           '/The-TensorFlow-Workshop/master/Chapter05/dataset'\
           '/shuttle.tst'
```

7. Load the test dataset into a DataFrame called **X_test** using **read_table()**, provide the URL to the CSV file, set **header=None** as the dataset doesn't provide column names, and use **sep=' '** as each column is separated by a space in this dataset. Print the first five rows using **head()** method.

```
X_test = pd.read_table(test_url, header=None, sep=' ')
X_test.head()
```

The expected output will be as follows:

	0	1	2	3	4	5	6	7	8	9
0	55	0	81	0	-6	11	25	88	64	4
1	56	0	96	0	52	-4	40	44	4	4
2	50	-1	89	-7	50	0	39	40	2	1
3	53	9	79	0	42	-2	25	37	12	4
4	55	2	82	0	54	-6	26	28	2	1

Figure 5.35: The first five rows of the test set

You can see that the test set is very similar to the training one.

8. Extract the target variable (column **9**) using the **pop()** method and save it in a variable called **y_test**:

```
y_test = X_test.pop(9)
```

9. Import the TensorFlow library and use **tf** as the alias:

```
import tensorflow as tf
```

10. Set the seed for TensorFlow as **8** using **tf.random.set_seed()** to get reproducible results:

```
tf.random.set_seed(8)
```

11. Instantiate a sequential model using **tf.keras.Sequential()** and store it in a variable called **model**:

```
model = tf.keras.Sequential()
```

12. Import the **Dense()** class from **tensorflow.keras.layers**:

```
from tensorflow.keras.layers import Dense
```

13. Create a fully connected layer of **512** units with **Dense()** and specify ReLu as the activation function and the input shape as **(9,)**, which corresponds to the number of features from the dataset. Save it in a variable called **fc1**:

```
fc1 = Dense(512, input_shape=(9,), activation='relu')
```

14. Create a fully connected layer of **512** units with **Dense()** and specify ReLu as the activation function. Save it in a variable called **fc2**:

```
fc2 = Dense(512, activation='relu')
```

15. Create a fully connected layer of **128** units with **Dense()** and specify ReLu as the activation function. Save it in a variable called **fc3**:

```
fc3 = Dense(128, activation='relu')
```

16. Again, create a fully connected layer of **128** units with **Dense()** and specify ReLu as the activation function. Save it in a variable called **fc4**:

```
fc4 = Dense(128, activation='relu')
```

17. Create a fully connected layer of 128 units with **Dense()** and specify softmax as the activation function. Save it in a variable called **fc5**:

```
fc5 = Dense(8, activation='softmax')
```

18. Sequentially add all five fully connected layers to the model using **add()** method.

```
model.add(fc1)
model.add(fc2)
model.add(fc3)
model.add(fc4)
model.add(fc5)
```

19. Print the summary of the model using **summary()** method:

```
model.summary()
```

The expected output will be as follows:

```
Model: "sequential"
```

Layer (type)	Output Shape	Param #
dense (Dense)	(None, 512)	5120
dense_1 (Dense)	(None, 512)	262656
dense_2 (Dense)	(None, 128)	65664
dense_3 (Dense)	(None, 128)	16512
dense_4 (Dense)	(None, 8)	1032

```
Total params: 350,984
Trainable params: 350,984
Non-trainable params: 0
```

Figure 5.36: Summary of the model architecture

The preceding output shows that there are five layers in your model (as expected) and tells you the number of parameters at each layer. For example, the first layer contains **5,120** parameters and the total number of parameters for this model is **350,984**. All these parameters will be trained while fitting the model.

20. Instantiate **SparseCategoricalCrossentropy()** from **tf.keras.losses** and save it in a variable called **loss**:

```
loss = tf.keras.losses.SparseCategoricalCrossentropy()
```

21. Instantiate **Adam()** from **tf.keras.optimizers** with **0.001** as the learning rate and save it in a variable called **optimizer**:

```
optimizer = tf.keras.optimizers.Adam(0.001)
```

22. Compile the model using the **compile()** method and specify the optimizer and loss parameters, with accuracy as the metric to be reported:

```
model.compile(optimizer=optimizer, loss=loss, \
              metrics=['accuracy'])
```

23. Start the model training process using **fit()** method on the training set for five epochs:

```
model.fit(X_train, y_train, epochs=5)
```

The expected output will be as follows:

```
Epoch 1/5
1360/1360 [==============================] - 11s 7ms/step - loss: 0.4522 - accuracy: 0.9785
Epoch 2/5
1360/1360 [==============================] - 10s 7ms/step - loss: 0.2207 - accuracy: 0.9937
Epoch 3/5
1360/1360 [==============================] - 11s 8ms/step - loss: 0.0831 - accuracy: 0.9954
Epoch 4/5
1360/1360 [==============================] - 10s 7ms/step - loss: 0.0863 - accuracy: 0.9960
Epoch 5/5
1360/1360 [==============================] - 11s 8ms/step - loss: 0.0737 - accuracy: 0.9963
<tensorflow.python.keras.callbacks.History at 0x2431133a948>
```

Figure 5.37: Logs of the training process

The preceding output shows the logs of each epoch during the training of the model. Note that it took around 7 seconds to process a single epoch, and the loss value decreased from **0.5859** (first epoch) to **0.0351** (fifth epoch).

24. Evaluate the performance of the model on the test set using the **evaluate()** method:

```
model.evaluate(X_test, y_test)
```

The expected output will be as follows:

```
454/454 [==============================] - 2s 3ms/step - loss: 0.1221 - accuracy: 0.9981
[0.12214773148298264, 0.9980689883232117]
```

Figure 5.38: Performance of the model on the test set

In this exercise, you learned how to build and train a multi-class classifier to predict an outcome composed of eight different classes. Your model achieved an accuracy score close to **0.997** on both the training and test sets, which is quite remarkable. This implies that your model correctly predicts the right class in the majority of cases.

Now, let's consolidate your learning in the following activity.

ACTIVITY 5.01: BUILDING A CHARACTER RECOGNITION MODEL WITH TENSORFLOW

In this activity, you are tasked with building and training a multi-class classifier that will recognize the 26 letters of the alphabet from images. In this dataset, the images have been converted into 16 different statistical measures that will constitute our features. The goal of this model is to determine which of the 26 characters each observation belongs to.

The original dataset was shared by David J. Slate of the Odesta Corporation, and can be found here: http://archive.ics.uci.edu/ml/datasets/Letter+Recognition.

The dataset can be accessed from here: https://packt.link/j8m3L.

The following steps will help you to complete the activity:

1. Load the data with **read_csv()** from pandas.

2. Extract the target variable with **pop()** method from pandas.

3. Split the data into training (the first 15,000 rows) and test (the last 5,000 rows) sets.

4. Build the multi-class classifier with five fully connected layers of **512, 512, 128, 128**, and **26** units, respectively.

5. Train this model on the training set.

6. Evaluate its performance on the test set with **evaluate()** method from TensorFlow.

7. Print the confusion matrix with **confusion_matrix()** from TensorFlow.

The expected output is as follows:

```
<tf.Tensor: shape=(26, 26), dtype=int32, numpy=
array([[155,    0,    0,    0,    0,    0,    0,    0,    0,    0,    0,    0,    0,
           1,    0,    0,    0,    1,    0,    0,   12,   15,    0,    0,    0,    0],
        [   0,  184,    2,    0,    0,    2,    0,    0,    1,    0,    0,    0,    0,
          12,    2,    0,    0,    0,    0,    1,    0,    0,    1,    0,    0,    0],
        [   0,    2,  214,    0,    0,    0,    0,    0,    0,    0,    0,    0,    0,
           0,    0,    0,    0,    0,    0,    0,    0,    0,    0,    0,    0,    0],
        [   0,    0,   10,  172,    0,    0,    0,    1,    0,    1,    0,    5,    0,
           0,    0,    2,    0,    1,    1,    0,    0,    3,    0,    2,    0,    0],
        [   0,    0,    2,    0,  152,    0,    1,    0,    0,    1,    0,    5,    1,
           1,    6,    3,    0,    5,    1,    1,    8,    0,   19,    0,    2,    0],
        [   0,    0,    0,    0,    2,  178,    2,    0,    0,    0,    1,    0,    0,
           3,    0,    0,    0,    0,    0,    2,    0,    0,    4,    0,    0,    6],
        [   1,    1,   13,    0,    1,    0,  153,    0,    0,    0,    0,    1,    0,
           0,    0,    2,    0,    0,    0,    0,    1,    0,    0,    0,    0,    0],
        [   0,    0,    2,    0,    0,    0,    0,  198,    0,    0,    0,    0,    0,
           0,    0,    2,    0,    0,    0,    0,    0,    4,    0,    0,    0,    0],
```

Figure 5.39: Confusion matrix of the test set

> **NOTE**
>
> The solution to this activity is available on page 483.

MULTI-LABEL CLASSIFICATION

Multi-label classification is another type of classification where you predict not only one target variable as in binary or multi-class classification, but several response variables at the same time. For instance, you can predict multiple outputs for the different objects present in an image (for instance, a model will predict whether there is a cat, a man, and a car in a given picture) or you can predict multiple topics for an article (such as whether the article is about the economy, international news, and manufacturing).

Implementing a multi-label classification with neural networks is extremely easy, and you have already learned everything required to build one. In TensorFlow, a multi-label classifier's architecture will look the same as for multi-class, with a final output layer with multiple units corresponding to the number of target variables you want to predict. But instead of using softmax as the activation function and categorical cross-entropy as the loss function, you will use sigmoid and binary cross-entropy as the activation and loss functions, respectively.

The sigmoid function will predict the probability of occurrence for each target variable:

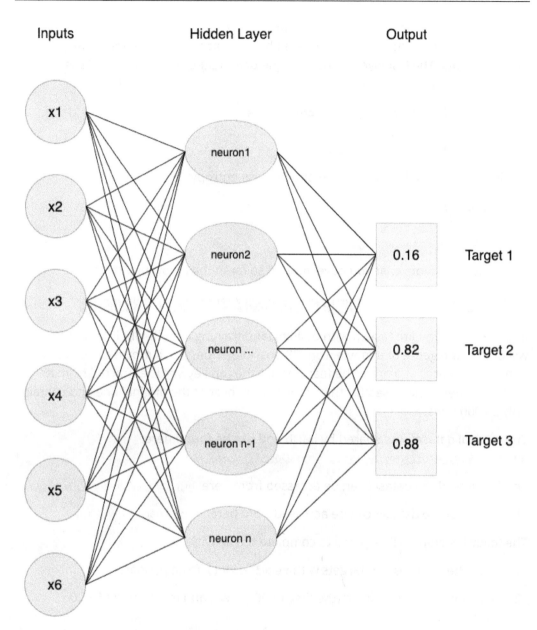

Figure 5.40: Architecture of the multi-label classifier

In the preceding example, you have three target variables and each of them has a probability of occurrence that is independent of the others (their sum will not equal 1). This model predicts that targets **2** and **3** are very likely to be the outputs for this observation.

Conceptually, multi-label classification combines several logistic regression models. They will share the same parameters (weights and biases) but with independent binary outputs. The last layer of the example of a multi-class classifier in TensorFlow will look like this:

```
from tensorflow.keras.layers import Dense

Dense(3, activation='sigmoid')
```

The loss function to be used will be binary cross-entropy:

```
from tensorflow.keras.losses import BinaryCrossentropy

bce = BinaryCrossentropy()
```

Now, put into action what you have learned so far in the following activity.

ACTIVITY 5.02: BUILDING A MOVIE GENRE TAGGING A MODEL WITH TENSORFLOW

In this activity, you are tasked with building and training a multi-label classifier that will predict the genre of a movie from 28 possible values. Each movie can be assigned to multiple genres at a time. The features are the top keywords extracted from its synopsis. The dataset used for this activity is a subset of the original one and contains only 20,000 rows.

The original dataset was shared by IMDb and can be found here: http://www.uco.es/kdis/mllresources/#ImdbDesc.

The features of the dataset can be accessed from here: https://packt.link/yW5ru.

The targets of the dataset can be accessed from here: https://packt.link/8f1mb.

The following steps will help you to complete the activity:

1. Load the features and targets with **read_csv()** from pandas.

2. Split the data into training (the first 15,000 rows) and test (the last 5,000 rows) sets.

3. Build the multi-class classifier with five fully connected layers of **512**, **512**, **128**, **128**, and **28** units, respectively.

4. Train this model on the training set.

5. Evaluate its performance on the test set with **evaluate()** method from TensorFlow.

The expected output is as follows:

```
157/157 [==============================] - 1s 5ms/step - loss: 0.9912 - accuracy: 0.1346
[0.9911884665489197, 0.13459999859333038]
```

Figure 5.41: Expected output of Activity 5.02

NOTE

The solution to this activity is available on page 489.

SUMMARY

You started your journey in this chapter with an introduction to classification models and their differences compared with regression models. You learned that the target variable for classifiers can only contain a limited number of possible values.

You then explored binary classification, wherein the response variable can only be from two possible values: **0** or **1**. You uncovered the specificities for building a logistic regression model with TensorFlow using the sigmoid activation function and binary cross-entropy as the loss function, and you built your own binary classifier for predicting the winning team on the video game Dota 2.

After this, you went through the different performance metrics that can be used to assess the performance of classifier models. You practiced calculating accuracy, precision, recall, and F1 scores with TensorFlow, and also plotted a confusion matrix, which is a visual tool to see where the model made correct and incorrect predictions.

Then you dove into the topic of multi-class classification. The difference between such models and binary classifiers is that their response variables can take more than two possible values. You looked at the softmax activation function and the categorical cross-entropy loss function, which are used for training such models in TensorFlow.

Finally, in the last section, you learned about multi-label classification, wherein the output can be multiple classes at the same time. In TensorFlow, such models can be easily built by constructing an architecture similar to multi-class classification but using sigmoid and binary cross-entropy, respectively, as the activation and loss functions.

In the next chapter, you will learn how to prevent model overfitting by applying some regularization techniques, which will help models to better generalize unseen data.

6

REGULARIZATION AND HYPERPARAMETER TUNING

OVERVIEW

In this chapter, you will be introduced to hyperparameter tuning. You will get hands-on experience in using TensorFlow to perform regularization on deep learning models to reduce overfitting. You will explore concepts such as L1, L2, and dropout regularization. Finally, you will look at the Keras Tuner package for performing automatic hyperparameter tuning.

By the end of the chapter, you will be able to apply regularization and tune hyperparameters in order to reduce the risk of overfitting your model and improve its performance.

INTRODUCTION

In the previous chapter, you learned how classification models can solve problems when the response variable is discrete. You also saw different metrics used to assess the performance of such classifiers. You got hands-on experience in building and training binary, multi-class, and multi-label classifiers with TensorFlow.

When evaluating a model, you will face three different situations: model overfitting, model underfitting, and model performing. The last one is the ideal scenario, in which a model is accurately predicting the right outcome and is generalizing to unseen data well.

If a model is underfitting, it means it is neither achieving satisfactory performance nor accurately predicting the target variable. In this case, a data scientist can try tuning different hyperparameters and finding the best combination that will boost the accuracy of the model. Another possibility is to improve the input dataset by handling issues such as the cleanliness of the data or feature engineering.

A model is overfitting when it can only achieve high performance on the training set and performs poorly on the test set. In this case, the model has only learned patterns from the data relevant to the data used for training. Regularization helps to lower the risk of overfitting.

REGULARIZATION TECHNIQUES

The main goal of a data scientist is to train a model that achieves high performance and generalizes to unseen data well. The model should be able to predict the right outcome on both data used during the training process and new data. This is the reason why a model is always assessed on the test set. This set of data serves as a proxy to evaluate the ability of the model to output correct results while in production.

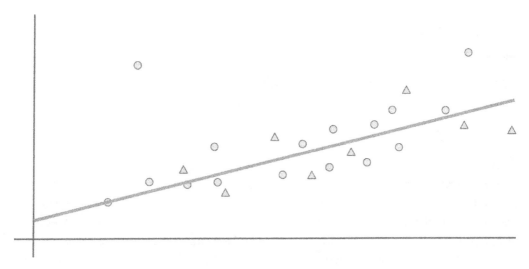

Figure 6.1: Model not overfitting or underfitting

In *Figure 6.1*, the linear model (line) seems to predict relatively accurate results for both the training (circles) and test (triangles) sets.

But sometimes a model fails to generalize well and will overfit the training set. In this case, the performance of the model will be very different between the training and test sets.

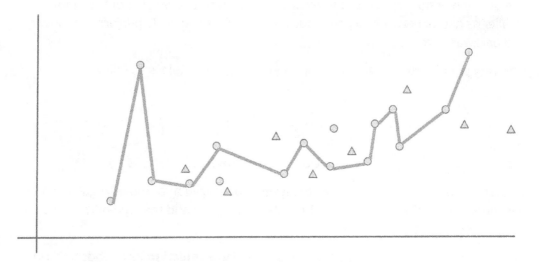

Figure 6.2: Model overfitting

Figure 6.2 shows the model (line) has only learned to predict accurately for the training set (circles) and is performing badly on the test set (triangles). This model is clearly overfitting.

Fortunately, there are **regularization techniques** that a data scientist can use to reduce and prevent overfitting, defined in the following sections.

L1 REGULARIZATION

For deep learning models, overfitting happens when some of the features have higher weights than they should. The model puts too much emphasis on these features as it believes they are extremely important for predicting the training set. Unfortunately, these features are less relevant for the test set or any new unseen data. Regularization techniques try to penalize such weights and reduce their importance to the model predictions.

There are multiple ways to perform regularization. One of them is to add a regularization component to the cost function:

$$cost\ function = loss + regularization$$

Figure 6.3: Adding a regularization component to the cost function

The addition of this regularization component will lead the weights of the model to be smaller as neural networks try to reduce the cost function while performing forward and backward propagations.

One very popular regularization component is L1. Its formula is as follows:

$$\lambda \sum_{i=1}^{n} |W_i|$$

Figure 6.4: L1 regularization

λ is a hyperparameter that defines the level of penalization of the L1 regularization. $\mathbf{W}$ is the weight of the model. With L1 regularization, you add the sum of the absolute value of the weights to the model loss.

L1 regularization is sometimes referred to as **feature selection** as it tends to push the weights of non-relevant features to **0**. Therefore, only the relevant features are used for making predictions.

In TensorFlow, you can define L1 regularization with the following code snippet:

```
from tensorflow.keras.regularizers import l1

l1_reg = l1(l=0.01)
```

The **l** parameter corresponds to the λ hyperparameter. The instantiated L1 regularization can then be added to any layer from TensorFlow Keras:

```
from tensorflow.keras.layers import Dense

Dense(10, kernel_regularizer=l1_reg)
```

In the preceding example, you added the L1 regularizer that you defined earlier to a fully connected layer of **10** units.

L2 REGULARIZATION

L2 regularization is similar to *L1* in that it adds a regularization component to the cost function, but its formula is different:

$$\lambda \sum_{i=1}^{n} W_i^2$$

Figure 6.5: L2 regularization

L2 regularization tends to decrease the weights of the non-relevant features. They will be close to **0**, but not exactly **0**. So, it reduces the impact of these features but does not disable them as L1 does.

In TensorFlow, you can define L2 regularization as follows:

```
from tensorflow.keras.regularizers import l2
from tensorflow.keras.layers import Dense

l2_reg = l2(l=0.01)
Dense(20, kernel_regularizer=l2_reg)
```

In the preceding example, you defined an L2 regularizer and added it to a fully connected layer of **20** units.

TensorFlow provides another regularizer class that combines both L1 and L2 regularizers. You can instantiate it with the following code snippet:

```
from tensorflow.keras.regularizers
import l1_l2
l1_l2_reg = l1_l2(l1=0.01, l2=0.001)
```

In the preceding example, you instantiated L1 and L2 regularizers and specified the factors for L1 and L2 as **0.01** and **0.001**, respectively. You can observe that more weights are put on the L1 regularization compared to L2. These values are hyperparameters that can be fine-tuned depending on the dataset.

In the next exercise, you will put this into practice as you apply L2 regularization to a model.

EXERCISE 6.01: PREDICTING A CONNECT-4 GAME OUTCOME USING THE L2 REGULARIZER

In this exercise, you will build and train two multi-class models in TensorFlow that will predict the class outcome for player one in the game Connect-4.

Each observation of this dataset contains different situations of the game with different positions. For each of these situations, the model tries to predict the outcome for the first player: win, loss, or draw. The first model will not have any regularization, while the second will have L2 regularization:

> **NOTE**
>
> The dataset can be accessed here: https://packt.link/xysRc.
>
> The original dataset can be found here:
> http://archive.ics.uci.edu/ml/datasets/Connect-4.

1. Open a new Jupyter notebook.

2. Import the pandas library and use **pd** as the alias:

   ```
   import pandas as pd
   ```

3. Create a variable called **file_url** that contains the URL to the dataset:

   ```
   file_url = 'https://raw.githubusercontent.com/PacktWorkshops'\
              '/The-TensorFlow-Workshop/master/Chapter06/dataset'\
              '/connect-4.csv'
   ```

4. Load the dataset into a DataFrame called **data** using the **read_csv()** function and provide the URL to the CSV file. Print the first five rows using the **head()** function:

```
data = pd.read_csv(file_url)
data.head()
```

The expected output will be as follows:

	a1	a2	a3	a4	a5	a6	b1	b2	b3	b4	b5	b6	c1	c2	c3	c4	c5	c6	d1	d2	d3	d4	d5	d6	e1	e2	e3	e4
0	0	0	0	0	0	0	0	0	0	0	0	0	2	1	0	0	0	0	2	1	2	1	2	1	0	0	0	0
1	0	0	0	0	0	0	0	0	0	0	0	0	2	0	0	0	0	0	2	1	2	1	2	1	1	0	0	0
2	0	0	0	0	0	0	1	0	0	0	0	0	2	0	0	0	0	0	2	1	2	1	2	1	0	0	0	0
3	0	0	0	0	0	0	0	0	0	0	0	0	2	0	0	0	0	0	2	1	2	1	2	1	0	0	0	0
4	1	0	0	0	0	0	0	0	0	0	0	0	2	0	0	0	0	0	2	1	2	1	2	1	0	0	0	0

Figure 6.6: First five rows of the dataset

The preceding figure shows the first five rows of the dataset.

5. Extract the target variable (the **class** column) using the **pop()** method and save it in a variable named **target**:

```
target = data.pop('class')
```

6. Import the TensorFlow library and use **tf** as the alias. Then, import the **Dense** class from **tensorflow.keras.layers**:

```
import tensorflow as tf
from tensorflow.keras.layers import Dense
```

7. Set the seed as **8** to get reproducible results:

```
tf.random.set_seed(8)
```

8. Instantiate a sequential model using **tf.keras.Sequential()** and store it in a variable called **model**:

```
model = tf.keras.Sequential()
```

9. Create a fully connected layer of **512** units with **Dense()** and specify ReLu as the activation function and the input shape as **(42,)**, which corresponds to the number of features from the dataset. Save it in a variable called **fc1**:

```
fc1 = Dense(512, input_shape=(42,), activation='relu')
```

10. Create three fully connected layers of **512**, **128**, and **128** units with **Dense ()** and specify ReLu as the activation function. Save them in three variables, called **fc2**, **fc3**, and **fc4**, respectively:

```
fc2 = Dense(512, activation='relu')
fc3 = Dense(128, activation='relu')
fc4 = Dense(128, activation='relu')
```

11. Create a fully connected layer of three units (corresponding to the number of classes) with **Dense ()** and specify softmax as the activation function. Save it in a variable called **fc5**:

```
fc5 = Dense(3, activation='softmax')
```

12. Sequentially add all five fully connected layers to the model using the **add ()** method:

```
model.add(fc1)
model.add(fc2)
model.add(fc3)
model.add(fc4)
model.add(fc5)
```

13. Print the summary of the model using the **summary ()** method:

```
model.summary()
```

The expected output will be as follows:

```
Model: "sequential"
_____
Layer (type)                 Output Shape              Param #
=================================================================
dense (Dense)                (None, 512)               22016

dense_1 (Dense)              (None, 512)               262656

dense_2 (Dense)              (None, 128)               65664

dense_3 (Dense)              (None, 128)               16512

dense_4 (Dense)              (None, 3)                 387

=================================================================
Total params: 367,235
Trainable params: 367,235
Non-trainable params: 0
_____
```

Figure 6.7: Summary of the model architecture

14. Instantiate a **SparseCategoricalCrossentropy()** function from **tf.keras.losses** and save it in a variable called **loss**:

```
loss = tf.keras.losses.SparseCategoricalCrossentropy()
```

15. Instantiate **Adam()** from **tf.keras.optimizers** with **0.001** as the learning rate and save it in a variable called **optimizer**:

```
optimizer = tf.keras.optimizers.Adam(0.001)
```

16. Compile the model using the **compile()** method, and specify the optimizer and loss you created in *steps 14* and *15* and **accuracy** as the metric to be displayed:

```
model.compile(optimizer=optimizer, loss=loss, \
              metrics=['accuracy'])
```

17. Start the model training process using the **fit()** method for five epochs and split the data into a validation set with 20% of the data:

```
model.fit(data, target, epochs=5, validation_split=0.2)
```

The expected output will be as follows:

```
Epoch 1/5
1689/1689 [==============================] - 14s 8ms/step - loss: 0.6394 - accuracy: 0.7450 - val_loss: 1.0408 - val_accuracy:
0.5797
Epoch 2/5
1689/1689 [==============================] - 14s 8ms/step - loss: 0.4950 - accuracy: 0.8055 - val_loss: 1.2903 - val_accuracy:
0.5477
Epoch 3/5
1689/1689 [==============================] - 13s 8ms/step - loss: 0.4351 - accuracy: 0.8270 - val_loss: 1.2402 - val_accuracy:
0.5800
Epoch 4/5
1689/1689 [==============================] - 14s 8ms/step - loss: 0.4021 - accuracy: 0.8417 - val_loss: 1.6979 - val_accuracy:
0.5504
Epoch 5/5
1689/1689 [==============================] - 14s 8ms/step - loss: 0.3729 - accuracy: 0.8521 - val_loss: 1.3584 - val_accuracy:
0.5873

<keras.callbacks.History at 0x7f3cce8dd250>
```

Figure 6.8: Logs of the training process

The preceding output reveals that the model is overfitting. It achieved an accuracy score of **0.85** on the training set and only **0.58** on the validation set. Now, train another model with L2 regularization.

18. Create five fully connected layers similar to the previous model's and specify the L2 regularizer for the **kernel_regularizer** parameters. Use the value **0.001** for the regularizer factor. Save the layers in five variables, called **reg_fc1**, **reg_fc2**, **reg_fc3**, **reg_fc4**, and **reg_fc5**:

```
reg_fc1 = Dense(512, input_shape=(42,), activation='relu', \
                kernel_regularizer=tf.keras.regularizers\
                           .l2(l=0.1))
reg_fc2 = Dense(512, activation='relu', \
                kernel_regularizer=tf.keras.regularizers\
                           .l2(l=0.1))
reg_fc3 = Dense(128, activation='relu', \
                kernel_regularizer=tf.keras.regularizers\
                           .l2(l=0.1))
reg_fc4 = Dense(128, activation='relu', \
                kernel_regularizer=tf.keras.regularizers\
                           .l2(l=0.1))
reg_fc5 = Dense(3, activation='softmax')
```

19. Instantiate a sequential model using **tf.keras.Sequential()**, store it in a variable called **model2**, and add sequentially all five fully connected layers to the model using the **add()** method:

```
model2 = tf.keras.Sequential()
model2.add(reg_fc1)
model2.add(reg_fc2)
model2.add(reg_fc3)
model2.add(reg_fc4)
model2.add(reg_fc5)
```

20. Print the summary of the model:

```
model2.summary()
```

The expected output will be as follows:

```
Model: "sequential_1"
```

Layer (type)	Output Shape	Param #
dense_10 (Dense)	(None, 512)	22016
dense_11 (Dense)	(None, 512)	262656
dense_12 (Dense)	(None, 128)	65664
dense_13 (Dense)	(None, 128)	16512
dense_14 (Dense)	(None, 3)	387

```
Total params: 367,235
Trainable params: 367,235
Non-trainable params: 0
```

Figure 6.9: Summary of the model architecture

21. Compile the model using the **compile()** method, and specify the optimizer and loss you created in *steps 14* and *15* and **accuracy** as the metric to be displayed:

```
model2.compile(optimizer=optimizer, loss=loss, \
               metrics=['accuracy'])
```

22. Start the model training process using the **fit()** method for five epochs and split the data into a validation set with 20% of the data:

```
model2.fit(data, target, epochs=5, validation_split=0.2)
```

The expected output will be as follows:

```
Epoch 1/5
1689/1689 [==============================] - 16s 9ms/step - loss: 1.1634 - accuracy: 0.6780 - val_loss: 0.9639 - val_accuracy:
0.5796
Epoch 2/5
1689/1689 [==============================] - 15s 9ms/step - loss: 0.8169 - accuracy: 0.6780 - val_loss: 0.9698 - val_accuracy:
0.5796
Epoch 3/5
1689/1689 [==============================] - 15s 9ms/step - loss: 0.8172 - accuracy: 0.6780 - val_loss: 0.9504 - val_accuracy:
0.5796
Epoch 4/5
1689/1689 [==============================] - 15s 9ms/step - loss: 0.8170 - accuracy: 0.6780 - val_loss: 0.9610 - val_accuracy:
0.5796
Epoch 5/5
1689/1689 [==============================] - 15s 9ms/step - loss: 0.8170 - accuracy: 0.6780 - val_loss: 0.9694 - val_accuracy:
0.5796

<keras.callbacks.History at 0x7f3cce80d3d0>
```

Figure 6.10: Logs of the training process

With the addition of L2 regularization, the model now has similar accuracy scores between the training (**0.68**) and test (**0.58**) sets. The model is not overfitting as much as before, but its performance is not great.

Now that you know how to apply L1 and L2 regularization to neural networks, the next section will introduce another regularization technique, called **dropout**.

DROPOUT REGULARIZATION

Unlike L1 and L2 regularization, dropout is a regularization technique specific to neural networks. The logic behind it is very simple: the networks will randomly change the weights of some features to **0**. This will force the model to rely on other features that would have been ignored and, therefore, bump up their weights.

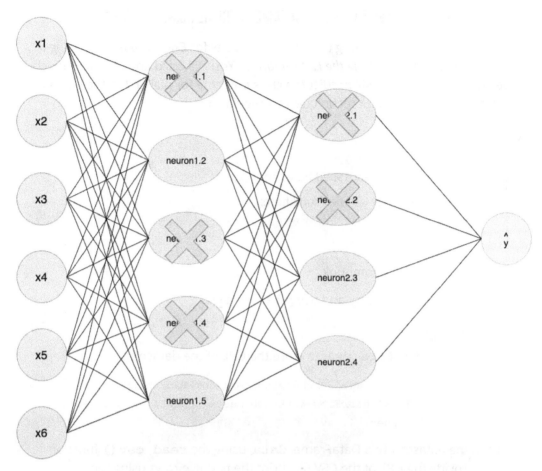

Figure 6.11: Dropout of neural networks

The preceding example shows an architecture with a dropout of 50%. This means that 50% of the units of the model are turned off at each iteration. The following code snippet shows you how to create a dropout layer of 50% in TensorFlow:

```
from tensorflow.keras.layers import Dropout

do = Dropout(0.5)
```

In the next exercise, you will extend the previous model by applying dropout.

EXERCISE 6.02: PREDICTING A CONNECT-4 GAME OUTCOME USING DROPOUT

In this exercise, you will be using the same dataset as for *Exercise 6.01, Predicting a Connect-4 Game Outcome Using the L2 Regularizer*. You will build and train a multi-class model in TensorFlow that will predict the class outcome for player 1 in the game Connect-4 using the dropout technique as a regularizer:

> **NOTE**
>
> The dataset can be accessed here: https://packt.link/0Bo1B.
>
> The original dataset can be found here:
> http://archive.ics.uci.edu/ml/datasets/Connect-4.

1. Open a new Jupyter notebook.

2. Import the pandas library and use **pd** as the alias:

```
import pandas as pd
```

3. Create a variable, **file_url**, to store the URL of the dataset:

```
file_url = 'https://raw.githubusercontent.com/PacktWorkshops'\
          '/The-TensorFlow-Workshop/master/Chapter06/dataset'\
          '/connect-4.csv'
```

4. Load the dataset into a DataFrame, **data**, using the **read_csv()** function and provide the URL of the CSV file. Print the first five rows using the **head()** function:

```
data = pd.read_csv(file_url)
data.head()
```

The expected output will be as follows:

	a1	a2	a3	a4	a5	a6	b1	b2	b3	b4	b5	b6	c1	c2	c3	c4	c5	c6	d1	d2	d3	d4	d5	d6	e1	e2	e3	e4
0	0	0	0	0	0	0	0	0	0	0	0	0	2	1	0	0	0	0	2	1	2	1	2	1	0	0	0	0
1	0	0	0	0	0	0	0	0	0	0	0	0	2	0	0	0	0	0	2	1	2	1	2	1	1	0	0	0
2	0	0	0	0	0	0	1	0	0	0	0	0	2	0	0	0	0	0	2	1	2	1	2	1	0	0	0	0
3	0	0	0	0	0	0	0	0	0	0	0	0	2	0	0	0	0	0	2	1	2	1	2	1	0	0	0	0
4	1	0	0	0	0	0	0	0	0	0	0	0	2	0	0	0	0	0	2	1	2	1	2	1	0	0	0	0

Figure 6.12: First five rows of the dataset

5. Extract the target variable (the column called **class**) using the **pop()** method, and save it in a variable called **target**:

```
target = data.pop('class')
```

6. Import the TensorFlow library and use **tf** as the alias. Then, import the **Dense** class from **tensorflow.keras.layers**:

```
import tensorflow as tf
from tensorflow.keras.layers import Dense
```

7. Set the seed as **8** to get reproducible results:

```
tf.random.set_seed(8)
```

8. Instantiate a sequential model using **tf.keras.Sequential()** and store it in a variable called **model**:

```
model = tf.keras.Sequential()
```

9. Create a fully connected layer of **512** units with **Dense()** and specify ReLu as the activation function and the input shape as **(42,)**, which corresponds to the number of features from the dataset. Save it in a variable called **fc1**:

```
fc1 = Dense(512, input_shape=(42,), activation='relu')
```

10. Create three fully connected layers of **512**, **128**, and **128** units with **Dense()** and specify ReLu as the activation function. Save them in three variables, called **fc2**, **fc3**, and **fc4**, respectively:

```
fc2 = Dense(512, activation='relu')
fc3 = Dense(128, activation='relu')
fc4 = Dense(128, activation='relu')
```

11. Create a fully connected layer of three units (corresponding to the number of classes) with **Dense()** and specify softmax as the activation function. Save it in a variable called **fc5**:

```
fc5 = Dense(3, activation='softmax')
```

12. Sequentially add all five fully connected layers to the model with a dropout layer of **0.75** in between each of them using the **add()** method:

```
model.add(fc1)
model.add(Dropout(0.75))
model.add(fc2)
model.add(Dropout(0.75))
model.add(fc3)
model.add(Dropout(0.75))
model.add(fc4)
model.add(Dropout(0.75))
model.add(fc5)
```

13. Print the summary of the model:

```
model.summary()
```

The expected output will be as follows:

```
Model: "sequential"
_____
Layer (type)                 Output Shape              Param #
================================================================
dense (Dense)                (None, 512)               22016

dropout (Dropout)            (None, 512)               0

dense_1 (Dense)              (None, 512)               262656

dropout_1 (Dropout)          (None, 512)               0

dense_2 (Dense)              (None, 128)               65664

dropout_2 (Dropout)          (None, 128)               0

dense_3 (Dense)              (None, 128)               16512

dropout_3 (Dropout)          (None, 128)               0

dense_4 (Dense)              (None, 3)                 387

================================================================
Total params: 367,235
Trainable params: 367,235
Non-trainable params: 0
```

Figure 6.13: Summary of the model architecture

14. Instantiate a **SparseCategoricalCrossentropy()** function from **tf.keras.losses** and save it in a variable called **loss**:

```
loss = tf.keras.losses.SparseCategoricalCrossentropy()
```

15. Instantiate **Adam()** from **tf.keras.optimizers** with **0.001** as the learning rate and save it in a variable called **optimizer**:

```
optimizer = tf.keras.optimizers.Adam(0.001)
```

16. Compile the model using the **compile()** method, specify the optimizer and loss, and set **accuracy** as the metric to be displayed:

```
model.compile(optimizer=optimizer, loss=loss, \
              metrics=['accuracy'])
```

17. Start the model training process using the **fit()** method for five epochs and split the data into a validation set with 20% of the data:

```
model.fit(data, target, epochs=5, validation_split=0.2)
```

The output will be as follows:

```
Epoch 1/5
1689/1689 [==============================] - 14s 8ms/step - loss: 0.8570 - accuracy: 0.6674 - val_loss: 0.9373 - val_accuracy:
0.5796
Epoch 2/5
1689/1689 [==============================] - 13s 8ms/step - loss: 0.8077 - accuracy: 0.6780 - val_loss: 0.9368 - val_accuracy:
0.5796
Epoch 3/5
1689/1689 [==============================] - 13s 8ms/step - loss: 0.7805 - accuracy: 0.6780 - val_loss: 0.9360 - val_accuracy:
0.5796
Epoch 4/5
1689/1689 [==============================] - 13s 8ms/step - loss: 0.7500 - accuracy: 0.6834 - val_loss: 0.9285 - val_accuracy:
0.5824
Epoch 5/5
1689/1689 [==============================] - 13s 8ms/step - loss: 0.7333 - accuracy: 0.6964 - val_loss: 0.9175 - val_accuracy:
0.5967

<keras.callbacks.History at 0x7f5b0d45db10>
```

Figure 6.14: Logs of the training process

With the addition of dropout, the model now has similar accuracy scores between the training (**0.69**) and test (**0.59**) sets. The model is not overfitting as much as before, but its performance is still less than ideal.

You have now seen how to apply L1, L2, or dropout as regularizers for a model. In deep learning, there is another very simple technique that you can apply to avoid overfitting—that is, early stopping.

EARLY STOPPING

Another reason why neural networks overfit is due to the training process. The more you train the model, the more it will try to improve its performance. By training the model for a longer duration (more epochs), it will at some point start finding patterns that are only relevant to the training set. In such a case, the difference between the scores of the training and test (or validation) sets will start increasing after a certain number of epochs.

To prevent this situation, you can stop the model training when the difference between the two sets starts to increase. This technique is called **early stopping**.

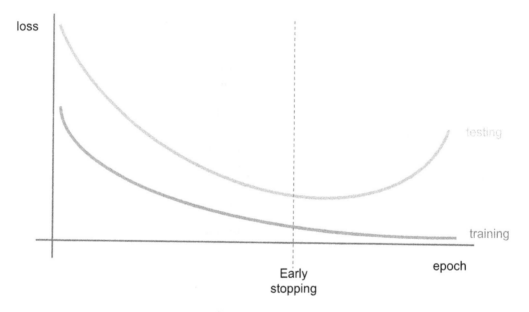

Figure 6.15: Early stopping to prevent overfitting

The preceding graph shows the loss value of a model on the training and test (or validation) sets according to the number of epochs. In early epochs, the loss value is quite different between the two sets. As the training goes on, the models start learning the relevant patterns for making predictions and both losses converge. But after a while, they start diverging. The loss of the training set keeps decreasing while the one for the test (or validation) set is increasing. You can observe that the model is overfitting and is optimizing only for the training set. Stopping the training at the point when the difference between the two losses starts to increase prevents the model from overfitting.

In TensorFlow, you can achieve this by setting up callbacks that analyze the performance of the models at each epoch and compare its score between the training and test sets. To define an early stopping callback, you will do the following:

```
from tensorflow.keras.callbacks import EarlyStopping

EarlyStopping(monitor='val_accuracy', patience=5)
```

The preceding code shows you how to instantiate an **EarlyStopping** class that will monitor the accuracy score of the validation set and wait for five successive epochs with no improvement before stopping the training process.

In the next activity, you will practice applying both L1 and L2 regularization to a model.

ACTIVITY 6.01: PREDICTING INCOME WITH L1 AND L2 REGULARIZERS

The **census-income-train.csv** dataset contains weighted census data extracted from the 1994 and 1995 current population surveys conducted by the US Census Bureau. The dataset is the subset of the original dataset shared by the US Census Bureau. In this activity, you are tasked with building and training a regressor to predict the income of a person based on their census data. The dataset can be accessed here: https://packt.link/G8xFd.

The following steps will help you to complete the activity:

1. Open a new Jupyter notebook.

2. Import the required libraries.

3. Create a list called **usecols** containing the column names **AAGE, ADTIND, ADTOCC, SEOTR, WKSWORK**, and **PTOTVAL**.

4. Load the data using the **read_csv()** method.

5. Split the data into training (the first 15,000 rows) and test (the last 5,000 rows) sets.

6. Build the multi-class classifier with five fully connected layers of, respectively, **512, 512, 128, 128**, and **26** units.

7. Train the model on the training set.

    The expected output will be as follows:

```
Epoch 1/5
4989/4989 [==============================] - 61s 12ms/step - loss: 13579405.0000 - mse: 13579149.0000 - val_loss: 3970202.7500
- val_mse: 3969996.2500
Epoch 2/5
4989/4989 [==============================] - 62s 12ms/step - loss: 4028319.2500 - mse: 4028123.2500 - val_loss: 3970189.5000 -
val_mse: 3969996.2500
Epoch 3/5
4989/4989 [==============================] - 60s 12ms/step - loss: 4028304.7500 - mse: 4028129.5000 - val_loss: 3970171.0000 -
val_mse: 3969996.2500
Epoch 4/5
4989/4989 [==============================] - 61s 12ms/step - loss: 4028274.5000 - mse: 4028119.5000 - val_loss: 3970124.2500 -
val_mse: 3969996.2500
Epoch 5/5
4989/4989 [==============================] - 61s 12ms/step - loss: 4028207.0000 - mse: 4028115.5000 - val_loss: 3970040.5000 -
val_mse: 3969996.2500

<keras.callbacks.History at 0x7f6e2267d3d0>
```

Figure 6.16: Logs of the training process

> **NOTE**
>
> The solution to this activity is available on page 494.

In the section ahead, you will see how to tune hyperparameters to achieve better results.

HYPERPARAMETER TUNING

Previously, you saw how to deal with a model that is overfitting by using different regularization techniques. These techniques help the model to better generalize to unseen data but, as you have seen, they can also lead to inferior performance and make the model underfit.

With neural networks, data scientists have access to different hyperparameters they can tune to improve the performance of a model. For example, you can try different learning rates and see whether one leads to better results, you can try different numbers of units for each hidden layer of a network, or you can test to see whether different ratios of dropout can achieve a better trade-off between overfitting and underfitting.

However, the choice of one hyperparameter can impact the effect of another one. So, as the number of hyperparameters and values you want to tune grows, the number of combinations to be tested will increase exponentially. It will also take a lot of time to train models for all these combinations—especially if you have to do it manually. There are some packages that can automatically scan the hyperparameter search space you defined and find the best combination overall for you. In the section ahead, you will see how to use one of them: Keras Tuner.

KERAS TUNER

Unfortunately, this package is not included in TensorFlow. You will need to install it manually by running the following command:

```
pip install keras-tuner
```

This package is very simple to use. There are two concepts to understand: **hyperparameters** and **tuners**.

Hyperparameters are the classes used to define a parameter that will be assessed by the tuner. You can use different types of hyperparameters. The main ones are the following:

- **hp.Boolean**: A choice between **True** and **False**

- **hp.Int**: A choice with a range of integers

- **hp.Float**: A choice with a range of decimals

- **hp.Choice**: A choice within a list of possible values

The following code snippet shows you how to define a hyperparameter called **learning_rate** that can only take one of four values—**0.1**, **0.01**, **0.001**, or **0.0001**:

```
hp.Choice('learning_rate', values = [0.1, 0.01, 0.001, 0.0001])
```

A tuner in the Keras Tuner package is an algorithm that will look at the hyperparameter search space, test some combinations, and find the one that gives the best result. The Keras Tuner package provides different tuners, and in the section ahead, you will look at three of them: **random search**, **Hyperband**, and **Bayesian optimization**.

Once defined with the algorithm of your choice, you can call the **search()** method to start the hyperparameter tuning process on the training and test sets, as follows:

```
tuner.search(X_train, y_train, validation_data=(X_test, y_test))
```

Once the search is complete, you can access the best combination with **get_best_hyperparameters()** and then look specifically at one of the hyperparameters you defined:

```
best_hps = tuner.get_best_hyperparameters()[0]
best_hps.get('learning_rate')
```

Finally, the **hypermodel.build()** method will instantiate a TensorFlow Keras model with the best hyperparameters found:

```
model = tuner.hypermodel.build(best_hps)
```

It's as simple as that. Now, let's have a look at the random search tuner.

RANDOM SEARCH

Random search is one of the available algorithms in this package. As its name implies, it randomly defines the combinations to be tested by sampling through the search space. Even though this algorithm doesn't test every single possible combination, random search provides very good results.

> **NOTE**
>
> The algorithm that tests every single combination of the search space is called grid search.

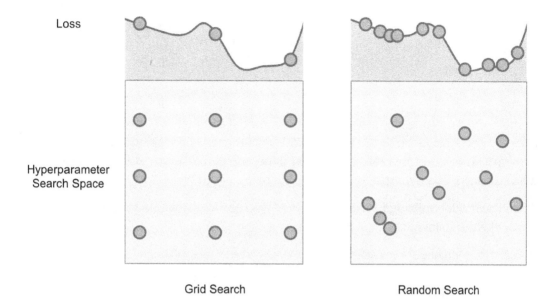

Figure 6.17: Comparison between grid search and random search

The preceding figure shows an example of the difference between grid search and random search. You can see that grid search splits the search space into a grid and tests each of the combinations, but some may lead to the same loss value, which makes it less efficient. On the other side, random search covers the search space more efficiently and helps find the optimal solution.

In Keras Tuner, before instantiating a tuner, you need to define a model-building function that will define the architecture of the TensorFlow Keras model to be trained with the hyperparameters you want to test. Here is an example of such a function:

```
def model_builder(hp):
    model = tf.keras.Sequential()

    hp_lr = hp.Choice('learning_rate', \
                      values = [0.1, 0.01, 0.001, 0.0001])

    model.add(Dense(512, input_shape=(100,), activation='relu'))
    model.add(Dense(128, activation='relu'))
    model.add(Dense(10, activation='softmax'))
```

```
loss = tf.keras.losses.SparseCategoricalCrossentropy()
optimizer = tf.keras.optimizers.Adam(hp_lr)
model.compile(optimizer=optimizer, loss=loss, \
              metrics=['accuracy'])


return model
```

In the preceding code snippet, you created a model composed of three fully connected layers of **512**, **128**, and **10** units that will be trained with a categorical cross-entropy loss function and the Adam optimizer. You defined the **learning_rate** hyperparameter that will be assessed by Keras Tuner.

Once the model-building function is defined, you can instantiate a random search tuner like the following:

```
import kerastuner as kt


tuner = kt.RandomSearch(model_builder, objective='val_accuracy', \
                        max_trials=10)
```

In the preceding code, you instantiated a **RandomSearch** tuner that will look at the model and hyperparameters defined in the **model_builder** function using the validation accuracy as the **objective** metric and will run for a maximum of **10** trials.

In the next exercise, you will use random search to find the best set of hyperparameters for a model.

EXERCISE 6.03: PREDICTING A CONNECT-4 GAME OUTCOME USING RANDOM SEARCH FROM KERAS TUNER

In this exercise, you will be using the same dataset as for *Exercise 6.01, Predicting a Connect-4 Game Outcome Using the L2 Regularizer*. You will build and train a multi-class model in TensorFlow that will predict the class outcome for player 1 in the game Connect-4 using the Keras Tuner package to find the best regularization factor for L2 regularization through random search:

> **NOTE**
>
> The dataset can be accessed here: https://packt.link/aTSbC.
>
> The original dataset can be found here:
> http://archive.ics.uci.edu/ml/datasets/Connect-4.

1. Open a new Jupyter notebook.

2. Import the pandas library and use **pd** as the alias:

```
import pandas as pd
```

3. Create a variable called **file_url** that contains the URL to the dataset:

```
file_url = 'https://raw.githubusercontent.com/PacktWorkshops'\
           '/The-TensorFlow-Workshop/master/Chapter06/dataset'\
           '/connect-4.csv'
```

4. Load the dataset into a DataFrame called **data** using the **read_csv()** method and provide the URL to the CSV file. Print the first five rows using the **head()** method:

```
data = pd.read_csv(file_url)
data.head()
```

The output will be as follows:

	a1	a2	a3	a4	a5	a6	b1	b2	b3	b4	b5	b6	c1	c2	c3	c4	c5	c6	d1	d2	d3	d4	d5	d6	e1	e2	e3	e4	e5	e6	f1	f2	f3	f4	f5	f6	g1	g2
0	0	0	0	0	0	0	0	0	0	0	0	0	2	1	0	0	0	0	2	1	2	1	2	1	0	0	0	0	0	0	0	0	0	0	0	0	0	0
1	0	0	0	0	0	0	0	0	0	0	0	0	2	0	0	0	0	0	2	1	2	1	2	1	1	0	0	0	0	0	0	0	0	0	0	0	0	0
2	0	0	0	0	0	0	1	0	0	0	0	0	2	0	0	0	0	0	2	1	2	1	2	1	0	0	0	0	0	0	0	0	0	0	0	0	0	0
3	0	0	0	0	0	0	0	0	0	0	0	0	2	0	0	0	0	0	2	1	2	1	2	1	0	0	0	0	0	0	1	0	0	0	0	0	0	0
4	1	0	0	0	0	0	0	0	0	0	0	0	2	0	0	0	0	0	2	1	2	1	2	1	0	0	0	0	0	0	0	0	0	0	0	0	0	0

Figure 6.18: First five rows of the dataset

5. Extract the target variable (the column called **class**) using the **pop()** method and save it in a variable called **target**:

```
target = data.pop('class')
```

6. Import **train_test_split** from **sklearn.model_selection**:

```
from sklearn.model_selection import train_test_split
```

7. Split the data into training and test sets using **train_test_split()**, with 20% of the data for testing and **42** for **random_state**:

```
X_train, X_test, y_train, y_test = train_test_split\
                                   (data, target, \
                                   test_size=0.2, \
                                   random_state=42)
```

8. Install the **kerastuner** package and then import it and assign it the **kt** alias:

```
!pip install keras-tuner
import kerastuner as kt
```

9. Import the TensorFlow library and use **tf** as the alias. Then, import the **Dense** class from **tensorflow.keras.layers**:

```
import tensorflow as tf
from tensorflow.keras.layers import Dense
```

10. Set the seed as **8** using **tf.random.set_seed()** to get reproducible results:

```
tf.random.set_seed(8)
```

11. Define a function called **model_builder** that will create a sequential model with the same architecture as *Exercise 6.02*, *Predicting a Connect-4 Game Outcome Using Dropout*, with L2 regularization, but this time, provide an **hp.Choice** hyperparameter for the regularization factor:

```
def model_builder(hp):
    model = tf.keras.Sequential()

    p_l2 = hp.Choice('l2', values = [0.1, 0.01, 0.001, 0.0001])

    reg_fc1 = Dense(512, input_shape=(42,), activation='relu', \
                    kernel_regularizer=tf.keras.regularizers\
                    .l2(l=hp_l2))
    reg_fc2 = Dense(512, activation='relu', \
                    kernel_regularizer=tf.keras.regularizers\
                    .l2(l=hp_l2))
    reg_fc3 = Dense(128, activation='relu', \
                    kernel_regularizer=tf.keras.regularizers\
                    .l2(l=hp_l2))
    reg_fc4 = Dense(128, activation='relu', \
                    kernel_regularizer=tf.keras.regularizers\
                    .l2(l=hp_l2))
    reg_fc5 = Dense(3, activation='softmax')

    model.add(reg_fc1)
    model.add(reg_fc2)
    model.add(reg_fc3)
    model.add(reg_fc4)
```

```
model.add(reg_fc5)

loss = tf.keras.losses.SparseCategoricalCrossentropy()
optimizer = tf.keras.optimizers.Adam(0.001)
model.compile(optimizer = optimizer, loss = loss, \
                metrics = ['accuracy'])

return model
```

12. Instantiate a **RandomSearch** tuner and assign **val_accuracy** to **objective** and **10** to **max_trials**:

```
tuner = kt.RandomSearch(model_builder, objective='val_accuracy', \
                        max_trials=10)
```

13. Launch the hyperparameter search with the **search()** method on the training and test sets:

```
tuner.search(X_train, y_train, validation_data=(X_test, y_test))
```

14. Extract the best hyperparameter combination (index **0**) with **get_best_hyperparameters()** and save it in a variable called **best_hps**:

```
best_hps = tuner.get_best_hyperparameters()[0]
```

15. Extract the best value for the **l2** regularization hyperparameter, save it in a variable called **best_l2**, and print its value:

```
best_l2 = best_hps.get('l2')
best_l2
```

You should get the following result:

```
0.0001
```

The best value for the **l2** hyperparameter found by random search is **0.0001**.

16. Start the model training process using the **fit()** method for five epochs and use the test set for **validation_data**:

```
model = tuner.hypermodel.build(best_hps)
model.fit(X_train, y_train, epochs=5, \
          validation_data=(X_test, y_test))
```

You will get the following output:

```
Epoch 1/5
1689/1689 [==============================] - 13s 8ms/step - loss: 0.5187 - accuracy: 0.8080 - val_loss: 0.5342 - val_accuracy:
0.7994
Epoch 2/5
1689/1689 [==============================] - 13s 8ms/step - loss: 0.4979 - accuracy: 0.8188 - val_loss: 0.5166 - val_accuracy:
0.8120
Epoch 3/5
1689/1689 [==============================] - 13s 8ms/step - loss: 0.4788 - accuracy: 0.8265 - val_loss: 0.5131 - val_accuracy:
0.8102
Epoch 4/5
1689/1689 [==============================] - 13s 8ms/step - loss: 0.4678 - accuracy: 0.8304 - val_loss: 0.5199 - val_accuracy:
0.8130
Epoch 5/5
1689/1689 [==============================] - 13s 8ms/step - loss: 0.4546 - accuracy: 0.8367 - val_loss: 0.5033 - val_accuracy:
0.8162

<keras.callbacks.History at 0x7fc5e52fe050>
```

Figure 6.19: Logs of the training process

Using a random search tuner, you found the best value for L2 regularization (**0.0001**), which helped the model to achieve an accuracy of **0.83** on the training set and **0.81** on the test set. These scores are quite an improvement on those from *Exercise 6.01*, *Predicting a Connect-4 Game Outcome Using the L2 Regularizer* (**0.69** for the training set and **0.59** for the test set).

In the next section, you will use another Keras tuner, called Hyperband.

HYPERBAND

Hyperband is another tuner available in the Keras Tuner package. Like random search, it randomly picks candidates from the search space, but more efficiently. The idea behind it is to test a set of combinations for just one or two iterations, keeping only the best performers and training them for longer. So, the algorithm doesn't waste time in training non-performing combinations as with random search. Instead, it simply discards them from the next run. Only the ones that achieve higher performance are kept for longer training. To instantiate a Hyperband tuner, execute the following command:

```
tuner = kt.Hyperband(model_builder, objective='val_accuracy', \
                     max_epochs=5)
```

This tuner takes a model-building function and an objective metric as input parameters, as for random search. But it requires an additional one, **max_epochs**, corresponding to the maximum number of epochs a model is allowed to train for during the hyperparameter search.

EXERCISE 6.04: PREDICTING A CONNECT-4 GAME OUTCOME USING HYPERBAND FROM KERAS TUNER

In this exercise, you will be using the same dataset as for *Exercise 6.01, Predicting a Connect-4 Game Outcome Using the L2 Regularizer*. You will build and train a multi-class model in TensorFlow that will predict the class outcome for player 1 in the game Connect-4 using the Keras Tuner package to find the best learning rate and the number of units for the input layer through Hyperband:

> **NOTE**
>
> The dataset can be accessed here: https://packt.link/WLgen.
>
> The original dataset can be found here:
> http://archive.ics.uci.edu/ml/datasets/Connect-4.

1. Open a new Jupyter notebook.

2. Import the pandas library and use **pd** as the alias:

```
import pandas as pd
```

3. Create a variable called **file_url** that contains the URL to the dataset:

```
file_url = 'https://raw.githubusercontent.com/PacktWorkshops'\
           '/The-TensorFlow-Workshop/master/Chapter06/dataset'\
           '/connect-4.csv'
```

4. Load the dataset into a DataFrame called **data** using the **read_csv()** method and provide the URL to the CSV file. Print the first five rows using the **head()** method:

```
data = pd.read_csv(file_url)
data.head()
```

The output will be as follows:

	a1	a2	a3	a4	a5	a6	b1	b2	b3	b4	b5	b6	c1	c2	c3	c4	c5	c6	d1	d2	d3	d4	d5	d6	e1	e2
0	0	0	0	0	0	0	0	0	0	0	0	0	2	1	0	0	0	0	2	1	2	1	2	1	0	0
1	0	0	0	0	0	0	0	0	0	0	0	0	2	0	0	0	0	0	2	1	2	1	2	1	1	0
2	0	0	0	0	0	0	1	0	0	0	0	0	2	0	0	0	0	0	2	1	2	1	2	1	0	0
3	0	0	0	0	0	0	0	0	0	0	0	0	2	0	0	0	0	0	2	1	2	1	2	1	0	0
4	1	0	0	0	0	0	0	0	0	0	0	0	2	0	0	0	0	0	2	1	2	1	2	1	0	0

Figure 6.20: First five rows of the dataset

5. Extract the target variable (**class**) using the **pop()** method, and save it in a variable called **target**:

```
target = data.pop('class')
```

6. Import **train_test_split** from **sklearn.model_selection**:

```
from sklearn.model_selection import train_test_split
```

7. Split the data into training and test sets using **train_test_split()**, with 20% of the data for testing and **42** for **random_state**:

```
X_train, X_test, y_train, y_test = train_test_split\
                                   (data, target, \
                                   test_size=0.2, \
                                   random_state=42)
```

8. Install the **keras-tuner** package, and then import it and assign it the **kt** alias:

```
!pip install keras-tuner
import kerastuner as kt
```

9. Import the TensorFlow library and use **tf** as the alias, and then import the **Dense** class from **tensorflow.keras.layers**:

```
import tensorflow as tf
from tensorflow.keras.layers import Dense
```

10. Set the seed as **8** using **tf.random.set_seed()** to get reproducible results:

```
tf.random.set_seed(8)
```

11. Define a function called **model_builder** to create a sequential model with the same architecture as *Exercise 6.02, Predicting a Connect-4 Game Outcome Using Dropout*, with L2 regularization and a **0.0001** regularization factor. But, this time, provide a hyperparameter, **hp.Choice**, for the learning rate (**0.01**, **0.001**, or **0.0001**) and an **hp.Int** function for the number of units (between **128** and **512** with a step of **64**) for the input fully connected layer:

```
def model_builder(hp):
    model = tf.keras.Sequential()

    hp_units = hp.Int('units', min_value=128, max_value=512, \
                      step=64)

    reg_fc1 = Dense(hp_units, input_shape=(42,), \
                    activation='relu', \
                    kernel_regularizer=tf.keras.regularizers\
                                .l2(l=0.0001))
    reg_fc2 = Dense(512, activation='relu', \
                    kernel_regularizer=tf.keras.regularizers\
                                .l2(l=0.0001))
    reg_fc3 = Dense(128, activation='relu', \
                    kernel_regularizer=tf.keras.regularizers\
                                .l2(l=0.0001))
    reg_fc4 = Dense(128, activation='relu', \
                    kernel_regularizer=tf.keras.regularizers\
                                .l2(l=0.0001))
    reg_fc5 = Dense(3, activation='softmax')
    model.add(reg_fc1)
    model.add(reg_fc2)
    model.add(reg_fc3)
    model.add(reg_fc4)
    model.add(reg_fc5)
    loss = tf.keras.losses.SparseCategoricalCrossentropy()
    hp_learning_rate = hp.Choice('learning_rate', \
                            values = [0.01, 0.001, 0.0001])

    optimizer = tf.keras.optimizers.Adam(hp_learning_rate)
    model.compile(optimizer = optimizer, loss = loss, \
                  metrics = ['accuracy'])
    return model
```

12. Instantiate a Hyperband tuner, and assign **val_accuracy** to the **objective** metric and **5** to **max_epochs**:

```
tuner = kt.Hyperband(model_builder, objective='val_accuracy', \
                     max_epochs=5)
```

13. Launch the hyperparameter search with **search()** on the training and test sets:

```
tuner.search(X_train, y_train, validation_data=(X_test, y_test))
```

14. Extract the best hyperparameter combination (index **0**) with **get_best_hyperparameters()** and save it in a variable called **best_hps**:

```
best_hps = tuner.get_best_hyperparameters()[0]
```

15. Extract the best value for the number of units for the input layer, save it in a variable called **best_units**, and print its value:

```
best_units = best_hps.get('units')
best_units
```

You will get the following output:

```
192
```

The best value for the number of units of the input layer found by Hyperband is **192**.

16. Extract the best value for the learning rate, save it in a variable called **best_lr**, and print its value:

```
best_lr = best_hps.get('learning_rate')
best_lr
```

17. The output will be the following:

```
0.001
```

The best value for the learning rate hyperparameter found by Hyperband is **0.001**.

18. Start the model training process using the **fit()** method for five epochs and use the test set for **validation_data**:

```
model.fit(X_train, y_train, epochs=5, \
          validation_data=(X_test, y_test))
```

You will get the following output:

```
Epoch 1/5
1689/1689 [==============================] - 12s 7ms/step - loss: 0.7449 - accuracy: 0.7147 - val_loss: 0.6485 - val_accuracy:
0.7485
Epoch 2/5
1689/1689 [==============================] - 12s 7ms/step - loss: 0.6084 - accuracy: 0.7719 - val_loss: 0.5957 - val_accuracy:
0.7753
Epoch 3/5
1689/1689 [==============================] - 11s 7ms/step - loss: 0.5551 - accuracy: 0.7939 - val_loss: 0.5493 - val_accuracy:
0.7943
Epoch 4/5
1689/1689 [==============================] - 11s 7ms/step - loss: 0.5214 - accuracy: 0.8089 - val_loss: 0.5685 - val_accuracy:
0.7938
Epoch 5/5
1689/1689 [==============================] - 11s 7ms/step - loss: 0.4970 - accuracy: 0.8193 - val_loss: 0.5335 - val_accuracy:
0.8043

<keras.callbacks.History at 0x7ff8a7663490>
```

Figure 6.21: Logs of the training process

Using Hyperband as the tuner, you found the best number of units for the input layer (**192**) and learning rate (**0.001**). With these hyperparameters, the final model achieved an accuracy of **0.81** on both the training and test sets. It is not overfitting much and achieved a satisfactory accuracy score.

Another very popular tuner is Bayesian optimization, which you will learn about in the following section.

BAYESIAN OPTIMIZATION

Bayesian optimization is another very popular algorithm used for automatic hyperparameter tuning. It uses probabilities to determine the best combination of hyperparameters. The objective is to iteratively build a probability model that optimizes the objective function from a set of hyperparameters. At each iteration, the probability model is updated from the results obtained. Therefore, unlike random search and Hyperband, Bayesian optimization takes past results into account to improve new ones. The following code snippet will show you how to instantiate a Bayesian optimizer in Keras Tuner:

```
tuner = kt.BayesianOptimization(model_builder, \
                                objective='val_accuracy', \
                                max_trials=10)
```

The expected parameters are similar to random search, including the model-building function, the **objective** metric, and the maximum number of trials.

In the following activity, you will use Bayesian optimization to predict the income of a person.

ACTIVITY 6.02: PREDICTING INCOME WITH BAYESIAN OPTIMIZATION FROM KERAS TUNER

In this activity, you will use the same dataset as used in *Activity 6.01, Predicting Income with L1 and L2 Regularizers*. You are tasked with building and training a regressor to predict the income of a person based on their census data. You will perform automatic hyperparameter tuning with Keras Tuner and find the best combination of hyperparameters for the learning rate, the number of units for the input layer, and L2 regularization with Bayesian optimization.

The following steps will help you to complete the activity:

1. Load the data with **read_csv()** from pandas.

2. Extract the target variable with the **pop()** method.

3. Split the data into training (the first 15,000 rows) and test (the last 5,000 rows) sets.

4. Create the model-building function multi-class classifier with five fully connected layers of **512**, **512**, **128**, **128**, and **26** units and the three different hyperparameters to be tuned: the learning rate (between **0.01** and **0.001**), the number of units for the input layer (between **128** and **512** and a step of **64**), and L2 regularization (between **0.1**, **0.01**, and **0.001**).

5. Find the best combination of hyperparameters with Bayesian optimization.

6. Train the model on the training set with the best hyperparameters found.

   The expected output will be as follows:

```
Epoch 1/5
6236/6236 [==============================] - 27s 4ms/step - loss: 1057466.0000 - mse: 1057465.1250 - val_loss: 1018290.7500 - v
al_mse: 1018289.8750
Epoch 2/5
6236/6236 [==============================] - 27s 4ms/step - loss: 1004625.5625 - mse: 1004624.7500 - val_loss: 999515.1250 - va
l_mse: 999514.6250
Epoch 3/5
6236/6236 [==============================] - 28s 5ms/step - loss: 1000156.8750 - mse: 1000156.5000 - val_loss: 995369.2500 - va
l_mse: 995368.3750
Epoch 4/5
6236/6236 [==============================] - 26s 4ms/step - loss: 997135.5000 - mse: 997134.9375 - val_loss: 1004317.0000 - val
_mse: 1004316.2500
Epoch 5/5
6236/6236 [==============================] - 27s 4ms/step - loss: 994731.3125 - mse: 994730.7500 - val_loss: 989958.5000 - val_
mse: 989957.7500

<keras.callbacks.History at 0x7f8c2be79b50>
```

Figure 6.22: Logs of the training process

NOTE

The solution to this activity is available on page 500.

SUMMARY

You started your journey in this chapter with an introduction to the different scenarios of training a model. A model is overfitting when its performance is much better on the training set than the test set. An underfitting model is one that can achieve good results only after training. Finally, a good model achieves good performance on both the training and test sets.

Then, you encountered several regularization techniques that can help prevent a model from overfitting. You first looked at the L1 and L2 regularizations, which add a penalty component to the cost function. This additional penalty helps to simplify the model by reducing the weights of some features. Then, you went through two different techniques specific to neural networks: dropout and early stopping. Dropout randomly drops some units in the model architecture and forces it to consider other features to make predictions. Early stopping is a mechanism that automatically stops the training of a model once the performance of the test set starts to deteriorate.

After this, you learned how to use the Keras Tuner package for automatic hyperparameter tuning. You considered three specific types of tuners: random search, Hyperband, and Bayesian optimization. You saw how to instantiate them, perform a hyperparameter search, and extract the best values and model. This process helped you to achieve better performance on the models trained for the exercises and activities.

In the next chapter, you will learn more about **Convolutional Neural Networks** (**CNNs**). Such architecture has led to groundbreaking results in computer vision in the past few years. The following chapter will show you how to use this architecture to recognize objects in images.

7

CONVOLUTIONAL NEURAL NETWORKS

OVERVIEW

In this chapter, you will learn how **convolutional neural networks (CNNs)** process image data. You will also learn how to correctly use a CNN on image data.

By the end of the chapter, you will be able to create your own CNN for classification and object identification on any image dataset using TensorFlow.

INTRODUCTION

This chapter covers CNNs. CNNs use convolutional layers that are well-suited to extracting features from images. They use learning filters that correlate with the task at hand. Simply put, they are very good at finding patterns in images.

In the previous chapter, you explored regularization and hyperparameter tuning. You used L1 and L2 regularization and added dropout to a classification model to prevent overfitting on the **connect-4** dataset.

You will now be shifting gears quite a bit as you dive into deep learning with CNNs. In this chapter, you will learn the fundamentals of how CNNs process image data and how to apply those concepts to your own image classification problem. This is truly where TensorFlow shines.

CNNS

CNNs share many common components with the ANNs you have built so far. The key difference is the inclusion of one or more convolutional layers within the network. Convolutional layers apply convolutions of input data with filters, also known as kernels. Think of a **convolution** as an **image transformer**. You have an input image, which goes through the CNN and gives you an output label. Each layer has a unique function or special ability to detect patterns such as curves or edges in an image. CNNs combine the power of deep neural networks and kernel convolutions to transform images and make these image edges or curves easy for the model to see. There are three key components in a CNN:

- **Input image**: The raw image data
- **Filter/kernel**: The image transformation mechanism
- **Output label**: The image classification

The following figure is an example of a CNN in which the image is input into the network on the left-hand side and the output is generated on the right-hand side. The image components are identified throughout the hidden layers with more basic components, such as edges, identified in earlier hidden layers. Image components combine in the hidden layers to form recognizable features from the dataset. For example, in a CNN to classify images into planes or cars, the recognizable features may be filters that resemble a wheel or propellor. Combinations of these features will be instrumental in determining whether the image is a plane or a car.

Finally, the output layer is a dense layer used to determine the specific output of the model. For a binary classification model, this may be a dense layer with one unit with a sigmoid activation function. For a more complex multi-class classification, it may be a dense layer with many units, determined by the number of classes, and a softmax activation function to determine one output label for each image presented to the model.

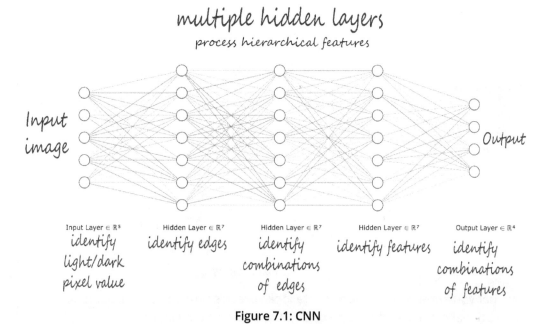

multiple hidden layers
process hierarchical features

Input image

Output

| Input Layer ∈ ℝ⁵ | Hidden Layer ∈ ℝ⁷ | Hidden Layer ∈ ℝ⁷ | Hidden Layer ∈ ℝ⁷ | Output Layer ∈ ℝ⁴ |

identify light/dark pixel value   *identify edges*   *identify combinations of edges*   *identify features*   *identify combinations of features*

Figure 7.1: CNN

A common CNN configuration includes a convolutional layer followed by a pooling layer. These layers are often used together in this order, as pairs (convolution and pooling). We'll get into the reason for this later in the chapter, but for now, think of these pooling layers as decreasing the size of input images by summarizing the filter results.

Before you move deeper into convolutional layers, you first need to understand what the data looks like from the computer's perspective.

IMAGE REPRESENTATION

First, consider how a computer processes an image. To a computer, images are numbers. To be able to work with images for classification or object identification, you need to understand how a model transforms an image input into data. A **pixel** in an image file is just a piece of data.

In the following figure, you can see an example of pixel values for a grayscale image of the number eight. For the **28x28**-pixel image, there are a total of **784** pixels. Each pixel has a value between **0** and **255** identifying how light or dark the pixel is. On the right side, there is one large column vector with each pixel value listed. This is used by the model to identify the image.

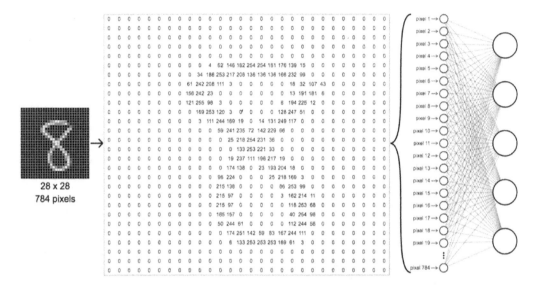

Figure 7.2: Pixel values

Now that you know what the input data looks like, it's time to get a closer look at the convolutional process—more specifically, the convolutional layer.

THE CONVOLUTIONAL LAYER

Think of a convolution as nothing more than an image transformer with three key elements. First, there is an input image, then a filter, and finally, a feature map.

This section will cover each of these in turn to give you a solid idea of how images are filtered in a convolutional layer. The convolution is the process of passing a filter window over the input data, which will result in a map of activations known as a **feature map**. The input data may be the input image to the model or the output of a prior, intermediary layer of the model. The filter is generally a much smaller array, such as **3x3** for two-dimensional data, in which the specific values of the filter are learned during the training process. The filter passes across the input data with a window size equal to the size of the filter, then, the scalar product of the filter and section of the input data is applied, producing what's known as an **activation**. As this process continues across the entire input data using the same filter, the map of activations is produced, also known as the **feature map**.

This concept is illustrated in the following figure, which has two convolutional layers, producing two sets of feature maps. After the feature maps are produced from the first convolutional layer, they are passed into the second convolutional layer. The feature map of the second convolutional layer is passed into a classifier:

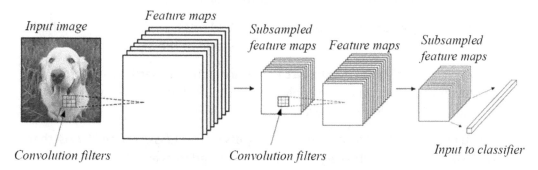

Figure 7.3: Convolution for classification

The distance, or number of steps, the filter moves with each operation is known as the **stride**. If the filter goes off the edge, you can do what's called **padding with zeros**. This way, the output map size is the same as the input map size. This is called **same padding**. However, if the filter cannot take its required stride without leaning over the edge somewhat, it will count any value over the edge as **0**. This is known as **valid padding**.

Let's recap some keywords. There's a **kernel**, which is a small matrix that is used to apply an effect, and what you saw in the example was a **2x2** kernel. There's **stride**, which is the number of pixels that you move the kernel by. Lastly, there's **padding with zeros** around the image, whether or not you add pixels. This ensures that the output is the same size as the input.

CREATING THE MODEL

From the very first chapter, you encountered different types of dimensional tensors. One important thing to note is that you will only be working with `Conv2D`. The layer name `Conv2D` refers only to the movement of a **filter** or **kernel**. So, if you recall the description of what the convolutional process is doing, it's simply sliding a kernel across a 2D space. So, for a flat, square image, the kernel only slides in two dimensions.

When you implement **Conv2D**, you need to pass in certain parameters:

1. The first parameter is **filter**. The filters are the dimensionality of the output space.

2. Specify **strides**, which is how many pixels will move the kernel across.

3. Then, specify **padding**, which is usually **valid** or **same** depending on whether you want an output that is of the same dimension as the input.

4. Finally, you can also have **activation**. Here, you will specify what sort of activation you would like to apply to the outputs. If you don't specify an activation, it's simply a linear activation.

Before you continue, recall from *Chapter 4, Regression and Classification Models*, that a dense layer is one in which every neuron is connected to every neuron in the previous layer. As you can see in the following code, you can easily add a dense layer with **model.add(Dense(32))**. **32** is the number of neurons, followed by the input shape. **AlexNet** is an example of a CNN with multiple convolution kernels that extracts interesting information from an image.

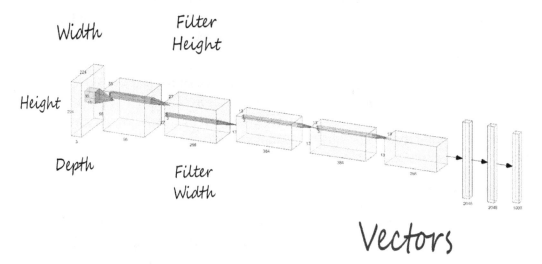

Figure 7.4: AlexNet consists of five convolution layers and three connected layers

> **NOTE**
>
> AlexNet is the name of a CNN designed by Alex Krizhevsky.

A sequential model can be used to build a CNN. Different methods can be used to add a layer; here, we will use the framework of sequentially adding layers to the model using the model's **add** method or passing in a list of all layers when the model is instantiated:

```
model = models.Sequential()
model.add(Dense(32, input_shape=(250,)))
```

The following is a code block showing the code that you'll be using later in the chapter:

```
our_cnn_model = models.Sequential([layers.Conv2D\
                        (filters = 32, \
                         kernel_size = (3,3),
                         input_shape=(28, 28, 1)), \
                        layers.Activation('relu'), \
                        layers.MaxPool2D\
                        (pool_size = (2, 2)), \
                        layers.Conv2D\
                        (filters = 64, \
                         kernel_size = (3,3)), \
                        layers.Activation('relu'), \
                        layers.MaxPool2D\
                        (pool_size = (2,2)), \
                        layers.Conv2D\
                        (filters = 64, \
                         kernel_size = (3,3)), \
                        layers.Activation('relu')])
```

Use the **Conv2D** layer when working with data that you want to convolve in two dimensions, such as images. For parameters, set the number of filters to **32**, followed by the kernel size of **3x3** pixels (**(3, 3)** in the example). In the first layer, you will always need to specify the **input_shape** dimensions, the height, width, and depth. **input_shape** is the size of the images you will be using. You can also select the activation function to be applied at the end of the layer.

Now that you have learned how to build a CNN layer in your model, you will practice doing so in your first exercise. In this exercise, you will build the first constructs of a CNN, initialize the model, and add a single convolutional layer to the model.

EXERCISE 7.01: CREATING THE FIRST LAYER TO BUILD A CNN

As a TensorFlow freelancer, you've been asked to show your potential employer a few lines of code that demonstrate how you might build the first layer in a CNN. They ask that you keep it simple but provide the first few steps to create a CNN layer. In this exercise, you will complete the first step in creating a CNN—that is, adding the first convolutional layer.

Follow these steps to complete this exercise:

1. Open a new Jupyter notebook.

2. Import the TensorFlow library and the **models** and **layers** classes from **tensorflow.keras**:

```
import tensorflow as tf
from tensorflow.keras import models, layers
```

3. Check the TensorFlow version:

```
print(tf.__version__)
```

You should get the following output:

```
2.6.0
```

4. Now, use **models.Sequential** to create your model. The first layer (**Conv2D**) will require the number of nodes (**filters**), the filter size (**3,3**), and the shape of the input. **input_shape** for your first layer will determine the shape of your input images. Add a ReLU activation layer:

```
image_shape = (300, 300, 3)

our_first_layer = models.Sequential([layers.Conv2D\
                             (filters = 16, \
                             kernel_size = (3,3), \
                             input_shape = image_shape), \
                             layers.Activation('relu')])
```

Simple enough. You have just taken the first steps in creating your first CNN.

You will now move on to the type of layer that usually follows a convolutional layer—the pooling layer.

POOLING LAYER

Pooling is an operation that is commonly added to a CNN to reduce the dimensionality of an image by reducing the number of pixels in the output from the convolutional layer it follows. **Pooling layers** shrink the input image to increase computational efficiency and reduce the number of parameters to limit the risk of **overfitting**.

A **pooling layer** immediately follows a convolution layer and is considered another important part of the CNN structure. This section will focus on two types of pooling:

- Max pooling

- Average pooling

MAX POOLING

With max pooling, a filter or kernel only retains the largest pixel value from an input matrix. To get a clearer idea of what is happening, consider the following example. Say you have a **4x4** input. This first step in max pooling would be to divide the **4x4** matrix into four quadrants. Each quadrant will be of the size **2x2**. Apply a filter of size 2. This means that your filter will look exactly like a **2x2** matrix.

Begin by placing the filter on top of your input. For max pooling, this filter will look at all values within the **2x2** area that it covers. It will find the largest value, send that value to your output, and store it there in the upper-left corner of the feature map.

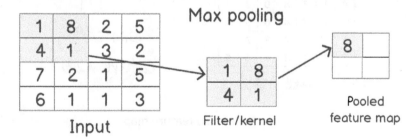

Figure 7.5: Max pooling

Then, the filter will move over to the right and repeat the same process, storing the value in the upper-right corner of the **2x2** matrix. Once this operation is complete, the filter will slide down and start at the far left, again repeating the same process, looking for the largest (or maximum) value, and then storing it in the correct place on the **2x2** matrix.

Recall that the sliding movement is referred to as **stride**. So, the filter was moving over two places. This would mean it has a stride value of **2**. This process is repeated until the maximum values in each of the four quadrants are **8, 5, 7,** and **5,** respectively. Again, to get these numbers, you used a filter of **2x2** and filtered for the largest number within that **2x2** matrix.

So, in this case, you had a stride of two because you moved two pixels. These are the **hyperparameters** for max pooling. The values of **filter** and **stride** are **2**. *Figure 7.6* shows what an implementation of max pooling might look like with a filter size of 3 x 3 and a **stride** of **1**.

There are two steps shown in *Figure 7.6*. Start at the upper left of the feature map. With the **3x3** filter, you would look at the following numbers, **2, 8, 2, 5, 4, 9, 8, 4,** and **6**, and choose the largest value, **9**. The **9** would be placed in the upper-left box of our pooled feature map. With a stride of **1**, you would slide the filter one place to the right, as shown in gray.

Now, look for the largest values from **8, 2, 1, 4, 9, 6, 4, 6,** and **4**. Again, **9** is the largest value, so add a **9** to the middle place in the top row of the pooled feature map (shown in gray).

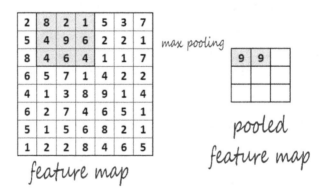

Figure 7.6: Pooled feature map

The preceding pool size is **(2, 2)**. It specifies factors that you will downscale with. Here's a more detailed look at what you could do to implement **MaxPool2D**:

```
layers.MaxPool2D(pool_size=(2, 2), strides=None, \
                 padding='valid')
```

MaxPool2D: The preceding code snippet introduces a **MaxPool2D** instance. The code snippet initializes a max pooling layer with a pool size of **2x2** and the **stride** value is not specified, so it will default to the pool size value. The **padding** parameter is set to **valid**, meaning there is no padding added. The following code snippet demonstrates its use within a CNN:

```
image_shape = (300, 300, 3)

our_first_model = models.Sequential([
    layers.Conv2D(filters = 16, kernel_size = (3,3), \
                  input_shape = image_shape), \
    layers.Activation('relu'), \
    layers.MaxPool2D(pool_size = (2, 2)), \
    layers.Conv2D(filters = 32, kernel_size = (3,3)), \
    layers.Activation('relu')])
```

In the preceding example, a sequential model is created with two convolutional layers, after each layer is a ReLU activation function, and after the activation function of the first convolutional layer is a max pooling layer.

Now that you have explored max pooling, let's look at the other type of pooling: average pooling.

AVERAGE POOLING

Average pooling operates in a similar way to max pooling, but instead of extracting the largest weight value within the filter, it calculates the average. It then passes along that value to the feature map. *Figure 7.7* highlights the difference between max pooling and average pooling.

In *Figure 7.7*, consider the **4x4** matrix on the left. The average of the numbers in the upper-left quadrant is **13**. This would be the average pooling value. The same upper-left quadrant would output **20** to its feature map if it were max pooled because **20** is the largest value within the filter frame. This is a comparison between max pooling and average pooling with hyperparameters, with the **filter** and **stride** parameters both set to **2**:

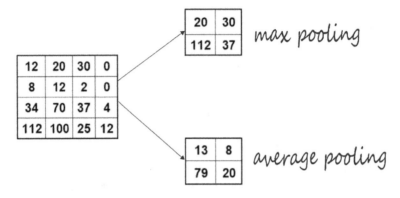

Figure 7.7: Max versus average pooling

For average pooling, you would use **AveragePooling2D** in place of **MaxPool2D**.

To implement the average pooling code, you could use the following:

```
layers.AveragePooling2D(pool_size=(2, 2), strides=None, \
                        padding='valid')
```

AveragePooling2D: The preceding code snippet demonstrates how to invoke an **AveragePooling2D** layer. In a similar manner to max pooling, the **pool_size**, **strides**, and **padding** parameters can be modified. The following code snippet demonstrates its use within a CNN:

```
image_shape = (300, 300, 3)

our_first_model = models.Sequential([
    layers.Conv2D(filters = 16, kernel_size = (3,3), \
                  input_shape = image_shape), \
    layers.Activation('relu'), \
    layers.AveragePooling2D(pool_size = (2, 2)), \
    layers.Conv2D(filters = 32, kernel_size = (3,3)), \
    layers.Activation('relu')])
```

It's a good idea to keep in mind the benefits of using pooling layers. One of these benefits is that if you down-sample the image, the *image shrinks*. This means that you have *less data to process* and fewer multiplications to do, which, of course, speeds things up.

Up to this point, you've created your first CNN layer and learned how to use pooling layers. Now you'll use what you've learned so far to build a pooling layer for the CNN in the following exercise.

EXERCISE 7.02: CREATING A POOLING LAYER FOR A CNN

You receive an email from your potential employer for the TensorFlow freelancing job that you applied for in *Exercise 7.01*, *Creating the First Layer to Build a CNN*. The email asks whether you can show how you would code a pooling layer for a CNN. In this exercise, you will build your base model by adding a pooling layer, as requested by your potential employer:

1. Open a new Jupyter notebook and import the TensorFlow library:

```
import tensorflow as tf
from tensorflow.keras import models, layers
```

2. Create your model using **models.Sequential**. The first layer, **Conv2D**, will require the number of nodes, the filter size, and the shape of the tensor, as in the previous exercise. It will be followed by an activation layer, a node at the end of the neural network:

```
image_shape = (300, 300, 3)

our_first_model = models.Sequential([
    layers.Conv2D(filters = 16, kernel_size = (3,3), \
                  input_shape = image_shape), \
    layers.Activation('relu')])
```

3. Now, add a **MaxPool2D** layer by using the model's **add** method:

```
our_first_model.add(layers.MaxPool2D(pool_size = (2, 2))
```

In this model, you have created a CNN with a convolutional layer, followed by a ReLU activation function then a max pooling layer. The models take images of size **300x300** with three color channels.

Now that you have successfully added a **MaxPool2D** layer to your CNN, the next step is to add a **flattening layer** so that your model can use all the data.

FLATTENING LAYER

Adding a flattening layer is an important step as you will need to provide the neural network with data in a form that it can process. Remember that after you perform the convolution operation, it will still be multi-dimensional. So, to change your data back into one-dimensional form, you will use a flattening layer. To achieve this, you take the pooled feature map and flatten it into a column, as shown in the following figure. In *Figure 7.8*, you can see that you start with the input matrix on the left side of the diagram, use a final pooled feature map, and stretch it out into a single column vector:

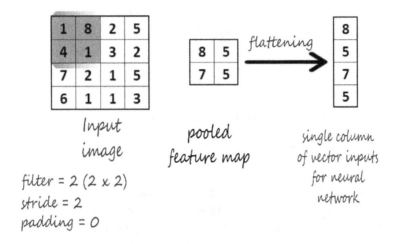

Figure 7.8: Flattening layer

The following is an implemented flattening layer:

```
image_shape = (300, 300, 3)

our_first_model = models.Sequential([
    layers.Conv2D(filters = 16, kernel_size = (3,3), \
                input_shape = image_shape), \
    layers.Activation('relu'), \
    layers.MaxPool2D(pool_size = (2, 2)), \
    layers.Conv2D(filters = 32, kernel_size = (3,3)), \
    layers.Activation('relu'), \
    layers.MaxPool2D(pool_size = (2, 2)), \
    layers.Flatten()])
```

Here, a flatten layer is added as the final layer to this model. Now that you've created your first CNN and pooling layers, you will put all the pieces together and build a CNN in the upcoming exercise.

EXERCISE 7.03: BUILDING A CNN

You were hired as a freelancer from your work in *Exercise 7.01, Creating the First Layer to Build a CNN*, and *Exercise 7.02, Creating a Pooling Layer for a CNN*. Now that you've got the job, your first assignment is to help your start-up company build its prototype product to show to investors and raise capital. The company is trying to develop a horse or human classifier app, and they want you to get started right away. They tell you that they just need the classifier to work for now and that there will be room for improvements on it soon.

In this exercise, you will build a convolutional base layer for your model using the **horses_or_humans** dataset. In this dataset, the images aren't centered. The target images are displayed at all angles and at different positions in the frame. You will continue to build on this foundation throughout the chapter, adding to it piece by piece.

> **NOTE**
>
> The dataset can be downloaded using the **tensorflow_datasets** package.

1. Import all the necessary libraries:

```
import numpy as np
import matplotlib.pyplot as plt
import matplotlib.image as mpimg
import tensorflow as tf
import tensorflow_datasets as tfds
from tensorflow.keras import models, layers
from tensorflow.keras.optimizers import RMSprop
from keras_preprocessing import image as kimage
```

First, you need to import the TensorFlow library. You will use **tensorflow_datasets** to load your dataset, **tensorflow.keras.models** to build a sequential TensorFlow model, **tensorflow.keras.layers** to add layers to your CNN model, **RMSprop** as your optimizer, and **matplotlib.pyplot** and **matplotlib.image** for some quick visualizations.

2. Load your dataset from the **tensorflow_datasets** package:

```
(our_train_dataset, our_test_dataset), \
dataset_info = tfds.load('horses_or_humans',\
                    split = ['train', 'test'],\
                    data_dir = 'content/',\
                    shuffle_files = True,\
                    with_info = True)
assert isinstance(our_train_dataset, tf.data.Dataset)
```

Here, you used the **tensorflow_datasets** package imported as **tfds**. You used the **tfds.load()** function to load the **horses_or_humans** dataset. It is a binary image classification dataset with two classes: horses and humans.

> **NOTE**
>
> More information on the dataset can be found at
> https://laurencemoroney.com/datasets.html.
>
> More information on the **tensorflow_datasets** package can be found
> at https://www.tensorflow.org/datasets.

The **split = ['train', 'test']** argument specifies which split of the data you want to load. In this example, you are loading the train and test splits into **our_train_dataset** and **our_test_dataset**, respectively. Specify **with_info = True** to load the metadata about the dataset into the **dataset_info** variable. After loading, use **assert** to make sure that the loaded dataset is an instance of the **tf.data.Dataset** object class.

3. View information about the dataset using the loaded metadata in **dataset_info**:

```
image_shape = dataset_info.features["image"].shape
print(f'Shape of Images in the Dataset: \t{image_shape}')
print(f'Number of Classes in the Dataset: \
```

```
        \t{dataset_info.features["label"].num_classes}')

names_of_classes = dataset_info.features["label"].names

for name in names_of_classes:
    print(f'Label for class "{name}": \
        \t\t{dataset_info.features["label"].str2int(name)}')
```

You should get the following output:

```
Shape of Images in the Dataset:        (300, 300, 3)
Number of Classes in the Dataset:               2
Label for class "horses":                       0
Label for class "humans":                       1
```

Figure 7.9: horses_or_humans dataset information

4. Now, view the number of images in the dataset and its distribution of classes:

```
print(f'Total examples in Train Dataset: \
    \t{len(our_train_dataset)}')
pos_tr_samples = sum(i['label'] for i in our_train_dataset)
print(f'Horses in Train Dataset: \t\t{len(our_train_dataset) \
                            - pos_tr_samples}')
print(f'Humans in Train Dataset: \t\t{pos_tr_samples}')

print(f'\nTotal examples in Test Dataset: \
    \t{len(our_test_dataset)}')
pos_ts_samples = sum(i['label'] for i in our_test_dataset)
print(f'Horses in Test Dataset: \t\t{len(our_test_dataset) \
                            - pos_ts_samples}')
print(f'Humans in Test Dataset: \t\t{pos_ts_samples}')
```

You should get the following output:

```
Total examples in Train Dataset:        1027
Horses in Train Dataset:                500
Humans in Train Dataset:                527

Total examples in Test Dataset:         256
Horses in Test Dataset:                 128
Humans in Test Dataset:                 128
```

Figure 7.10: horses_or_humans dataset distribution

5. Now, view some sample images in the training dataset, using the **tfds.show_examples()** function:

```
fig = tfds.show_examples(our_train_dataset, dataset_info)
```

This function is for interactive use, and it displays and returns a plot of images from the training dataset.

Your output should be something like the following:

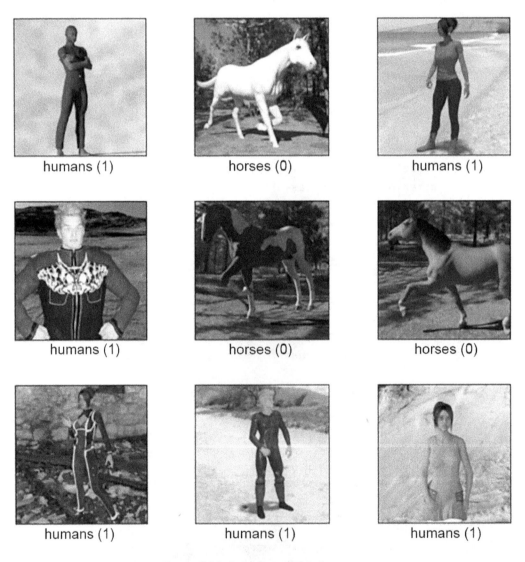

Figure 7.11: Sample training images

6. View some sample images in the test dataset:

```
fig = tfds.show_examples(our_test_dataset, dataset_info)
```

You will get the following output:

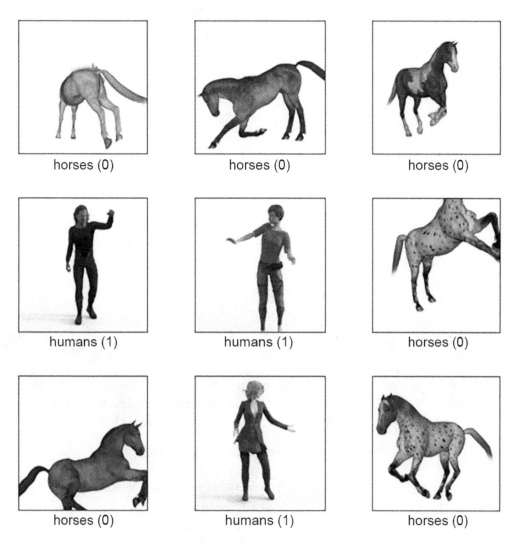

Figure 7.12: Sample test images

7. Finally, create your model with **our_model = models.Sequential**. Set up the first **Conv2D** layer and set **filters** to **16**. The kernel is **3x3**. Use ReLU activation. Because this is the first convolutional layer, you also need to set **input_shape** to **image_shape**, the dimensions of the color images you're working with. Now, add the **MaxPool2D** pooling layer. Then, add another **Conv2D** and **MaxPool2D** pair for more model depth, followed by the flatten and dense layers:

```
our_cnn_model = models.Sequential([
    layers.Conv2D(filters = 16, kernel_size = (3,3), \
                  input_shape = image_shape),\
    layers.Activation('relu'),\
    layers.MaxPool2D(pool_size = (2, 2)),\

    layers.Conv2D(filters = 32, kernel_size = (3,3)),\
    layers.Activation('relu'),\
    layers.MaxPool2D(pool_size = (2, 2)),\

    layers.Flatten(),\
    layers.Dense(units = 512),\
    layers.Activation('relu'),\
    layers.Dense(units = 1),\
    layers.Activation('sigmoid')
])
```

8. Compile the model with **RMSProp** for **optimizer** set to the recommended default of **0.001**, **loss** as **binary_crossentropy**, and **metrics** set to **acc** for accuracy. Print the model summary using the **summary()** method:

```
our_cnn_model.compile(optimizer=RMSprop(learning_rate=0.001), \
                      loss='binary_crossentropy',\
                      metrics=['acc'], loss_weights=None,\
                      weighted_metrics=None, run_eagerly=None,\
                      steps_per_execution=None)

print(our_cnn_model.summary())
```

This will print the model summary with details on the layer type, output shape, and parameters:

```
Model: "sequential"

_____
Layer (type)                 Output Shape              Param #
=================================================================
conv2d (Conv2D)              (None, 298, 298, 16)      448

activation (Activation)      (None, 298, 298, 16)      0

max_pooling2d (MaxPooling2D  (None, 149, 149, 16)      0
)

conv2d_1 (Conv2D)            (None, 147, 147, 32)      4640

activation_1 (Activation)    (None, 147, 147, 32)      0

max_pooling2d_1 (MaxPooling  (None, 73, 73, 32)        0
2D)

flatten (Flatten)            (None, 170528)            0

dense (Dense)                (None, 512)               87310848

activation_2 (Activation)    (None, 512)               0

dense_1 (Dense)              (None, 1)                 513

activation_3 (Activation)    (None, 1)                 0

=================================================================
Total params: 87,316,449
Trainable params: 87,316,449
Non-trainable params: 0
```

Figure 7.13: Model summary

In the preceding screenshot, you can see that there are layers and types listed on the left side. The layers are listed in order from first to last, top to bottom. The output shape is shown in the middle. There are several parameters for each layer listed alongside the assigned layer. At the bottom, you'll see a count of the total parameters, trainable parameters, and non-trainable parameters.

You've been able to explore the convolutional layer and pooling layers quite a bit. Let's now dive into another important component when using image data: image augmentation.

IMAGE AUGMENTATION

Augmentation is defined as making something better by making it greater in size or amount. This is exactly what data or image augmentation does. You use augmentation to provide the model with more versions of your image training data. Remember that the more data you have, the better the model's performance will be. By *augmenting* your data, you can transform your images in a way that makes the model generalize better on real data. To do this, you *transform* the images that you have at your disposal so that you can use your augmented images alongside your original image dataset to train with a greater variation and variety than you would have otherwise. This improves results and prevents overfitting. Take a look at the following three images:

Figure 7.14: Augmented leopard images

It's clear that this is the same leopard in all three images. They're just in different positions. Neural networks can still make sense of this due to convolution. However, with the use of image augmentation, you can improve the model's ability to learn **translational invariance**.

Unlike most other types of data with images, you can shift, rotate, and move the images around to make variations of the original image. This creates more data, and with CNNs, more data and data variation will create a better-performing model. To be able to create these image augmentations, take a look at how you would do this in TensorFlow with the loaded **tf.data.Dataset** object. You will use the **dataset.map()** function to map preprocessing image augmentation functions to your dataset, that is, **our_train_dataset**:

```
from tensorflow import image as tfimage
from tensorflow.keras.preprocessing import image as kimage
```

You will use the **tensorflow.image** and **tensorflow.keras.preprocessing.image** packages for this purpose. These packages have a lot of image manipulation functions that can be used for image data augmentation:

```
augment_dataset(image, label):
    image = kimage.random_shift(image, wrg = 0.1, hrg = 0.1)
    image = tfimage.random_flip_left_right(image)
    return image, label
```

Additional functions include the following:

- **kimage.random_rotation**: This function allows you to rotate an image randomly between specified degrees.

- **kimage.random_brightness**: This function randomly adjusts the brightness level.

- **kimage.random_shear**: This function applies shear transformations.

- **kimage.random_zoom**: This function randomly zooms images.

- **tfimage.random_flip_left_right**: This function randomly flips images horizontally.

- **tfimage.random_flip_up_down**: This function randomly flips images vertically.

In the next step, you will pass in the data that you want to augment with the **tf.data.Dataset.map()** function:

```
augment_dataset(image, label):
    image = kimage.random_shift(image, wrg = 0.1, hrg = 0.1)
    image = tfimage.random_flip_left_right(image)
    return image, label


our_train_dataset = our_train_dataset.map(augment_dataset)


model.fit(our_train_dataset, \
          epochs=50, \
          validation_data=our_test_dataset)
```

In the preceding code block, with **fit()**, you just need to pass the generator that you have already created. You need to pass in the **epochs** value. If you don't do this, the generator will never stop. The **fit()** function returns the history (plots loss per iteration and so on).

You need some more functions to add to **our_train_dataset** before you can train the model on it. With **batch()** function, you specify how many images per batch you will train. With **cache()** function, you fit your dataset in memory to improve performance. With **shuffle()** function, you set the shuffle buffer of your dataset to the entire length of the dataset, for true randomness. **prefetch()** function is also used for good performance:

```
our_train_dataset = our_train_dataset.cache()
our_train_dataset = our_train_dataset.map(augment_dataset)
our_train_dataset = our_train_dataset.shuffle\
                    (len(our_train_dataset))
our_train_dataset = our_train_dataset.batch(128)
our_train_dataset = our_train_dataset.prefetch\
                    (tf.data.experimental.AUTOTUNE)
```

Now that you've seen how you would implement augmentation in your training model, take a closer look at what some of those transformations are doing.

Here's an example of **random_rotation**, **random_shift**, and **random_brightnes** implementation. Use the following code to randomly rotate an image up to an assigned value:

```
image = kimage.random_rotation(image, rg = 135)
```

In *Figure 7.15*, you can see the outcome of **random_rotation**.

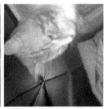

Figure 7.15: Rotation range

The images were randomly rotated up to 135 degrees.

random_shift is used to randomly shift the pixels width-wise. Notice the **.15** in the following code, which means the image can be randomly shifted up to 15 pixels:

```
image = kimage.random_shift(image, wrg = 0.15, hrg = 0)
```

The following figure shows the random adjustment of an image's width by up to 15 pixels:

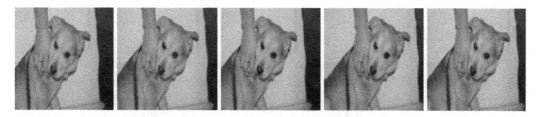

Figure 7.16: Width shift range

Again, **random_shift** is used here, which randomly adjusts the height by 15 pixels:

```
image = kimage.random_shift(image, wrg = 0, hrg = 0.15)
```

Figure 7.17 shows the random adjustment of an image's height by up to 15 pixels:

Figure 7.17: Height shift range

For random brightness levels using **random_brightness**, you will use a float value range to lighten or darken the image by percentage. Anything below **1.0** will darken the image. So, in this example, the images are being darkened randomly between 10% and 90%:

```
image = kimage.random_brightness(image, brightness_range=(0.1,0.9))
```

In the following figure, you've adjusted the brightness with **random_brightness**:

Figure 7.18: Brightness range

Now that you've been exposed to some of the image augmentation options, take a look at how you can use batch normalization to drive performance improvement in models.

BATCH NORMALIZATION

In 2015, **batch normalization**, also called **batch norm**, was introduced by *Christian Szegedy* and *Sergey Ioffe*. Batch norm is a technique that reduces the number of training epochs to improve performance. Batch norm standardizes the inputs for a mini-batch and "normalizes" the input layer. It is most commonly used following a convolutional layer, as shown in the following figure:

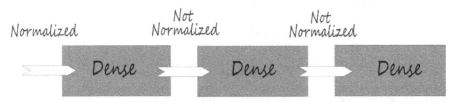

Figure 7.19: Batch norm

The following figure shows one common way that batch normalization is implemented. In the following example, you can see that you have a batch norm layer following a convolutional layer three times. Then you have a flattening layer, followed by two dense layers:

Figure 7.20: Layer sequences

Batch norm helps the model generalize better. With each batch that batch norm trains, the model has a different mean and standard deviation. Because the batch means and standard deviations each vary slightly from the true overall mean and standard deviation, these changes act as noise that you are training with, making the model perform better overall.

The following is an example of **BatchNormalization** implementation. You can simply add a batch norm layer, followed by an activation layer:

```
model.add(layers.Conv2D(filters = 64, kernel_size = (3, 3), use_
bias=False))
model.add(layers.BatchNormalization())
model.add(layers.Activation("relu"))
```

So far, you've created a CNN model and learned how to utilize image augmentation. Now you will bring everything together and build a CNN with some additional convolutional layers in the following exercise.

EXERCISE 7.04: BUILDING A CNN WITH ADDITIONAL CONVOLUTIONAL LAYERS

Your new employers were happy with what you were able to make in *Exercise 7.03*, *Building a CNN*. Now that the **Minimal Viable Product** (**MVP**), or prototype, is complete, it's time to build a better model.

In this exercise, you will add additional ANN layers to your model. You will be adding additional layers to your convolutional base layer that you created earlier. You will be using the **horses_or_humans** dataset again.

Let's get started.

Because you're expanding on *Exercise 7.03*, *Building a CNN*, and using the same data, begin from where you left off with the last step in the previous exercise:

1. Create a function to rescale the images then apply the function to the train and test datasets using the **map** method. Continue building your train and test dataset pipelines using the **cache**, **shuffle**, **batch**, and **prefetch** methods of the dataset:

```
normalization_layer = layers.Rescaling(1./255)

our_train_dataset = our_train_dataset.map\
                (lambda x: (normalization_layer(x['image']), \
                                            x['label']), \
                num_parallel_calls = \
                tf.data.experimental.AUTOTUNE)
our_train_dataset = our_train_dataset.cache()
our_train_dataset = our_train_dataset.shuffle\
                (len(our_train_dataset))
our_train_dataset = our_train_dataset.batch(128)
our_train_dataset = \
our_train_dataset.prefetch(tf.data.experimental.AUTOTUNE)

our_test_dataset = our_test_dataset.map\
                (lambda x: (normalization_layer(x['image']), \
                                            x['label']),\
                num_parallel_calls = \
                tf.data.experimental.AUTOTUNE)
```

```
our_test_dataset = our_test_dataset.cache()
our_test_dataset = our_test_dataset.batch(32)
our_test_dataset = our_test_dataset.prefetch\
                   (tf.data.experimental.AUTOTUNE)
```

2. Fit the model. Specify the values of **epochs** and **validation_steps** and set **verbose** equal to **1**:

```
history = our_cnn_model.fit\
           (our_train_dataset, \
           validation_data = our_test_dataset, \
           epochs=15, \
           validation_steps=8, \
           verbose=1)
```

The output looks like this:

```
Epoch 1/15
9/9 [==============================] - 129s 13s/step - loss: 28.2280 - acc: 0.5443 - val_loss: 5.4822 - val_acc: 0.5000
Epoch 2/15
9/9 [==============================] - 111s 12s/step - loss: 0.9930 - acc: 0.7352 - val_loss: 0.4075 - val_acc: 0.8672
Epoch 3/15
9/9 [==============================] - 106s 12s/step - loss: 0.1377 - acc: 0.9513 - val_loss: 1.1562 - val_acc: 0.7578
Epoch 4/15
9/9 [==============================] - 112s 13s/step - loss: 0.0624 - acc: 0.9883 - val_loss: 1.1676 - val_acc: 0.7812
Epoch 5/15
9/9 [==============================] - 111s 12s/step - loss: 0.0190 - acc: 0.9990 - val_loss: 1.5149 - val_acc: 0.7695
Epoch 6/15
9/9 [==============================] - 116s 13s/step - loss: 0.3292 - acc: 0.9133 - val_loss: 1.5085 - val_acc: 0.5742
Epoch 7/15
9/9 [==============================] - 113s 13s/step - loss: 0.8383 - acc: 0.8754 - val_loss: 0.4985 - val_acc: 0.8945
Epoch 8/15
9/9 [==============================] - 113s 13s/step - loss: 0.1069 - acc: 0.9591 - val_loss: 1.4484 - val_acc: 0.8438
Epoch 9/15
9/9 [==============================] - 115s 13s/step - loss: 0.0553 - acc: 0.9786 - val_loss: 1.9581 - val_acc: 0.7930
Epoch 10/15
9/9 [==============================] - 114s 13s/step - loss: 0.0052 - acc: 1.0000 - val_loss: 1.7870 - val_acc: 0.8281
Epoch 11/15
9/9 [==============================] - 115s 13s/step - loss: 0.0022 - acc: 1.0000 - val_loss: 2.2920 - val_acc: 0.8047
Epoch 12/15
9/9 [==============================] - 109s 12s/step - loss: 0.0011 - acc: 1.0000 - val_loss: 1.8607 - val_acc: 0.8398
Epoch 13/15
9/9 [==============================] - 113s 12s/step - loss: 7.2231e-04 - acc: 1.0000 - val_loss: 2.3023 - val_acc: 0.8281
Epoch 14/15
9/9 [==============================] - 111s 12s/step - loss: 2.8262e-04 - acc: 1.0000 - val_loss: 2.5894 - val_acc: 0.8242
Epoch 15/15
9/9 [==============================] - 116s 13s/step - loss: 1.5618e-04 - acc: 1.0000 - val_loss: 2.9473 - val_acc: 0.8047
```

Figure 7.21: Model fitting process

3. Take a batch from the test dataset and plot the first image from the batch. Convert the image to an array, then use the model to predict what the image shows:

```
from matplotlib.pyplot import imshow

for images, lables in our_test_dataset.take(1):
    imshow(np.asarray(images[0]))
```

```
image_to_test = kimage.img_to_array(images[0])
image_to_test = np.array([image_to_test])


prediction = our_cnn_model.predict(image_to_test)
print(prediction)
if prediction > 0.5:
    print("Image is a human")
    else:
    print("Image is a horse")
```

The output will have the following details:

Choose Files   musk.jpg
• **musk.jpg**(image/jpeg) - 171941 bytes, last modified: 12/4/2021 - 100% done
```
Saving musk.jpg to musk.jpg
[1.]
musk.jpg predicted to be a human
```

Figure 7.22: Output of image test with its metadata

For prediction, you have a picture of a person from the test set to see what the classification would be.

4. Take a look at what's happening with each successive layer. Do this by creating a list containing all names of the layers within the CNN and another list containing predictions on a random sample from each of the layers in the list created previously. Next, iterate through the list of names of the layers and their respective predictions and plot the features:

```
layer_outputs = []
for layer in our_cnn_model.layers[1:]:
    layer_outputs.append(layer.output)
layer_names = []
for layer in our_cnn_model.layers:
    layer_names.append(layer.name)


features_model = models.Model(inputs = our_cnn_model.input, \
                              outputs = layer_outputs)
random_sample = our_train_dataset.take(1)


layer_predictions = features_model.predict(random_sample)


for layer_name, prediction in zip(layer_names, \
```

```
                                    layer_predictions):
    if len(prediction.shape) != 4:
        continue
    num_features = prediction.shape[-1]
    size = prediction.shape[1]
    grid = np.zeros((size, size * num_features))

    for i in range(num_features):
        img = prediction[0, :, :, i]
        img = ((((img - img.mean()) / img.std()) * 64) + 128)
        img = np.clip(img, 0, 255).astype('uint8')
        grid[:, i * size : (i + 1) * size] = img

    scale = 20. / num_features
    plt.figure(figsize=(scale * num_features, scale))
    plt.title(layer_name)
    plt.imshow(grid)
```

You should get something like the following:

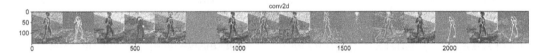

Figure 7.23: Transformation at different layers

Now that you have created your own CNN model and used it to determine whether an image was a horse or a human, you're now going to focus on how you can classify whether an image is or isn't a specific class.

BINARY IMAGE CLASSIFICATION

Binary classification is the simplest approach for classification models as it classifies images into just two categories. In this chapter, we started with the convolutional operation and discussed how you use it as an image transformer. Then, you learned what a pooling layer does and the differences between max and average pooling. Next, we also looked at how a flattening layer converts a pooled feature map into a single column. Then, you learned how and why to use image augmentation, and how to use batch normalization. These are the key components that differentiate CNNs from other ANNs.

After convolutional base layers, pooling, and normalization layers, CNNs are often structured like many ANNs you've built thus far, with a series of one or more dense layers. Much like other binary classifiers, binary image classifiers terminate with a dense layer with one unit and a sigmoid activation function. To provide more utility, image classifiers can be outfitted to classify more than two objects. Such classifiers are known generally as object classifiers, which you will learn about in the next section.

OBJECT CLASSIFICATION

In this section, you will learn about object detection and classification. The next step involves image classification for a dataset with more than two classes. The three different types of models for object classification we will cover are **image classification**, **classification with localization**, and **detection**:

- **Image classification**: This involves training with a set number of classes and then trying to determine which of those classes is shown in the image. Think of the MNIST handwriting dataset. For these problems, you'll use a traditional CNN.

- **Classification with localization**: With this type, the model tries to predict where the object is in the image space. For these models, you use a simplified **You Only Look Once** (**YOLO**) or R-CNN.

- **Detection**: The last type is detection. This is where your model can detect several different objects and where they are located. For this, you use YOLO or an R-CNN:

Figure 7.24: Object classification types

Now, you'll take a brief look at image classification with the **Fashion-MNIST** dataset. **Fashion-MNIST** was compiled from a dataset of Zalando's article images. Zalando is a fashion-focused e-commerce company based in Berlin, Germany. The dataset consists of 10 classes with a training set of 60,000 **28x28** grayscale images and 10,000 test images.

1. Import TensorFlow:

```
import tensorflow as tf
```

2. Next, make some additional imports, such as for NumPy, Matplotlib, and of course, layers and models. You'll notice here that you will be using additional dropout layers. If you recall, dropout layers help prevent overfitting:

```
import numpy as np
import matplotlib.pyplot as plt
import tensorflow_datasets as tfds
from tensorflow.keras.layers import Input, Conv2D, Dense, Flatten, \
    Dropout, GlobalMaxPooling2D, Activation, Rescaling
from tensorflow.keras.models import Model
from sklearn.metrics import confusion_matrix, ConfusionMatrixDisplay
import itertools
import matplotlib.pyplot as plt
```

3. Load the **Fashion-MNIST** dataset using **tdfs** in any one of the datasets that they have decided to include. Others include **CIFAR-10** and **CIFAR-100**, just to name a couple:

```
(our_train_dataset, our_test_dataset), \
dataset_info = tfds.load(\
                    'fashion_mnist'
                    , split = ['train', 'test']
                    , data_dir = 'content/FashionMNIST/'
                    , shuffle_files = True
                    , as_supervised = True
                    , with_info = True)
assert isinstance(our_train_dataset, tf.data.Dataset)
```

4. Check the data for its properties:

```
image_shape = dataset_info.features["image"].shape
print(f'Shape of Images in the Dataset: \t{image_shape}')
num_classes = dataset_info.features["label"].num_classes
```

```
print(f'Number of Classes in the Dataset: \t{num_classes}')


names_of_classes = dataset_info.features["label"].names
print(f'Names of Classes in the Dataset: \t{names_of_classes}\n')


for name in names_of_classes:
    print(f'Label for class \
        "{name}":  \t\t{dataset_info.features["label"].\
        str2int(name)}')
```

This will give you the following output:

```
Shape of Images in the Dataset:       (28, 28, 1)
Number of Classes in the Dataset:     10
Names of Classes in the Dataset:      ['T-shirt/top', 'Trouser', 'Pullover', 'Dress', 'Coat', 'Sandal', 'Shirt', 'Sneaker', 'Bag', 'Ankle boot']

Label for class      "T-shirt/top":              0
Label for class      "Trouser":          1
Label for class      "Pullover":         2
Label for class      "Dress":            3
Label for class      "Coat":             4
Label for class      "Sandal":           5
Label for class      "Shirt":            6
Label for class      "Sneaker":          7
Label for class      "Bag":              8
Label for class      "Ankle boot":       9
```

Figure 7.25: Details of properties for data

5. Now, print the total examples of the train and test data:

```
print(f'Total examples in Train Dataset: \
    \t{len(our_train_dataset)}')
print(f'Total examples in Test Dataset: \
    \t{len(our_test_dataset)}')
```

This will give you the following output:

```
Total examples in Train Dataset:          60000
Total examples in Test Dataset:           10000
```

Figure 7.26: Details of train and test datasets

6. Build your model with the functional API:

```
input_layer = Input(shape=image_shape)
x = Conv2D(filters = 32, kernel_size = (3, 3), \
          strides=2)(input_layer)
x = Activation('relu')(x)


x = Conv2D(filters = 64, kernel_size = (3, 3), strides=2)(x)
```

```
x = Activation('relu')(x)


x = Conv2D(filters = 128, kernel_size = (3, 3), strides=2)(x)
x = Activation('relu')(x)


x = Flatten()(x)
x = Dropout(rate = 0.2)(x)


x = Dense(units = 512)(x)
x = Activation('relu')(x)
x = Dropout(rate = 0.2)(x)


x = Dense(units = num_classes)(x)
output = Activation('softmax')(x)


our_classification_model = Model(input_layer, output)
```

7. Compile and fit your model. With **compile()** method, use **adam** as your optimizer, set the loss to **sparse_categorical_crossentropy**, and set the **accuracy** metric. Then, call **model.fit()** on your training and validation sets:

```
our_classification_model.compile(
                    optimizer='adam', \
                    loss='sparse_categorical_crossentropy',
                    metrics=['accuracy'], loss_weights=None,
                    weighted_metrics=None, run_eagerly=None,
                    steps_per_execution=None
)
history = our_classification_model.fit(our_train_dataset, validation_
data=our_test_dataset, epochs=15)
```

This will give the following as output:

```
Epoch 1/15
469/469 [==============================] - 8s 5ms/step - loss: 0.8613 - accuracy: 0.6934 - val_loss: 0.4498 - val_accuracy: 0.8280
Epoch 2/15
469/469 [==============================] - 1s 3ms/step - loss: 0.4180 - accuracy: 0.8422 - val_loss: 0.3870 - val_accuracy: 0.8532
Epoch 3/15
469/469 [==============================] - 1s 3ms/step - loss: 0.3626 - accuracy: 0.8624 - val_loss: 0.3587 - val_accuracy: 0.8637
Epoch 4/15
469/469 [==============================] - 1s 3ms/step - loss: 0.3256 - accuracy: 0.8756 - val_loss: 0.3329 - val_accuracy: 0.8758
Epoch 5/15
469/469 [==============================] - 1s 3ms/step - loss: 0.3025 - accuracy: 0.8871 - val_loss: 0.3068 - val_accuracy: 0.8858
Epoch 6/15
469/469 [==============================] - 1s 3ms/step - loss: 0.2707 - accuracy: 0.8974 - val_loss: 0.2965 - val_accuracy: 0.8918
Epoch 7/15
469/469 [==============================] - 1s 3ms/step - loss: 0.2584 - accuracy: 0.9012 - val_loss: 0.3041 - val_accuracy: 0.8900
Epoch 8/15
469/469 [==============================] - 1s 3ms/step - loss: 0.2405 - accuracy: 0.9097 - val_loss: 0.2830 - val_accuracy: 0.8958
Epoch 9/15
469/469 [==============================] - 1s 3ms/step - loss: 0.2252 - accuracy: 0.9153 - val_loss: 0.2794 - val_accuracy: 0.9003
Epoch 10/15
469/469 [==============================] - 1s 3ms/step - loss: 0.2115 - accuracy: 0.9195 - val_loss: 0.2906 - val_accuracy: 0.9014
Epoch 11/15
469/469 [==============================] - 1s 3ms/step - loss: 0.1976 - accuracy: 0.9260 - val_loss: 0.2771 - val_accuracy: 0.9018
Epoch 12/15
469/469 [==============================] - 1s 3ms/step - loss: 0.1787 - accuracy: 0.9339 - val_loss: 0.2787 - val_accuracy: 0.9045
Epoch 13/15
469/469 [==============================] - 1s 3ms/step - loss: 0.1699 - accuracy: 0.9350 - val_loss: 0.2797 - val_accuracy: 0.8998
Epoch 14/15
469/469 [==============================] - 1s 3ms/step - loss: 0.1544 - accuracy: 0.9412 - val_loss: 0.2818 - val_accuracy: 0.9030
Epoch 15/15
469/469 [==============================] - 1s 3ms/step - loss: 0.1497 - accuracy: 0.9437 - val_loss: 0.2841 - val_accuracy: 0.9044
```

Figure 7.27: Function returning history

8. Use **matplotlib.pyplot** to plot the loss and accuracy:

```
def plot_trend_by_epoch(tr_values, val_values, title):
    epoch_number = range(len(tr_values))
    plt.plot(epoch_number, tr_values, 'r')
    plt.plot(epoch_number, val_values, 'b')
    plt.title(title)
    plt.xlabel('epochs')
    plt.legend(['Training '+title, 'Validation '+title])
    plt.figure()


hist_dict = history.history
tr_accuracy, val_accuracy = hist_dict['accuracy'], \
                            hist_dict['val_accuracy']
plot_trend_by_epoch(tr_accuracy, val_accuracy, "Accuracy")
```

This will give the following plot as output:

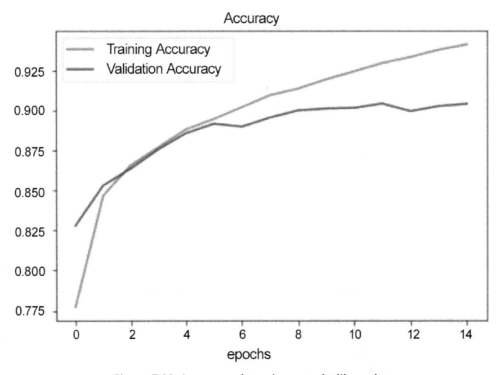

Figure 7.28: Accuracy plot using matplotlib.pyplot

9. Plot the validation loss and training loss. Use the following code:

```
tr_loss, val_loss = hist_dict['loss'], hist_dict['val_loss']
plot_trend_by_epoch(tr_loss, val_loss, "Loss")
```

This will give the following plot as output:

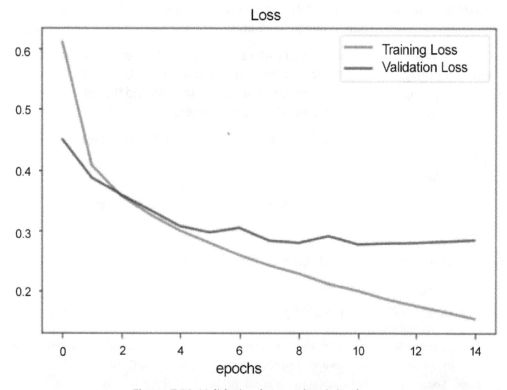

Figure 7.29: Validation loss and training loss

As you can see from the accuracy and loss curves as a function of epochs, the accuracy increases, and loss decreases. On the validation set, both begin to plateau, which is a good signal to stop training to prevent overfitting to the training dataset.

In the next exercise, you will build a CNN to classify images into 10 distinct classes from the **CIFAR-10** dataset.

EXERCISE 7.05: BUILDING A CNN

The start-up now wants to expand its capabilities and to work with more classes and larger image datasets. Your challenge is to accurately predict the class of an image.

The dataset you will be using is the **CIFAR-10** dataset, a dataset containing 60,000 **32x32** color images across 10 classes: airplanes, automobiles, birds, cats, deer, dogs, frogs, horses, ships, and trucks. Each class has 6,000 images and the entire dataset contains 50,000 training images and 10,000 test images.

More info on the dataset can be found at *Learning Multiple Layers of Features from Tiny Images* (http://www.cs.toronto.edu/~kriz/learning-features-2009-TR.pdf), *Alex Krizhevsky, 2009*:

1. Start a new Jupyter notebook and import the TensorFlow library:

```
import tensorflow as tf
```

2. Import the other additional libraries that are needed:

```
import numpy as np
import matplotlib.pyplot as plt
import tensorflow_datasets as tfds
from tensorflow.keras.layers import Input, Conv2D, Dense, Flatten, \
    Dropout, GlobalMaxPooling2D, Activation, Rescaling
from tensorflow.keras.models import Model
from sklearn import metrics import confusion_matrix, \
    ConfusionMatrixDisplay
import itertools
import matplotlib.pyplot as plt
```

3. Load the **CIFAR-10** dataset directly from **tfds** as follows:

```
(our_train_dataset, our_test_dataset), \
dataset_info = tfds.load('cifar10',\
                      split = ['train', 'test'],\
                      data_dir = 'content/Cifar10/',\
                      shuffle_files = True,\
                      as_supervised = True,\
                      with_info = True)
assert isinstance(our_train_dataset, tf.data.Dataset)
```

4. Print the properties of your dataset using the following code:

```
image_shape = dataset_info.features["image"].shape
print(f'Shape of Images in the Dataset: \t{image_shape}')
num_classes = dataset_info.features["label"].num_classes
print(f'Number of Classes in the Dataset: \t{num_classes}')


names_of_classes = dataset_info.features["label"].names
print(f'Names of Classes in the Dataset: \t{names_of_classes}\n')


for name in names_of_classes:
    print(f'Label for class "{name}": \
            \t\t{dataset_info.features["label"].str2int(name)}')


print(f'Total examples in Train Dataset: \
        \t{len(our_train_dataset)}')
print(f'Total examples in Test Dataset: \
        \t{len(our_test_dataset)}')
```

This will give the following output with the properties and the number of classes:

```
Shape of Images in the Dataset:          (32, 32, 3)
Number of Classes in the Dataset:        10
Names of Classes in the Dataset:         ['airplane', 'automobile', 'bird', 'cat', 'deer', 'dog', 'frog', 'horse', 'ship', 'truck']

Label for class "airplane":              0
Label for class "automobile":                1
Label for class "bird":                  2
Label for class "cat":                   3
Label for class "deer":                  4
Label for class "dog":                   5
Label for class "frog":                  6
Label for class "horse":                 7
Label for class "ship":                  8
Label for class "truck":                 9
Total examples in Train Dataset:    50000
Total examples in Test Dataset:     10000
```

Figure 7.30: Number of classes

5. Build the train and test data pipelines, as shown in *Exercise 7.03, Building a CNN*:

```
normalization_layer = Rescaling(1./255)
our_train_dataset = our_train_dataset.map\
                    (lambda x, y: (normalization_layer(x), y),\
                     num_parallel_calls = \
                     tf.data.experimental.AUTOTUNE)
our_train_dataset = our_train_dataset.cache()
our_train_dataset = our_train_dataset.shuffle\
                    (len(our_train_dataset))
```

```
our_train_dataset = our_train_dataset.batch(128)
our_train_dataset = our_train_dataset.prefetch\
                    (tf.data.experimental.AUTOTUNE)

our_test_dataset = our_test_dataset.map\
                   (lambda x, y: (normalization_layer(x), y),\
                    num_parallel_calls = \
                    tf.data.experimental.AUTOTUNE)
our_test_dataset = our_test_dataset.cache()
our_test_dataset = our_test_dataset.batch(1024)
our_test_dataset = our_test_dataset.prefetch\
                   (tf.data.experimental.AUTOTUNE)
```

6. Build the model using the functional API. Set the shape, layer types, strides, and activation functions:

```
input_layer = Input(shape=image_shape)
x = Conv2D(filters = 32, \
           kernel_size = (3, 3), strides=2)(input_layer)
x = Activation('relu')(x)

x = Conv2D(filters = 64, kernel_size = (3, 3), strides=2)(x)
x = Activation('relu')(x)

x = Conv2D(filters = 128, kernel_size = (3, 3), strides=2)(x)
x = Activation('relu')(x)

x = Flatten()(x)
x = Dropout(rate = 0.5)(x)

x = Dense(units = 1024)(x)
x = Activation('relu')(x)
x = Dropout(rate = 0.2)(x)

x = Dense(units = num_classes)(x)
output = Activation('softmax')(x)

our_classification_model = Model(input_layer, output)
```

7. Compile and fit your model. Be sure to use your GPU for this, if possible, as it will speed up the process quite a bit. If you decide not to use the GPU and your machine has difficulty in terms of computation, you can decrease the number of epochs accordingly:

```
our_classification_model.compile(
                    optimizer='adam', \
                    loss='sparse_categorical_crossentropy',
                    metrics=['accuracy'], loss_weights=None,
                    weighted_metrics=None, run_eagerly=None,
                    steps_per_execution=None

)


print(our_classification_model.summary())


history = our_classification_model.fit(our_train_dataset, validation_
data=our_test_dataset, epochs=15)
```

The function will return the following history:

```
Epoch 1/15
391/391 [==============================] - 41s 81ms/step - loss: 1.6524 - accuracy: 0.3997 - val_loss: 1.4263 - val_accuracy: 0.4851
Epoch 2/15
391/391 [==============================] - 29s 75ms/step - loss: 1.3798 - accuracy: 0.4994 - val_loss: 1.2617 - val_accuracy: 0.5467
Epoch 3/15
391/391 [==============================] - 29s 75ms/step - loss: 1.2604 - accuracy: 0.5470 - val_loss: 1.1560 - val_accuracy: 0.5897
Epoch 4/15
391/391 [==============================] - 29s 75ms/step - loss: 1.1666 - accuracy: 0.5835 - val_loss: 1.0646 - val_accuracy: 0.6221
Epoch 5/15
391/391 [==============================] - 29s 75ms/step - loss: 1.0889 - accuracy: 0.6106 - val_loss: 1.0072 - val_accuracy: 0.6446
Epoch 6/15
391/391 [==============================] - 29s 75ms/step - loss: 1.0153 - accuracy: 0.6399 - val_loss: 0.9468 - val_accuracy: 0.6669
Epoch 7/15
391/391 [==============================] - 29s 75ms/step - loss: 0.9644 - accuracy: 0.6547 - val_loss: 0.9358 - val_accuracy: 0.6696
Epoch 8/15
391/391 [==============================] - 29s 75ms/step - loss: 0.9142 - accuracy: 0.6763 - val_loss: 0.8945 - val_accuracy: 0.6814
Epoch 9/15
391/391 [==============================] - 29s 75ms/step - loss: 0.8714 - accuracy: 0.6903 - val_loss: 0.8812 - val_accuracy: 0.6948
Epoch 10/15
391/391 [==============================] - 29s 75ms/step - loss: 0.8222 - accuracy: 0.7085 - val_loss: 0.8530 - val_accuracy: 0.7014
Epoch 11/15
391/391 [==============================] - 29s 75ms/step - loss: 0.7807 - accuracy: 0.7216 - val_loss: 0.8282 - val_accuracy: 0.7093
Epoch 12/15
391/391 [==============================] - 29s 75ms/step - loss: 0.7499 - accuracy: 0.7320 - val_loss: 0.8330 - val_accuracy: 0.7134
Epoch 13/15
391/391 [==============================] - 30s 75ms/step - loss: 0.7219 - accuracy: 0.7447 - val_loss: 0.8071 - val_accuracy: 0.7200
Epoch 14/15
391/391 [==============================] - 30s 76ms/step - loss: 0.6889 - accuracy: 0.7548 - val_loss: 0.8080 - val_accuracy: 0.7201
Epoch 15/15
391/391 [==============================] - 30s 76ms/step - loss: 0.6595 - accuracy: 0.7665 - val_loss: 0.7760 - val_accuracy: 0.7295
```

Figure 7.31: Fitting the model

8. Get a visual representation of the model's performance by plotting your loss and accuracy per epoch:

```
def plot_trend_by_epoch(tr_values, val_values, title):
    epoch_number = range(len(tr_values))
    plt.plot(epoch_number, tr_values, 'r')
    plt.plot(epoch_number, val_values, 'b')
    plt.title(title)
    plt.xlabel('epochs')
    plt.legend(['Training '+title, 'Validation '+title])
    plt.figure()

hist_dict = history.history
tr_loss, val_loss = hist_dict['loss'], hist_dict['val_loss']
plot_trend_by_epoch(tr_loss, val_loss, "Loss")
```

This will produce the following plot:

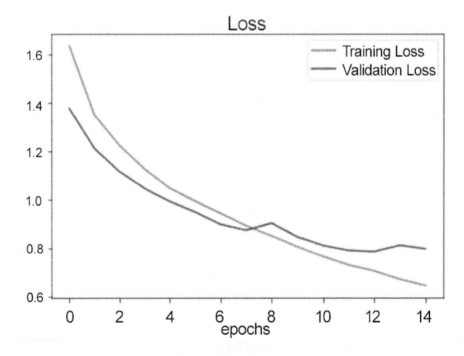

Figure 7.32: Loss plot

9. Next, get an accuracy plot by using the following code:

```
tr_accuracy, val_accuracy = hist_dict['accuracy'], \
                            hist_dict['val_accuracy']
plot_trend_by_epoch(tr_accuracy, val_accuracy, "Accuracy")
```

This will give the following plot:

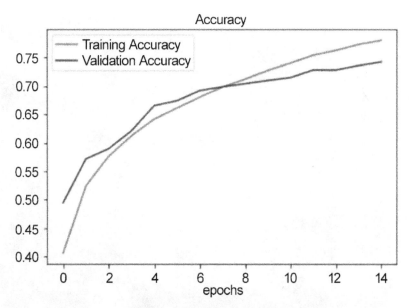

Figure 7.33: Accuracy plot

10. Plot the confusion matrix without normalization:

```
test_labels = []
test_images = []
for image, label in tfds.as_numpy(our_test_dataset.unbatch()):
    test_images.append(image)
    test_labels.append(label)
test_labels = np.array(test_labels)

predictions = our_classification_model.predict(our_test_dataset).
argmax(axis=1)

conf_matrix = confusion_matrix(test_labels, predictions)
```

```
disp = ConfusionMatrixDisplay(conf_matrix, \
                            display_labels = names_of_classes)
fig = plt.figure(figsize = (12, 12))
axis = fig.add_subplot(111)
disp.plot(values_format = 'd', ax = axis)
```

This will give the following output:

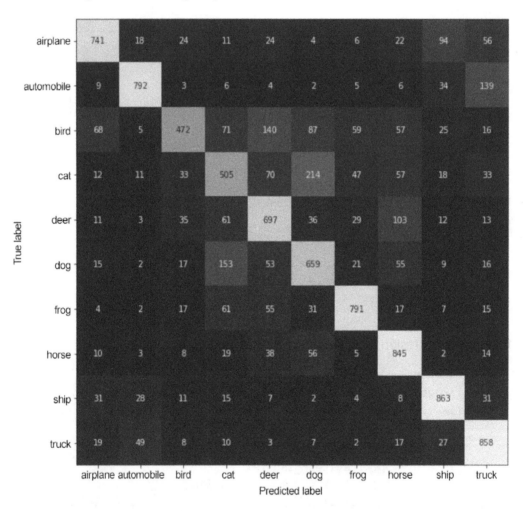

Figure 7.34: Confusion matrix without normalization

11. Use the following code to plot the confusion matrix with normalization:

```
conf_matrix = conf_matrix.astype\
            ('float') / conf_matrix.sum(axis=1) \
            [:, np.newaxis]

disp = ConfusionMatrixDisplay(\
        conf_matrix, display_labels = names_of_classes)
fig = plt.figure(figsize = (12, 12))
axis = fig.add_subplot(111)
disp.plot(ax = axis)
```

The output will look like this:

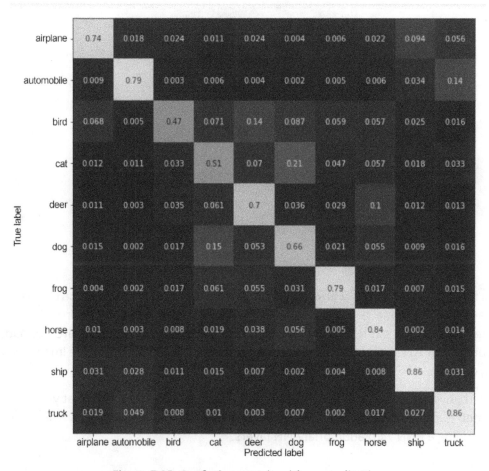

Figure 7.35: Confusion matrix with normalization

12. Take a look at one of the images that the model got wrong. Plot one of the incorrect predictions with the following code:

```
incorrect_predictions = np.where(predictions != test_labels)[0]
index = np.random.choice(incorrect_predictions)

plt.imshow(test_images[index])
print(f'True label: {names_of_classes[test_labels[index]]}')
print(f'Predicted label: {names_of_classes[predictions[index]]}')
```

The output will look like this:

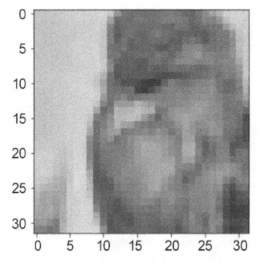

Figure 7.36: True versus predicted results

You'll notice it says **True label: bird** and **Predicted label: cat**. This means that the model predicted that this image was a cat, but it was a bird. The image is blurry since the resolution is only **32x32**; however, the results are not bad. It would be fair to say that it is difficult for a human to identify whether the image was a dog or a cat.

Now that you have completed this chapter, it's time to put everything that you've learned to the test with *Activity 7.01, Building a CNN with More ANN Layers*, where you'll be building a CNN with additional ANN layers.

ACTIVITY 7.01: BUILDING A CNN WITH MORE ANN LAYERS

The start-up that you've been working for has loved your work so far. They have tasked you with creating a new model that is capable of classifying images from 100 different classes.

In this activity, you'll be putting everything that you've learned to use as you build your own classifier with **CIFAR-100**. **CIFAR-100** is a more advanced version of the **CIFAR-10** dataset, with 100 classes, and is commonly used for benchmarking performance in machine learning research.

1. Start a new Jupyter notebook.

2. Import the TensorFlow library.

3. Import the additional libraries that you will need, including NumPy, Matplotlib, Input, Conv2D, Dense, Flatten, Dropout, GlobalMaxPooling2D, Activation, Model, confusion_matrix, and itertools.

4. Load the **CIFAR-100** dataset directly from **tensorflow_datasets** and view its properties from the metadata, and build a train and test data pipeline:

```
Shape of Images in the Dataset:        (32, 32, 3)
Number of Classes in the Dataset:      100
Names of Classes in the Dataset:       ['apple', 'aquarium_fish', 'baby', 'bear', 'beaver', 'bed', 'bee'

Total examples in Train Dataset:       50000
Total examples in Test Dataset:        10000
```

Figure 7.37: Properties of the CIFAR-100 dataset

5. Create a function to rescale images. Then, build a test and train data pipeline by rescaling, caching, shuffling, batching, and prefetching the images.

6. Build the model using the functional API using **Conv2D** and **Flatten**, among others.

7. Compile and fit the model using `model.compile` and `model.fit`:

```
Epoch 1/15
1563/1563 [==============================] - 13s 4ms/step - loss: 6.3395 - accuracy: 0.0375 - val_loss: 4.0599 - val_accuracy: 0.0707
Epoch 2/15
1563/1563 [==============================] - 5s 3ms/step - loss: 3.9944 - accuracy: 0.0829 - val_loss: 3.7596 - val_accuracy: 0.1258
Epoch 3/15
1563/1563 [==============================] - 5s 3ms/step - loss: 3.7720 - accuracy: 0.1189 - val_loss: 3.5724 - val_accuracy: 0.1675
Epoch 4/15
1563/1563 [==============================] - 4s 3ms/step - loss: 3.6113 - accuracy: 0.1491 - val_loss: 3.4189 - val_accuracy: 0.1909
Epoch 5/15
1563/1563 [==============================] - 4s 3ms/step - loss: 3.4475 - accuracy: 0.1772 - val_loss: 3.2566 - val_accuracy: 0.2149
Epoch 6/15
1563/1563 [==============================] - 4s 3ms/step - loss: 3.3013 - accuracy: 0.2020 - val_loss: 3.1660 - val_accuracy: 0.2416
Epoch 7/15
1563/1563 [==============================] - 4s 3ms/step - loss: 3.1800 - accuracy: 0.2207 - val_loss: 3.0630 - val_accuracy: 0.2589
Epoch 8/15
1563/1563 [==============================] - 4s 3ms/step - loss: 3.0794 - accuracy: 0.2403 - val_loss: 2.9940 - val_accuracy: 0.2709
Epoch 9/15
1563/1563 [==============================] - 4s 3ms/step - loss: 2.9897 - accuracy: 0.2561 - val_loss: 2.9154 - val_accuracy: 0.2869
Epoch 10/15
1563/1563 [==============================] - 5s 3ms/step - loss: 2.9122 - accuracy: 0.2728 - val_loss: 2.8597 - val_accuracy: 0.3027
Epoch 11/15
1563/1563 [==============================] - 5s 3ms/step - loss: 2.8356 - accuracy: 0.2852 - val_loss: 2.8523 - val_accuracy: 0.2980
Epoch 12/15
1563/1563 [==============================] - 4s 3ms/step - loss: 2.7869 - accuracy: 0.2950 - val_loss: 2.7875 - val_accuracy: 0.3176
Epoch 13/15
1563/1563 [==============================] - 4s 3ms/step - loss: 2.7299 - accuracy: 0.3097 - val_loss: 2.7684 - val_accuracy: 0.3194
Epoch 14/15
1563/1563 [==============================] - 5s 3ms/step - loss: 2.6782 - accuracy: 0.3192 - val_loss: 2.7655 - val_accuracy: 0.3211
Epoch 15/15
1563/1563 [==============================] - 5s 3ms/step - loss: 2.6242 - accuracy: 0.3305 - val_loss: 2.7470 - val_accuracy: 0.3285
```

Figure 7.38: Model fitting

8. Plot the loss with `plt.plot`. Remember to use the history collected during the `model.fit()` procedure:

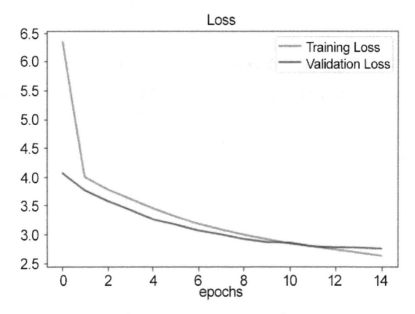

Figure 7.39: Loss versus epochs

9. Plot the accuracy with **plt.plot**:

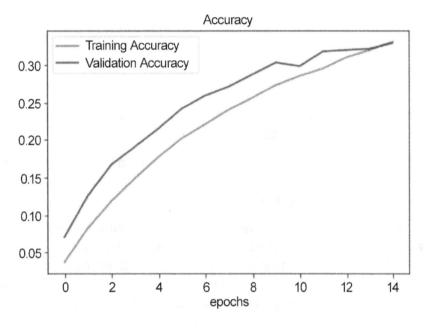

Figure 7.40: Accuracy versus epochs

10. Specify the labels for the different classes in your dataset.

11. Display a misclassified example with **plt.imshow**:

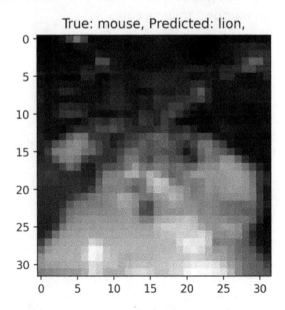

Figure 7.41: Wrong classification example

> **NOTE**
>
> The solution to this activity can be found on page 505.

SUMMARY

This chapter covered CNNs. We reviewed core concepts such as neurons, layers, model architecture, and tensors to understand how to create effective CNNs.

You learned about the convolution operation and explored kernels and feature maps. We analyzed how to assemble a CNN, and then explored the different types of pooling layers and when to apply them.

You then learned about the stride operation and how padding is used to create extra space around images if needed. Then, we delved into the flattening layer and how it is able to convert data into a 1D array for the next layer. You put everything that you learned to the test in the final activity, as you were presented with several classification problems, including **CIFAR-10** and even **CIFAR-100**.

In completing this chapter, you are now well on your way to being able to implement CNNs to confront image classification problems head-on and with confidence.

In the next chapter, you'll learn about pre-trained models and how to utilize them for your own applications by adding ANN layers on top of the pre-trained model and fine-tuning the weights given your own training data.

8

PRE-TRAINED NETWORKS

OVERVIEW

In this chapter, you will analyze pre-trained models. You will get hands-on
experience using the different state-of-the-art model architectures available
on TensorFlow. You will explore concepts such as transfer learning
and fine-tuning and look at TensorFlow Hub and its published deep
learning resources.

By the end of the chapter, you will be able to use pre-trained models directly
from TensorFlow and TensorFlow Hub.

INTRODUCTION

In the previous chapter, you learned how **convolution neural networks** (**CNNs**) analyze images and learn relevant patterns to classify their main subjects or identify objects within them. You also saw the different types of layers used for such models.

But rather than training a model from scratch, it would be more efficient if you could reuse existing models with pre-calculated weights. This is exactly what **transfer learning** and **fine-tuning** are about. You will learn how to apply these techniques to your own projects and datasets in this chapter.

You will also look at the ImageNet competition and the corresponding dataset that is used by deep learning researchers to benchmark their models against state-of-the-art algorithms. Finally, you will learn how to use TensorFlow Hub's resources to build your own model.

IMAGENET

ImageNet is a large dataset containing more than 14 million images annotated for image classification or object detection. It was first consolidated by Fei-Fei Li and her team in 2007. The goal was to build a dataset that computer vision researchers could benefit from.

The dataset was presented for the first time in 2009, and every year since 2010, an annual competition called the **ImageNet Large-Scale Visual Recognition Challenge** (**ILSVRC**) has been organized for image classification and object detection tasks.

Figure 8.1: Examples of images from ImageNet

Over the years, some of the most famous CNN architectures (such as AlexNet, Inception, VGG, and ResNet) have achieved amazing results in this ILSVRC competition. In the following graph, you can see how some of the most famous CNN architectures performed in this competition. In less than 10 years, performance increased from 50% accuracy to almost 90%.

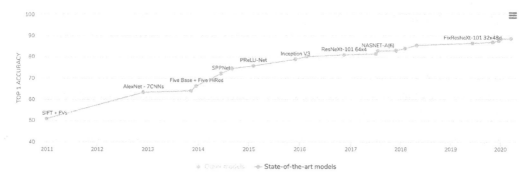

Figure 8.2: Model benchmarking from paperswithcode.com

You will see in the next section how you can use transfer learning with these models.

TRANSFER LEARNING

In the previous chapter, you got hands-on practice training different CNN models for image classification purposes. Even though you achieved good results, the models took quite some time to learn the relevant parameters. If you kept training the models, you could have achieved even better results. Using **graphical processing units** (**GPUs**) can shorten the training time, but it will still take a bit of time, especially for bigger or more complex datasets.

Deep learning researchers have published their work for the benefit of the community. Everyone can benefit by taking existing model architectures and customizing them, rather than designing architectures from scratch. More than this though, researchers also share the weights of their models. You can then not only reuse an architecture but also leverage all the training performed on it. This is what transfer learning is about. By reusing pre-trained models, you don't have to start from scratch. These models are trained on a large dataset such as ImageNet and have learned how to recognize thousands of different categories of objects. You can reuse these state-of-the-art models straight out of the box without having to train them. Isn't that amazing? Rather than training a model for weeks, you can now just use an existing model.

TensorFlow provides a list of state-of-the-art models pre-trained on the ImageNet dataset for transfer learning in its Keras API.

> **NOTE**
>
> You can find the full list of pre-trained models available in TensorFlow at the following link: https://www.tensorflow.org/api_docs/python/tf/keras/applications.

Importing a pre-trained model is quite simple in TensorFlow, as shown with the following example, where you load the **InceptionV3** model:

```
import tensorflow as tf
from tensorflow.keras.applications import InceptionV3
```

Now that you have imported the class for the pre-trained model, you need to instantiate it by specifying the dimensions of the input image and **imagenet** as the pre-trained weights to be loaded:

```
model = InceptionV3(input_shape=(224, 224, 3), \
                    weights='imagenet', include_top=True)
```

The **include_top=True** parameter specifies that you will be re-using the exact same top layer (which is the final layer) as for the original model trained on ImageNet. This means that the last layer is designed to predict the 1,000 classes that are in this dataset.

Now that you have instantiated your pre-trained model, you can make predictions from it:

```
model.predict(input_image)
```

If you want to use this pre-trained model to predict different categories than the ones from ImageNet, you will need to replace the top layer with another one that will be trained to recognize the specific categories of the input dataset.

First, you need to remove this layer by specifying **include_top=False**:

```
model = InceptionV3(input_shape=(224, 224, 3), \
                    weights='imagenet', include_top=False)
```

In the preceding example, you have loaded an **InceptionV3** model. The next step will be to *freeze* all the layers from this model so that their weights will not be updated:

```
model.trainable = False
```

After this, you will instantiate a new fully connected layer with the number of units and activation function of your choice. In the following example, you want to predict 50 different classes. To do this, you create a dense layer with **20** units and use softmax as the activation function:

```
top_layer = tf.keras.layers.Dense(20, activation='softmax')
```

Then you need to add this fully connected layer to your base model with the Sequential API from Keras:

```
new_model = tf.keras.Sequential([model, top_layer])
```

Now, you can train this model and only the top-layer weights will be updated. All the other layers have been frozen:

```
new_model.compile(loss='sparse_categorical_crossentropy', \
                  optimizer=tf.keras.optimizers.Adam(0.001))
```

```
new_model.fit(X_train, t_train, epochs=50)
```

In just a few lines of code, you have loaded the Inception V3 model, which is a state-of-the-art model that won the ILSVRC competition in 2016. You learned how to adapt it to your own project and dataset.

In the next exercise, you will have hands-on practice on transfer learning.

EXERCISE 8.01: CLASSIFYING CATS AND DOGS WITH TRANSFER LEARNING

In this exercise, you will use transfer learning to correctly classify images as either cats or dogs. You will use a pre-trained model, NASNet-Mobile, that is already available in TensorFlow. This model comes with pre-trained weights on ImageNet.

> **NOTE**
>
> The original dataset used in this exercise has been provided by Google. It contains 25,000 images of dogs and cats. It can be found here: https://storage.googleapis.com/mledu-datasets/cats_and_dogs_filtered.zip.

1. Open a new Jupyter notebook.

2. Import the TensorFlow library:

    ```
    import tensorflow as tf
    ```

3. Create a variable called **file_url** containing a link to the dataset:

    ```
    file_url = 'https://storage.googleapis.com'\
               '/mledu-datasets/cats_and_dogs_filtered.zip'
    ```

4. Download the dataset using **tf.keras.get_file**, with **'cats_and_dogs.zip'**, **origin=file_url**, and **extract=True** as parameters, and save the result to a variable called **zip_dir**:

    ```
    zip_dir = tf.keras.utils.get_file('cats_and_dogs.zip', \
                                      origin=file_url, extract=True)
    ```

5. Import the **pathlib** library:

    ```
    import pathlib
    ```

6. Create a variable called **path** containing the full path to the **cats_and_dogs_filtered** directory using **pathlib.Path(zip_dir).parent**:

    ```
    path = pathlib.Path(zip_dir).parent / 'cats_and_dogs_filtered'
    ```

7. Create two variables called **train_dir** and **validation_dir** that take the full path to the **train** and **validation** folders, respectively:

```
train_dir = path / 'train'
validation_dir = path / 'validation'
```

8. Create four variables called **train_cats_dir**, **train_dogs_dir**, **validation_cats_dir**, and **validation_dogs_dir** that take the full path to the **cats** and **dogs** folders for the train and validation sets, respectively:

```
train_cats_dir = train_dir / 'cats'
train_dogs_dir = train_dir /'dogs'
validation_cats_dir = validation_dir / 'cats'
validation_dogs_dir = validation_dir / 'dogs'
```

9. Import the **os** package. In the next step, you will need to count the number of images from a folder:

```
import os
```

10. Create two variables called **total_train** and **total_val** that get the number of images for the training and validation sets:

```
total_train = len(os.listdir(train_cats_dir)) \
             + len(os.listdir(train_dogs_dir))
total_val = len(os.listdir(validation_cats_dir)) \
             + len(os.listdir(validation_dogs_dir))
```

11. Import **ImageDataGenerator** from **tensorflow.keras.preprocessing**:

```
from tensorflow.keras.preprocessing.image
    import ImageDataGenerator
```

12. Instantiate two **ImageDataGenerator** classes and call them **train_image_generator** and **validation_image_generator**. These will rescale images by dividing by **255**:

```
train_image_generator = ImageDataGenerator(rescale=1./255)
validation_image_generator = ImageDataGenerator(rescale=1./255)
```

13. Create three variables called **batch_size**, **img_height**, and **img_width** that take the values **16**, **224**, and **224**, respectively:

```
batch_size = 16
img_height = 224
img_width = 224
```

14. Create a data generator called **train_data_gen** using **flow_from_directory()** method, and specify the batch size, the path to the training folder, the size of the target, and the mode of the class:

```
train_data_gen = train_image_generator.flow_from_directory\
                 (batch_size = batch_size, \
                  directory = train_dir, \
                  shuffle=True, \
                  target_size = (img_height, img_width), \
                  class_mode='binary')
```

15. Create a data generator called **val_data_gen** using **flow_from_directory()** method and specify the batch size, the path to the validation folder, the size of the target, and the mode of the class:

```
val_data_gen = validation_image_generator.flow_from_directory\
               (batch_size = batch_size, \
                directory = validation_dir, \
                target_size=(img_height, img_width), \
                class_mode='binary')
```

16. Import **numpy** as **np**, **tensorflow** as **tf**, and **layers** from **tensorflow.keras**:

```
import numpy as np
import tensorflow as tf
from tensorflow.keras import layers
```

17. Set **8** (this is totally arbitrary) as **seed** for NumPy and TensorFlow:

```
np.random.seed(8)
tf.random.set_seed(8)
```

18. Import the **NASNETMobile** model from **tensorflow.keras.applications**:

```
from tensorflow.keras.applications import NASNetMobile
```

19. Instantiate the model with the ImageNet weights, remove the top layer, and specify the correct input dimensions:

```
base_model = NASNetMobile(include_top=False, \
                          input_shape=(img_height, img_width, 3),\
                          weights='imagenet')
```

20. Freeze all the layers of this model:

```
base_model.trainable = False
```

21. Print a summary of the model using the **summary()** method:

```
base_model.summary()
```

The expected output will be as follows:

```
Model: "NASNet"
```

Layer (type)	Output Shape	Param #	Connected to
input_1 (InputLayer)	[(None, 224, 224, 3)]	0	[]
stem_conv1 (Conv2D)	(None, 111, 111, 32)	864	['input_1[0][0]']
stem_bn1 (BatchNormalization)	(None, 111, 111, 32)	128	['stem_conv1[0][0]']
activation (Activation)	(None, 111, 111, 32)	0	['stem_bn1[0][0]']
reduction_conv_1_stem_1 (Conv2 D)	(None, 111, 111, 11)	352	['activation[0][0]']
reduction_bn_1_stem_1 (BatchNo rmalization)	(None, 111, 111, 11)	44	['reduction_conv_1_stem_1[0][0]']

Figure 8.3: Summary of the model

22. Create a new model that combines the **NASNETMobile** model with two new top layers with **500** and **1** unit(s) and ReLu and sigmoid as the activation functions:

```
model = tf.keras.Sequential([base_model,\
                             layers.Flatten(),
                             layers.Dense(500, \
                                   activation='relu'),
                             layers.Dense(1, \
                                   activation='sigmoid')])
```

23. Compile the model by providing **binary_crossentropy** as the **loss** function, an Adam optimizer with a learning rate of **0.001**, and **accuracy** as the metric to be displayed:

```
model.compile(loss='binary_crossentropy', \
              optimizer=tf.keras.optimizers.Adam(0.001), \
              metrics=['accuracy'])
```

24. Fit the model, provide the train and validation data generators, and run it for five epochs:

```
model.fit(train_data_gen, \
          steps_per_epoch = total_train // batch_size, \
          epochs=5, \
          validation_data = val_data_gen, \
          validation_steps = total_val // batch_size)
```

The expected output is as follows:

```
Epoch 1/5
125/125 [==============================] - 46s 179ms/step - loss: 1.4444 - accuracy: 0.9675 - val_loss: 1.4239 - val_accuracy: 0.9798
Epoch 2/5
125/125 [==============================] - 18s 143ms/step - loss: 0.5265 - accuracy: 0.9890 - val_loss: 1.9579 - val_accuracy: 0.9738
Epoch 3/5
125/125 [==============================] - 18s 143ms/step - loss: 0.2033 - accuracy: 0.9965 - val_loss: 0.5904 - val_accuracy: 0.9889
Epoch 4/5
125/125 [==============================] - 18s 143ms/step - loss: 0.0766 - accuracy: 0.9970 - val_loss: 0.3927 - val_accuracy: 0.9889
Epoch 5/5
125/125 [==============================] - 18s 143ms/step - loss: 0.0525 - accuracy: 0.9975 - val_loss: 1.0821 - val_accuracy: 0.9758
<keras.callbacks.History at 0x7f110cb67550>
```

Figure 8.4: Model training output

You can observe that the model achieved an accuracy score of **0.99** on the training set and **0.98** on the validation set. This is quite a remarkable result given that you only trained the last two layers, and it took less than a minute. This is the benefit of applying transfer learning and using pre-trained state-of-the-art models.

In the next section, you will see how you can apply fine-tuning to a pre-trained model.

FINE-TUNING

Previously, you used transfer learning to leverage pre-trained models on your own dataset. You used the weights of state-of-the-art models that have been trained on large datasets such as ImageNet. These models learned the relevant parameters to recognize different patterns from images and helped you to achieve amazing results on different datasets.

But there is a catch with this approach. Transfer learning works well in general if the classes you are trying to predict belong to the same list as that of ImageNet. If this is the case, the weight learned from ImageNet will also be relevant to your dataset. For example, the **cats** and **dogs** classes from the preceding exercise are present in ImageNet, so its weights will also be relevant for this dataset.

However, if your dataset is very different from ImageNet, then the weights from these pre-trained models may not all be relevant. For example, if your dataset contains satellite images, and you are trying to determine whether a house has solar panels installed on its roof, this will be very different compared to ImageNet. The weights from the last layers will be very specific to the classes from ImageNet, such as cat whiskers or car wheels (which are not very useful for the satellite image dataset case), while the ones from earlier layers will be more generic, such as for detecting shapes, colors, or texture (which can be applied to the satellite image dataset).

So, it will be great to still leverage some of the weights from earlier layers but train the final layers so that your models can learn the specific patterns relevant to your dataset and improve its performance.

This technique is called fine-tuning. The idea behind it is quite simple: you freeze early layers and update the weights of the final layers only. Let's see how you can achieve this in TensorFlow:

1. First, instantiate a pre-trained **MobileNetV2** model without the top layer:

```
from tensorflow.keras.applications import MobileNetV2

base_model = MobileNetV2(input_shape=(224, 224, 3), \
                         weights='imagenet', include_top=False)
```

2. Next, iterate through the first layers and freeze them by setting them as non-trainable. In the following example, you will freeze only the first **100** layers:

```
for layer in base_model.layers[:100]:
    layer.trainable = False
```

3. Now you need to add your custom top layer to your base model. In the following example, you will be predicting 20 different classes, so you need to add a fully connected layer of **20** units with the softmax activation function:

```
prediction_layer = tf.keras.layers.Dense(20, activation='softmax')
model = tf.keras.Sequential([base_model, prediction_layer])
```

4. Finally, you will compile and then train this model:

```
model.compile(loss='sparse_categorical_crossentropy', \
              optimizer = tf.keras.optimizers.Adam(0.001))
model.fit(features_train, label_train, epochs=5)
```

This will display a number of logs, as seen in the following screenshot:

```
Epoch 1/5
712/712 [==============================] - 148s 207ms/step - loss: 2.7766 - accuracy: 0.4375 - val_loss: 3.0142 - val_accuracy: 0.4840
```

Figure 8.5: Fine-tuning results on a pre-trained MobileNetV2 model

That's it. You have just performed fine-tuning on a pre-trained MobileNetV2 model. You have used the first 100 pre-trained weights from ImageNet and only updated the weights from layer 100 onward according to your dataset.

In the next activity, you will put into practice what you have just learned and apply fine-tuning to a pre-trained model.

ACTIVITY 8.01: FRUIT CLASSIFICATION WITH FINE-TUNING

The **Fruits 360** dataset (https://arxiv.org/abs/1712.00580), which was originally shared by *Horea Muresan and Mihai Oltean, Fruit recognition from images using deep learning, Acta Univ. Sapientiae, Informatica Vol. 10, Issue 1, pp. 26-42, 2018*, contains more than 82,000 images of 120 different types of fruit. You will be using a subset of this dataset with more than 16,000 images. The numbers of images in the training and validation sets are **11398** and **4752** respectively.

In this activity, you are tasked with training a **NASNetMobile** model to recognize images of different varieties of fruits (classification into 120 different classes). You will use fine-tuning to train the final layers of this model.

> **NOTE**
>
> The dataset can be found here: http://packt.link/OFUJj.

The following steps will help you to complete this activity:

1. Import the dataset and unzip the file using TensorFlow.

2. Create a data generator with the following data augmentation:

```
Rescale = 1./255,
rotation_range = 40,
width_shift_range = 0.1,
height_shift_range = 0.1,
shear_range = 0.2,
zoom_range = 0.2,
horizontal_flip = True,
fill_mode = 'nearest
```

3. Load a pre-trained **NASNetMobile** model from TensorFlow.

4. Freeze the first **600** layers of the model.

5. Add two fully connected layers on top of **NASNetMobile**:

   – A fully connected layer with **Dense(1000, activation=relu)**

   – A fully connected layer with **Dense(120, activation='softmax')**

6. Specify an Adam optimizer with a learning rate of **0.001**.

7. Train the model.

8. Evaluate the model on the test set.

   The expected output is as follows:

```
Epoch 1/5
712/712 [==============================] - 148s 207ms/step - loss: 2.7766 - accuracy: 0.4375 - val_loss: 3.0142 - val_accuracy: 0.4840
Epoch 2/5
712/712 [==============================] - 144s 202ms/step - loss: 0.5059 - accuracy: 0.8475 - val_loss: 3.0363 - val_accuracy: 0.5149
Epoch 3/5
712/712 [==============================] - 142s 200ms/step - loss: 0.2538 - accuracy: 0.9220 - val_loss: 0.9776 - val_accuracy: 0.7919
Epoch 4/5
712/712 [==============================] - 142s 200ms/step - loss: 0.2049 - accuracy: 0.9380 - val_loss: 0.8568 - val_accuracy: 0.8523
Epoch 5/5
712/712 [==============================] - 142s 199ms/step - loss: 0.1554 - accuracy: 0.9549 - val_loss: 0.8855 - val_accuracy: 0.8264
<tensorflow.python.keras.callbacks.History at 0x7fa90f22c860>
```

Figure 8.6: Expected output of the activity

NOTE

The solution to this activity can be found on page 511.

Now that you know how to use pre-trained models from TensorFlow, you will learn how models can be accessed from TensorFlow Hub in the following section.

TENSORFLOW HUB

TensorFlow Hub is a repository of TensorFlow modules shared by publishers such as Google, NVIDIA, and Kaggle. TensorFlow modules are self-contained models built on TensorFlow that can be reused for different tasks. Put simply, it is an external collection of published TensorFlow modules for transfer learning and fine-tuning. With TensorFlow Hub, you can access different deep learning models or weights than the ones provided directly from TensorFlow's core API.

> **NOTE**
>
> You can find more information about TensorFlow Hub here:
> https://tfhub.dev/.

In order to use it, you first need to install it:

```
pip install tensorflow-hub
```

Once it's installed, you can load available classification models with the **load()** method by specifying the link to a module:

```
import tensorflow_hub as hub

MODULE_HANDLE = 'https://tfhub.dev/tensorflow/efficientnet'\
                '/b0/classification/1'
module = hub.load(MODULE_HANDLE)
```

In the preceding example, you have loaded the **EfficientNet B0** model, which was trained on ImageNet. You can find more details on this at the TensorFlow Hub page: https://tfhub.dev/tensorflow/efficientnet/b0/classification/1.

> **NOTE**
>
> TensorFlow Hub provides a search engine to find a specific module:
> https://tfhub.dev/s?subtype=module,placeholder.

By default, modules loaded from TensorFlow Hub contain the final layer of a model without an activation function. For classification purposes, you need to add an activation layer of your choice. To do so, you can use the Sequential API from Keras. You just need to convert your model into a Keras layer with the **KerasLayer** class:

```
import tensorflow as tf

model = tf.keras.Sequential([
    hub.KerasLayer(MODULE_HANDLE,input_shape=(224, 224, 3)),
    tf.keras.layers.Activation('softmax')
])
```

Then, you can use your final model to perform predictions:

```
model.predict(data)
```

You just performed transfer learning with a model from TensorFlow Hub. This is very similar to what you learned previously using the Keras API, where you loaded an entire model with **include_top=True**. With TensorFlow Hub, you can access a library of pre-trained models for object detection or image segmentation.

In the next section, you will learn how to extract features from TensorFlow Hub pre-trained modules.

FEATURE EXTRACTION

TensorFlow Hub provides the option of downloading a model without the final layer. In this case, you will be using a TensorFlow module as a feature extractor; you can design your custom final layers on top of it. In TensorFlow Hub, a module used for feature extraction is known as a feature vector:

```
import tensorflow_hub as hub

MODULE_HANDLE = 'https://tfhub.dev/google/efficientnet/b0'\
                '/feature-vector/1'
module = hub.load(MODULE_HANDLE)
```

> **NOTE**
>
> To find all the available feature vectors on TensorFlow Hub, you can use its search engine: https://tfhub.dev/s?module-type=image-feature-vector&tf-version=tf2.

Once loaded, you can add your own final layer to the feature vector with the Sequential API:

```
model = tf.keras.Sequential([
    hub.KerasLayer(MODULE_HANDLE, input_shape=(224, 224, 3)),
    tf.keras.layers.Dense(20, activation='softmax')
])
```

In the preceding example, you added a fully connected layer of **20** units with the softmax activation function. Next, you need to compile and train your model:

```
model.compile(optimizer=optimizer, \
              loss='sparse_categorical_crossentropy', \
              metrics=['accuracy'])
model.fit(X_train, epochs=5)
```

And with that, you just used a feature vector from TensorFlow Hub and added your custom final layer to train the final model on your dataset.

Now, test the knowledge you have gained so far in the next activity.

ACTIVITY 8.02: TRANSFER LEARNING WITH TENSORFLOW HUB

In this activity, you are required to correctly classify images of cats and dogs using transfer learning. Rather than training a model from scratch, you will benefit from the **EfficientNet B0** feature vector from TensorFlow Hub, which contains pre-computed weights that can recognize different types of objects.

You can find the dataset here: https://packt.link/RAAtm.

The following steps will help you to complete this activity:

1. Import the dataset and unzip the file using TensorFlow.

2. Create a data generator that will perform rescaling.

3. Load a pre-trained **EfficientNet B0** feature vector from TensorFlow Hub.

4. Add two fully connected layers on top of the feature vector:

   – A fully connected layer with **Dense(500, activation=relu)**

   – A fully connected layer with **Dense(1, activation='sigmoid')**

5. Specify an Adam optimizer with a learning rate of **0.001**.

6. Train the model.

7. Evaluate the model on the test set.

    The expected output is as follows:

```
Epoch 1/5
62/62 [==============================] - 28s 240ms/step - loss: 0.0554 - accuracy: 0.9776 - val_loss: 0.0247 - val_accuracy: 0.9899
Epoch 2/5
62/62 [==============================] - 14s 223ms/step - loss: 0.0116 - accuracy: 0.9959 - val_loss: 0.0253 - val_accuracy: 0.9919
Epoch 3/5
62/62 [==============================] - 14s 221ms/step - loss: 0.0032 - accuracy: 0.9995 - val_loss: 0.0245 - val_accuracy: 0.9909
Epoch 4/5
62/62 [==============================] - 14s 222ms/step - loss: 0.0011 - accuracy: 1.0000 - val_loss: 0.0265 - val_accuracy: 0.9919
Epoch 5/5
62/62 [==============================] - 14s 222ms/step - loss: 5.1736e-04 - accuracy: 1.0000 - val_loss: 0.0243 - val_accuracy: 0.9929
<keras.callbacks.History at 0x7f04bc0a6b10>
```

Figure 8.7: Expected output of the activity

The expected accuracy scores should be around **1.0** for the training and validation sets.

> **NOTE**
>
> The solution to this activity can be found on page 515.

SUMMARY

In this chapter, you learned two very important concepts: transfer learning and fine-tuning. Both help deep learning practitioners to leverage existing pre-trained models and adapt them to their own projects and datasets.

Transfer learning is the re-use of models that have been trained on large datasets such as ImageNet (which contains more than 14 million images). TensorFlow provides a list of such pre-trained models in its core API. You can also access other models from renowned publishers such as Google and NVIDIA through TensorFlow Hub.

Finally, you got some hands-on practice fine-tuning a pre-trained model. You learned how to freeze the early layers of a model and only train the last layers according to the specificities of the input dataset.

These two techniques were a major breakthrough for the community as they facilitated access to state-of-the-art models for anyone interested in applying deep learning models.

In the next chapter, you will look at another type of model architecture, **recurrent neural networks (RNNs)**. This type of architecture is well suited for sequential data such as time series or text.

9

RECURRENT NEURAL NETWORKS

OVERVIEW

In this chapter, you will learn how to handle real sequential data. You will extend your knowledge of **artificial neural network** (**ANN**) models and **recurrent neural network** (**RNN**) architecture for training sequential data. You will also learn how to build an RNN model with an LSTM layer for natural language processing.

By the end of this chapter, you will have gained hands-on experience of applying multiple LSTM layers to build RNNs for stock price predictions.

INTRODUCTION

Sequential data refers to datasets in which each data point is dependent on the previous ones. Think of it like a sentence, which is composed of a sequence of words that are related to each other. A verb will be linked to a subject and an adverb will be related to a verb. Another example is a stock price, where the price on a particular day is related to the price of the previous days. Traditional neural networks are not fit for processing this kind of data. There is a specific type of architecture that can ingest sequences of data. This chapter will introduce you to such models—known as **recurrent neural networks (RNNs)**.

An RNN model is a specific type of deep learning architecture in which the output of the model feeds back into the input. Models of this kind have their own challenges (known as vanishing and exploding gradients) that will be addressed later in the chapter.

In many ways, an RNN is a representation of how a brain might work. RNNs use memory to help them learn. But how can they do this if information only flows in one direction? To understand this, you'll need to first review sequential data. This is a type of data that requires a working memory to process data effectively. Until now, you have only explored non-sequential models, such as a perceptron or CNN. In this chapter, you will look at sequential models such as RNN, LSTM, or GRU.

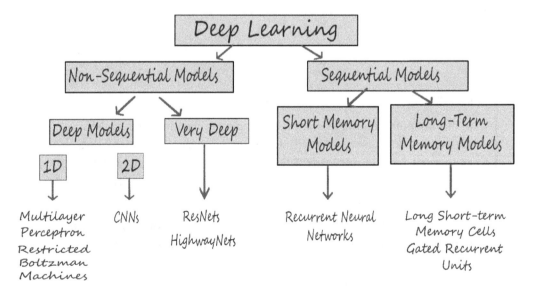

Figure 9.1: Sequential versus non-sequential models

SEQUENTIAL DATA

Sequential data is information that happens in a sequence and is related to past and future data. An example of sequential data is time series data; as you perceive it, time only travels in one direction.

Suppose you have a ball (as in *Figure 9.2*), and you want to predict where this ball will travel next. If you have no prior information about the direction from which the ball was thrown, you will simply have to guess. However, if in addition to the ball's current location, you also had information about its previous location, the problem would be much simpler. To be able to predict the ball's next location, you need the previous location information in a sequential (or ordered) form to make a prediction about future events.

Figure 9.2: Direction of the ball

RNNs function in a way that allows the sequence of the information to retain value with the help of internal memory.

You'll take a look at some examples of sequential data in the following section.

EXAMPLES OF SEQUENTIAL DATA

Sequential data is a specific type of data where the order of each piece of information is important, and they all depend on each other.

One example of sequential data is financial data, such as stock prices. If you want to predict future data values for a given stock, you need to use previous values in time. In fact, you will work on stock prediction in *Exercise 9.01, Training an ANN for Sequential Data – Nvidia Stock Prediction*.

Audio and text can also be considered sequential data. Audio can be split up into a sequence of sound waves, and text can be split up into sequences of either characters or words. The sound waves or sequences of characters or words should be processed in order to convey the desired result. Beyond these two examples that you encounter every day, there are many more examples in which sequential processing may be useful, from analyzing medical signals such as EEGs, projecting stock prices, and inferring and understanding genomic sequences. There are three categories of sequential data:

- **Many-to-One** produces one output from many inputs.

- **One-to-Many** produces many outputs from one input.

- **Many-to-Many** produces many outputs from many inputs.

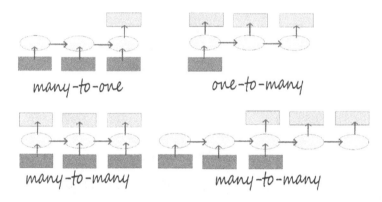

Figure 9.3: Categories of sequential data

Consider another example. Suppose you have a language model with a sentence or a phrase and you are trying to predict the word that comes next, as in the following figure:

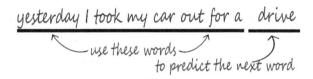

Figure 9.4: Sentence example

Say you're given the words **yesterday I took my car out for a**..., and you want to try to predict the next word, **drive**. One way you could do this is by building a deep neural network such as a feed-forward neural network. However, you would immediately run into a problem. A feed-forward network can only take a fixed-length input vector as its input; you have to specify the size of that input right from the start.

Because of this, your model needs a way to be able to handle variable-length inputs. One way you can do this is by using a fixed window. That means that you force your input vector to be just a certain length. For example, you can split the sentence into groups of two consecutive words (also called a **bi-gram**) and predict the next one. This means that no matter where you're trying to make that next prediction, your model will only be taking in the previous two words as its input. You need to consider how you can numerically represent this data. One way you can do this is by taking a fixed-length vector and allocating some space in that vector for the first word and some space in that vector for the second word. In those spaces, encode the identity of each word. However, this is problematic.

Why? Because you're using only a portion of the information available (that is, two consecutive words only). You have access to a limited window of data that doesn't give enough context to accurately predict what will be the next word. That means you cannot effectively model long-term dependencies. This is important in sentences like the one in *Figure 9.5* where you clearly need information from much earlier in the sentence to be able to accurately predict the next word.

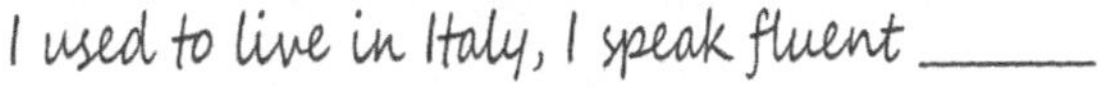

I used to live in Italy, I speak fluent _____

Figure 9.5: Sentence example

If you were only looking at the past two or three words, you wouldn't be able to make this next prediction, which you know is **Italian**. So, this means that you really need a way to integrate the information in the sentence from start to finish.

To do this, you could use a set of counts as a fixed-length vector and use the entire sentence. This method is known as **bag of words**.

You have a fixed-length vector regardless of the identity of the sentence, but what differs is adding the counts over this vocabulary. You can feed this into your model as an input to generate a prediction.

However, there's another big problem with this. Using just the counts means that you lose all sequential information and all information about the prior history.

Consider *Figure 9.6*. So, these two sentences, which have completely opposite semantic meanings would have the exact same representations in this bag of words format. This is because they have the exact same list of words, just in a different order. So, obviously, this isn't going to work. Another idea could be simply to extend the fixed window.

it is what it is
is it what it is

Figure 9.6: Bag of words example

Now, consider *Figure 9.7*. You can represent your sentence in this way, feed the sentence into your model, and generate your prediction. The problem is that if you were to feed this vector into a feed-forward neural network, each of these inputs, **yesterday I took my car**, would have a separate weight connecting it to the network. So, if you were to repeatedly see the word **yesterday** at the beginning of the sentence, the network may be able to learn that **yesterday** represents a time or a setting. However, if **yesterday** were to suddenly appear later in that fixed-length vector, at the end of a sentence, the network may have difficulty understanding the meaning of **yesterday**. This is because the parameters that are at the end of a vector may never have seen the term **yesterday** before, and the parameters from the beginning of the sentence weren't shared across the entire sequence.

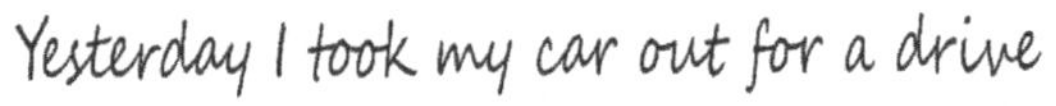

Yesterday I took my car out for a drive

Figure 9.7: Sentence example

So, you need to be able to handle variable-length input and long-term dependencies, track sequential order, and have parameters that can be shared across the entirety of your sequence. Specifically, you need to develop models that can do the following:

- Handle variable-length input sequences.
- Track long-term dependencies in the data.
- Maintain information about the sequence's order.
- Share parameters across the entirety of the sequence.

How can you do this with a model where information only flows in one direction? You need a different kind of neural network. You need a recursive model. You will practice processing sequential data in the following exercise.

EXERCISE 9.01: TRAINING AN ANN FOR SEQUENTIAL DATA — NVIDIA STOCK PREDICTION

In this exercise, you will build a simple ANN model to predict the Nvidia stock price. But unlike examples from previous chapters, this time the input data is sequential. So, you need to manually do some processing to create a dataset that will contain the price of the stock for a given day as the target variable and the price for the previous 60 days as features. You are required to split the data into training and testing sets before and after the date **2019-01-01**.

> **NOTE**
>
> You can find the **NVDA.csv** dataset here: https://packt.link/Mxi80.

1. Open a new Jupyter or Colab notebook.

2. Import the libraries needed. Use **numpy** for computation, **matplotlib** for plotting visualization, **pandas** to help work with your dataset, and **MinMaxScaler** to scale the dataset between zero and one:

```
import numpy as np
import matplotlib.pyplot as plt
import pandas as pd
from sklearn.preprocessing import StandardScaler, MinMaxScaler
```

3. Use the **read_csv()** function to read in the CSV file and store your dataset in a pandas DataFrame, **data**, for manipulation:

```
import io
data = pd.read_csv('NVDA.csv')
```

4. Call the **head()** function on your data to take a look at the first five rows of your DataFrame:

```
data.head()
```

You should get the following output:

	Date	Open	High	Low	Close	Adj Close	Volume
0	2015-07-22	19.650000	19.650000	19.17	19.410000	18.851749	8911800
1	2015-07-23	19.450001	19.940001	19.41	19.650000	19.084845	4247900
2	2015-07-24	19.790001	19.809999	19.34	19.420000	18.861464	4721100
3	2015-07-27	19.250000	19.530001	19.09	19.309999	18.754622	4810500
4	2015-07-28	19.360001	19.860001	19.16	19.730000	19.162542	4957700

Figure 9.8: First five rows of output

The preceding table shows the raw data. You can see that each row represents a day where you have information about the stock price when the market opened and closed, the highest price, the lowest price, and the adjusted close price of the stock (taking into account dividend or stock split, for instance).

5. Now, split the training data. Use all data that is older than **2019-01-01** using the **Date** column for your training data. Save it as **data_training**. Save this in a separate file by using the **copy()** method:

```
data_training = data[data['Date']<'2019-01-01'].copy()
```

6. Now, split the test data. Use all data that is more recent than or equal to **2019-01-01** using the **Date** column. Save it as **data_test**. Save this in a separate file by using the **copy()** method:

```
data_test = data[data['Date']>='2019-01-01'].copy()
```

7. Use **drop()** to remove your **Date** and **Adj Close** columns in your DataFrame. Remember that you used the **Date** column to split your training and test sets, so the date information is not needed. Use **axis = 1** to specify that you also want to drop labels from your columns. To make sure it worked, call the **head()** function to take a look at the first five rows of the DataFrame:

```
training_data = data_training.drop\
                (['Date', 'Adj Close'], axis = 1)
training_data.head()
```

You should get the following output:

	Open	High	Low	Close	Volume
0	19.650000	19.650000	19.17	19.410000	8911800
1	19.450001	19.940001	19.41	19.650000	4247900
2	19.790001	19.809999	19.34	19.420000	4721100
3	19.250000	19.530001	19.09	19.309999	4810500
4	19.360001	19.860001	19.16	19.730000	4957700

Figure 9.9: New training data

This is the output you should get after removing those two columns.

8. Create a scaler from **MinMaxScaler** to scale **training_data** to numbers between zero and one. Use the **fit_transform** function to fit the model to the data and then transform the data according to the fitted model:

```
scaler = MinMaxScaler()
training_data = scaler.fit_transform(training_data)
training_data
```

You should get the following output:

```
array([[1.48109745e-03, 4.39186751e-04, 3.00198896e-04, 3.70305518e-04,
        8.35120643e-02],
       [7.40552430e-04, 1.50056724e-03, 1.20079559e-03, 1.25902987e-03,
        3.22671736e-02],
       [1.99948527e-03, 1.02477031e-03, 9.38121551e-04, 4.07335700e-04,
        3.74664879e-02],
       ...,
       [4.13744593e-01, 4.13021997e-01, 3.98101261e-01, 4.14219607e-01,
        1.60582121e-01],
       [4.17484345e-01, 4.31358175e-01, 4.17351508e-01, 4.23403077e-01,
        1.58297807e-01],
       [4.30073651e-01, 4.28869458e-01, 4.24668845e-01, 4.22847646e-01,
        1.13361974e-01]])
```

Figure 9.10: Scaled training data

9. Split your data into **X_train** and **y_train** datasets:

```
X_train = []
y_train = []
```

10. Check the shape of **training_data**:

```
training_data.shape[0]
```

You should get the following output:

```
868
```

You can see there are 868 observations in the training set.

11. Create a training dataset that has the previous 60 days' stock prices so that you can predict the closing stock price for day 61. Here, **X_train** will have two columns. The first column will store the values from 0 to 59, and the second will store values from 1 to 60. In the first column of **y_train**, store the 61st value at index 60, and in the second column, store the 62nd value at index 61. Use a **for** loop to create data in 60 time steps:

```
for i in range(60, training_data.shape[0]):
    X_train.append(training_data[i-60:i])
    y_train.append(training_data[i, 0])
```

12. Convert **X_train** and **y_train** into NumPy arrays:

```
X_train, y_train = np.array(X_train), np.array(y_train)
```

13. Call the **shape()** function on **X_train** and **y_train**:

```
X_train.shape, y_train.shape
```

You should get the following output:

```
((808, 60, 5), (808,))
```

The preceding snippet shows that the prepared training set contains **808** observations with **60** days of data for the five features you kept (**Open**, **Low**, **High**, **Close**, and **Volume**).

14. Transform the data into a 2D matrix with the shape of the sample (the number of samples and the number of features in each sample). Stack the features for all 60 days on top of each other to get an output size of **(808, 300)**. Use the following code for this purpose:

```
X_old_shape = X_train.shape
X_train = X_train.reshape(X_old_shape[0], \
                          X_old_shape[1]*X_old_shape[2])
X_train.shape
```

You should get the following output:

```
(808, 300)
```

15. Now, build an ANN. You will need some additional libraries for this. Use **Sequential** to initialize the neural net, **Input** to add an input layer, **Dense** to add a dense layer, and **Dropout** to help prevent overfitting:

```
from tensorflow.keras import Sequential
from tensorflow.keras.layers import Input, Dense, Dropout
```

16. Initialize the neural network by calling **regressor_ann = Sequential()**.

```
regressor_ann = Sequential()
```

17. Add an input layer with **shape** as **300**:

```
regressor_ann.add(Input(shape = (300,)))
```

18. Then, add the first dense layer. Set it to **512** units, which will be your dimensionality for the output space. Use a ReLU activation function. Finally, add a dropout layer that will remove 20% of the units during training to prevent overfitting:

```
regressor_ann.add(Dense(units = 512, activation = 'relu'))
regressor_ann.add(Dropout(0.2))
```

19. Add another dense layer with **128** units, ReLU as the activation function, and a dropout of **0.3**:

```
regressor_ann.add(Dense(units = 128, activation = 'relu'))
regressor_ann.add(Dropout(0.3))
```

20. Add another dense layer with **64** units, ReLU as the activation function, and a dropout of **0.4**:

```
regressor_ann.add(Dense(units = 64, activation = 'relu'))
regressor_ann.add(Dropout(0.4))
```

21. Again, add another dense layer with **128** units, ReLU as the activation function, and a dropout of **0.3**:

```
regressor_ann.add(Dense(units = 16, activation = 'relu'))
regressor_ann.add(Dropout(0.5))
```

22. Add a final dense layer with one unit:

```
regressor_ann.add(Dense(units = 1))
```

23. Check the summary of the model:

```
regressor_ann.summary()
```

You will get valuable information about your model layers and parameters.

```
Model: "sequential"

_____
Layer (type)                 Output Shape              Param #
=================================================================
dense (Dense)                (None, 512)               154112

dropout (Dropout)            (None, 512)               0

dense_1 (Dense)              (None, 128)               65664

dropout_1 (Dropout)          (None, 128)               0

dense_2 (Dense)              (None, 64)                8256

dropout_2 (Dropout)          (None, 64)                0

dense_3 (Dense)              (None, 16)                1040

dropout_3 (Dropout)          (None, 16)                0

dense_4 (Dense)              (None, 1)                 17
=================================================================
Total params: 229,089
Trainable params: 229,089
Non-trainable params: 0
```

Figure 9.11: Model summary

24. Use the **compile()** method to configure your model for training. Choose Adam as your optimizer and mean squared error to measure your loss function:

```
regressor_ann.compile(optimizer='adam', \
                    loss = 'mean_squared_error')
```

25. Finally, fit your model and set it to run on **10** epochs. Set your batch size to **32**:

```
regressor_ann.fit(X_train, y_train, epochs=10, batch_size=32)
```

You should get the following output:

```
Epoch 1/10
26/26 [==============================] - 1s 3ms/step - loss: 0.3294
Epoch 2/10
26/26 [==============================] - 0s 2ms/step - loss: 0.1459
Epoch 3/10
26/26 [==============================] - 0s 3ms/step - loss: 0.0920
Epoch 4/10
26/26 [==============================] - 0s 3ms/step - loss: 0.0881
Epoch 5/10
26/26 [==============================] - 0s 3ms/step - loss: 0.0888
Epoch 6/10
26/26 [==============================] - 0s 3ms/step - loss: 0.0729
Epoch 7/10
26/26 [==============================] - 0s 3ms/step - loss: 0.0707
Epoch 8/10
26/26 [==============================] - 0s 4ms/step - loss: 0.0674
Epoch 9/10
26/26 [==============================] - 0s 3ms/step - loss: 0.0600
Epoch 10/10
26/26 [==============================] - 0s 3ms/step - loss: 0.0590
<keras.callbacks.History at 0x2a967d36388>
```

Figure 9.12: Training the model

26. Test and predict the stock price and prepare the dataset. Check your data by calling the **head()** method:

```
data_test.head()
```

You should get the following output:

	Date	Open	High	Low	Close	Adj Close	Volume
868	2019-01-02	130.639999	138.479996	130.050003	136.220001	135.547104	12718800
869	2019-01-03	133.789993	135.160004	127.690002	127.989998	127.357750	17638800
870	2019-01-04	130.940002	137.729996	129.699997	136.190002	135.517258	14640500
871	2019-01-07	138.500000	144.889999	136.429993	143.399994	142.691620	17729000
872	2019-01-08	146.690002	146.779999	136.899994	139.830002	139.139282	19650400

Figure 9.13: First five rows of a DataFrame

27. Use the **tail(60)** method to create a **past_60_days** variable, which consists of the last 60 days of data in the training set. Add the **past_60_days** variable to the test data with the **append()** function. Assign **True** to **ignore_index**:

```
past_60_days = data_training.tail(60)
df = past_60_days.append(data_test, ignore_index = True)
```

28. Now, prepare your test data for predictions by repeating what you did for the training data in *steps 8* to *15*:

```
df = df.drop(['Date', 'Adj Close'], axis = 1)

inputs = scaler.transform(df)

X_test = []
y_test = []

for i in range(60, inputs.shape[0]):
    X_test.append(inputs[i-60:i])
    y_test.append(inputs[i, 0])

X_test, y_test = np.array(X_test), np.array(y_test)

X_old_shape = X_test.shape
```

```
X_test = X_test.reshape(X_old_shape[0], \
                        X_old_shape[1] * X_old_shape[2])
```

```
X_test.shape, y_test.shape
```

You should get the following output:

```
((391, 300), (391,))
```

29. Test some predictions for your stock prices by calling the **predict()** method on **X_test**:

```
y_pred = regressor_ann.predict(X_test)
```

30. Before looking at the results, reverse the scaling you did earlier so that the number you get as output will be at the correct scale using the **StandardScaler** utility class that you imported with **scaler.scale_**:

```
scaler.scale_
```

You should get the following output:

```
array([3.70274364e-03, 3.65992009e-03, 3.75248621e-03, 3.70301815e-03,
       1.09875621e-08])
```

Figure 9.14: Using StandardScaler

31. Use the first value in the preceding array to set your scale in preparation for the multiplication of **y_pred** and **y_test**. Recall that you are converting your data back from your earlier scale, in which you converted all values to between zero and one:

```
scale = 1/3.70274364e-03
scale
```

You should get the following output:

```
270.0700067909643
```

32. Multiply **y_pred** and **y_test** by **scale** to convert your data back to the proper values:

```
y_pred = y_pred*scale
y_test = y_test*scale
```

33. Review the real Nvidia stock price and your predictions:

```
plt.figure(figsize=(14,5))
plt.plot(y_test, color = 'black', label = "Real NVDA Stock Price")
plt.plot(y_pred, color = 'gray',\
        label = 'Predicted NVDA Stock Price')
plt.title('NVDA Stock Price Prediction')
plt.xlabel('time')
plt.ylabel('NVDA Stock Price')
plt.legend()
plt.show()
```

You should get the following output:

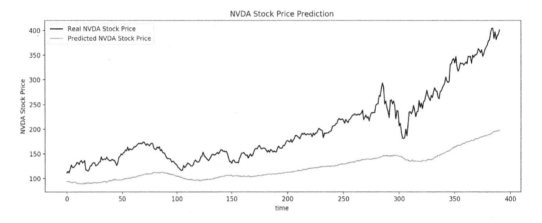

Figure 9.15: Real Nvidia stock price versus your predictions

In the preceding graph, you can see that your trained model is able to capture some of the trends of the Nvidia stock price. Observe that the predictions are quite different from the real values. It is evident from this result that ANNs are not suited for sequential data.

In this exercise, you saw the inability of simple ANNs to deal with sequential data. In the next section, you will learn about recurrent neural networks, which are designed to learn from the temporal dimensionality of sequential data. Then, in *Exercise 9.02, Building an RNN with LSTM Layer Nvidia Stock Prediction*, you will perform predictions on the same Nvidia stock price dataset using RNNs and compare your results.

RECURRENT NEURAL NETWORKS

The first formulation of a recurrent-like neural network was created by John Hopfield in 1982. He had two motivations for doing so:

- Sequential processing of data

- Modeling of neuronal connectivity

Essentially, an RNN processes input data at each time step and stores information in its memory that will be used for the next step. Information is first transformed into vectors that can be processed by machines. The RNN then processes the vector sequence one at a time. As it processes each vector, it passes the previous hidden state. The hidden state retains information from the previous step, acting as a type of memory. It does this by combining the input and the previous hidden state with a tanh function that compresses the values between **−1** and **1**.

Essentially, this is how the RNN functions. RNNs don't need a lot of computation and work well with short sequences.

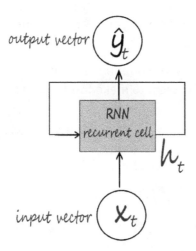

Figure 9.16: RNN data flow

Now turn your attention to applying neural networks to problems that involve sequential processing of data. You've already learned a bit about why these sorts of tasks require a fundamentally different type of network architecture from what you've seen so far.

RNN ARCHITECTURE

This section will go through the key principles behind RNNs, how they are fundamentally different from what you've learned so far, and how RNN computation actually works.

But before you do that, take one step back and consider the standard feed-forward neural network that was discussed previously.

In feed-forward neural networks, data propagates in one direction only, that is, from input to output.

Therefore, you need a different kind of network architecture to handle sequential data. RNNs are particularly well-suited to handling cases in which you have a sequence of inputs rather than a single input. These are great for problems in which a sequence of data is being propagated to give a single output.

For example, imagine that you are training a model that takes a sequence of words as input and outputs an emotion associated with that sequence. Similarly, consider cases in which, instead of returning a single output, you could have a sequence of inputs and propagate them through your network, where each time step in the sequence generates an output.

Simply put, RNNs are networks that offer a mechanism to persist previously processed data over time and use it to make future predictions.

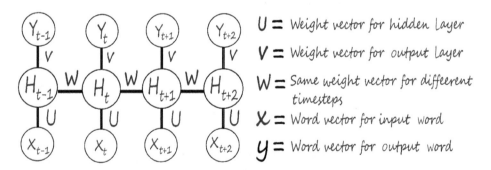

Figure 9.17: RNN computation

In the preceding diagram, at some time step denoted by t, the RNN takes in X_t as the input, and at that time step, it computes a prediction value, Y_t, which is the output of the network.

In addition to that output, it saved an internal state, called update, H_t. This internal state from time step **t** can then be used to complement the input of the next time step **t+1**. So, basically, it provides information about the previous step to the next one. This mechanism is called **recurrent** because information is being passed from one time step to the next within the network.

What's really happening here? This is done by using a simple recurrence relation to process the sequential data. RNNs maintain internal state, H_t, and combine it with the next input data, X_{t+1}, to make a prediction, Y_{t+1}, and store the new internal state, H_{t+1}. The key idea is that the state update is a combination of the previous state time step as well as the current input that the network is receiving.

It's important to note that, in this computation, it's the same function **f** of **W** and the same set of parameters that are used at every time step, and it's those sets of parameters that you learn during the course of training. To get a better sense of how these networks work, step through the RNN algorithm:

1. You begin by initializing your RNN and the hidden state of that network. You can denote a sentence for which you are interested in predicting the next word. The RNN computation simply consists of them looping through the words in this sentence.

2. At each time step, you feed both the current word that you're considering, as well as the previous hidden state of your RNN into the network. This can then generate a prediction for the next word in the sequence and use this information to update its hidden state.

3. Finally, after you've looped through all the words in the sentence, your prediction for that missing word is simply the RNN's output at that final time step.

As you can see in the following diagram, this RNN computation includes both the internal state update and the formal output vector.

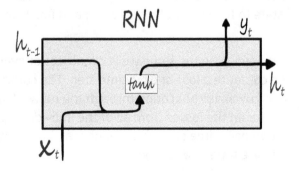

Figure 9.18: RNN data flow

Given the input vector, $\mathbf{X}_t$, the RNN applies a function to update its hidden state. This function is simply a standard neural net operation. It consists of multiplication by a weight matrix and the application of a non-linearity activation function. The key difference is that, in this case, you're feeding in both the input vector, $\mathbf{X}_t$, and the previous state as inputs to this function, $\mathbf{H}_{t-1}$.

Next, you apply a non-linearity activation function such as tanh to the previous step. You have these two weight matrices, and finally, your output, $\mathbf{y}_t$, at a given time step is then a modified, transformed version of this internal state.

After you've looped through all the words in the sentence, your prediction for that missing word is simply the RNN's output at that final time step, after all the words have been fed through the model. So, as mentioned, RNN computation includes both internal state updates and formal output vectors.

Another way you can represent RNNs is by unrolling their modules over time. You can think of RNNs as having multiple copies of the same network, where each passes a message on to its descendant.

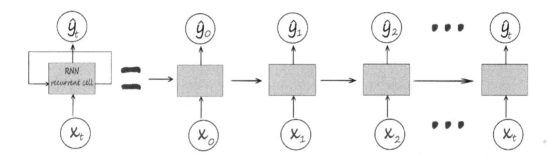

Figure 9.19: Computational graph with time

In this representation, you can make your weight matrices explicit, beginning with the weights that transform the input to the $\mathbf{H}$ weights that are used to transform the previous hidden state to the current hidden state, and finally the hidden state to the output.

It's important to note that you use the same weight matrices at every time step. From these outputs, you can compute a loss at each time step. The computation of the loss will then complete your forward propagation through the network. Finally, to define the total loss, you simply sum the losses from all of the individual time steps. Since your loss is dependent on each time step, this means that, in training the network, you will have to also involve time as a component.

Now that you've got a bit of a sense of how these RNNs are constructed and how they function, you can walk through a simple example of how to implement an RNN from scratch in TensorFlow.

The following snippet uses a simple RNN from **keras.models.Sequential**. You specify the number of units as **1** and set the first input dimension to **None** as an RNN can process any number of time steps. A simple RNN uses tanh activation by default:

```
model = keras.models.Sequential([

                              keras.layers.SimpleRNN\
                              (1, input_shape=[None, 1])
])
```

The preceding code creates a single layer with a single neuron.

That was easy enough. Now you need to stack some additional recurrent layers. The code is similar, but there is a key difference here. You will notice **return_sequences=True** on all but the last layer. This is to ensure that the output is a 3D array. As you can see, the first two layers each have **20** units:

```
model = keras.models.Sequential\
        ([Keras.layers.SimpleRNN\
           (20, return_sequences=True, input_shape=[None, 1]), \
           Keras.layers.SimpleRNN(20, return_sequences=True), \
           Keras.layers.SimpleRNN(1)])
```

The RNN is defined as a layer, and you can build it by inheriting it from the layer class. You can also initialize your weight matrices and the hidden state of your RNN cell to zero.

The key step here is defining the call function, which describes how you make a forward pass through the network given an input **X**. And, to break down this call function, you would first update the hidden state according to the equation discussed previously.

Take the previous hidden state and the input **X**, multiply them by the relevant weight matrices, add them together, and then pass them through a non-linearity, like a hyperbolic tangent (tanh).

Then, the output is simply a transformed version of the hidden state, and at each time step, you return both the current output and the updated hidden state.

TensorFlow has made it easy by having a built-in dense layer. The same applies to RNNs. TensorFlow has implemented these types of RNN cells with the simple RNN layer. But this type of layer has some limitations, such as vanishing gradients. You will look at this problem in the next section before exploring different types of recurrent layers.

VANISHING GRADIENT PROBLEM

If you take a closer look at how gradients flow in this chain of repeating modules, you can see that between each time step you need to perform matrix multiplication. That means that the computation of the gradient—that is, the derivative of the loss with respect to the parameters, tracing all the way back to your initial state—requires many repeated multiplications of this weight matrix, as well as repeated use of the derivative of your activation function.

You can have one of two scenarios that could be particularly problematic: the exploding gradient problem or the vanishing gradient problem.

The exploding gradients problem is when gradients become continuously larger and larger due to the matrix multiplication operation, and you can't optimize them anymore. One way you may be able to mitigate this is by performing what's called gradient clipping. This amounts to scaling back large gradients so that their values are smaller and closer to **1**.

You can also have the opposite problem where your gradients are too small. This is what is known as the vanishing gradient problem. This is when gradients become increasingly smaller (close to **0**) as you make these repeated multiplications, and you can no longer train the network. This is a very real problem when it comes to training RNNs.

For example, consider a scenario in which you keep multiplying a number by some number that's in between zero and one. As you keep doing this repeatedly, that number is constantly shrinking until, eventually, it vanishes and becomes 0. When this happens to gradients, it's hard to propagate errors further back into the past because the gradients are becoming smaller and smaller.

Consider the earlier example from the language model where you were trying to predict the next word. If you're trying to predict the last word in the following phrase, it's relatively clear what the next word is going to be. There's not that much of a gap between the key relevant information, such as the word "fish," and the place where the prediction is needed.

fish swim in the _____

Figure 9.20: Word prediction

However, there are other cases where more context is necessary, like in the following example. Information from early in the sentence, **She lived in Spain**, suggests that the next word of the sentence after **she speaks fluent** is most likely the name of a language, **Spanish**.

She lived in Spain...she speaks fluent _____

Figure 9.21: Sentence example

But you need the context of **Spain**, which is located at a much earlier position in this sentence, to be able to fill in the relevant gaps and identify which language is correct. As this gap between words that are semantically important grows, RNNs become increasingly unable to connect the dots and link these relevant pieces of information together. That is due to the vanishing gradient problem.

How can you alleviate this? The first trick is simple. You can choose either tanh or sigmoid as your activation function. Both of these functions have derivatives that are less than **1**.

Another simple trick you can use is to initialize the weights for the parameters of your network. It turns out that initializing the weights to the identity matrix helps prevent them shrinking to zero too rapidly during back-propagation.

But the final and most robust solution is to use a slightly more complex recurrent unit that can track long-term dependencies in the data more effectively. It can do this by controlling what information is passed through and what information is used to update its internal state. Specifically, this is the concept of a gated cell, like in the LSTM layer, which is the focus of the next section.

LONG SHORT-TERM MEMORY NETWORK

LSTMs are well-suited to learning long-term dependencies and overcoming the vanishing gradient problem. They are very performant models for sequential data, and they're widely used by the deep learning community.

LSTMs have a chain-like structure. In an LSTM, the repeating unit contains different interacting layers. The key point is that these layers interact to selectively control the flow of information within the cell.

The key building block of the LSTM is a structure called a gate, which functions to enable the LSTM to selectively add or remove information from its cell state. Gates consist of a neural net layer like a sigmoid.

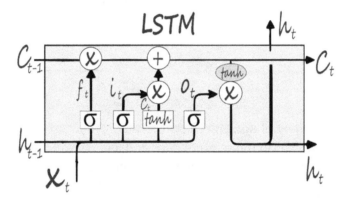

Figure 9.22: LSTM architecture

Take a moment to think about what a gate like this would do in an LSTM. In this case, the sigmoid function would force its input to be between **0** and **1**. You can think of this mechanism as capturing how much of the information that's passed through the gate should be retained. It's between zero and one. This effectively gates the flow of information.

LSTMs process information through four simple steps:

1. The first step in the LSTM is to decide what information is going to be thrown away from the cell state, to forget irrelevant history. This is a function of both the prior internal state, $\mathbf{H}_{t-1}$, and the input, $\mathbf{X}_t$, because some of that information may not be important.

2. Next, the LSTM decides what part of the new information is relevant and uses this to store this information in its cell state.

3. Then, it takes both the relevant parts of the prior information, as well as the current input, and uses this to selectively update its cell state.

4. Finally, it returns an output, and this is known as the output gate, which controls what information encoded in the cell state is sent to the network.

Figure 9.23: LSTM processing steps

The key takeaway here for LSTMs is the sequence of how they regulate information flow and storage. Once again, LSTMs operate as follows:

- Forgetting irrelevant history

- Storing what's new and what's important

- Using its internal memory to update the internal state

- Generating an output

An important property of LSTMs is that all these different gating and update mechanisms work to create an internal cell state, **C**, which allows the uninterrupted flow of gradients through time. You can think of it as sort of a highway of cell states where gradients can flow uninterrupted. This enables you to alleviate and mitigate the vanishing gradient problem that's seen with standard RNNs.

LSTMs are able to maintain this separate cell state independently of what is output, and they use gates to control the flow of information by forgetting irrelevant history, storing relevant new information, selectively updating their cell state, and then returning a filtered version as the output.

The key point in terms of training and LSTMs is that maintaining the separate independent cell state allows the efficient training of an LSTM to backpropagate through time, which is discussed later.

Now that you've gone through the fundamental workings of RNNs, the backpropagation through time algorithm, and a bit about the LSTM architecture, you can put some of these concepts to work in the following example.

Consider the following LSTM model:

```
regressor = Sequential()

regressor.add(LSTM(units= 50, activation = 'relu', \
                   return_sequences = True, \
                   input_shape = (X_train.shape[1], 5)))
regressor.add(Dropout(0.2))

regressor.add(LSTM(units= 60, activation = 'relu', \
                   return_sequences = True))
regressor.add(Dropout(0.3))

regressor.add(LSTM(units= 80, activation = 'relu', \
                   return_sequences = True))
regressor.add(Dropout(0.4))

regressor.add(LSTM(units= 120, activation = 'relu'))
regressor.add(Dropout(0.5))

regressor.add(Dense(units = 1))
```

First, you have initialized a neural network by calling **regressor = Sequential()**. Again, it's important to note that in the last line you omit **return_sequences = True** because it is the final output:

```
regressor = Sequential()
```

Then, the LSTM layer is added. In the first instance, set the LSTM layer to **50** units. Use a relu activation function and specify the shape of the training set. Finally, the dropout layer is added with **regressor.add(Dropout(0.2)**. The **0.2** means that 20% of the layers will be removed. Set **return_sequences = True**, which allows the return of the last output.

Similarly, add three more LSTM layers and one dense layer to the LSTM model.

Now that you are familiar with the basic concepts surrounding working with sequential data, it's time to complete the following exercise using some real data.

EXERCISE 9.02: BUILDING AN RNN WITH AN LSTM LAYER – NVIDIA STOCK PREDICTION

In this exercise, you will be working on the same dataset as for *Exercise 9.01, Training an ANN for Sequential Data – Nvidia Stock Prediction*. You will still try to predict the Nvidia stock price based on the data of the previous 60 days. But this time, you will be training an LSTM model. You will need to split the data into training and testing sets before and after the date **2019-01-01**.

> **NOTE**
>
> You can find the **NVDA.csv** dataset here: https://packt.link/Mxi80.

You will need to prepare the dataset like in *Exercise 9.01, Training an ANN for Sequential Data – Nvidia Stock Prediction* (*steps 1* to *15*) before applying the following code:

1. Start building the LSTM. You will need some additional libraries for this. Use **Sequential** to initialize the neural net, **Dense** to add a dense layer, **LSTM** to add an LSTM layer, and **Dropout** to help prevent overfitting:

```
from tensorflow.keras import Sequential
from tensorflow.keras.layers import Dense, LSTM, Dropout
```

2. Initialize the neural network by calling **regressor = Sequential()**. Add four LSTM layers with **50**, **60**, **80**, and **120** units each. Use a ReLU activation function and assign **True** to **return_sequences** for all but the last LSTM layer. Provide the shape of your training set to the first LSTM layer. Finally, add dropout layers with 20%, 30%, 40%, and 50% dropouts:

```
regressor = Sequential()

regressor.add(LSTM(units= 50, activation = 'relu',\
                   return_sequences = True,\
                   input_shape = (X_train.shape[1], 5)))
regressor.add(Dropout(0.2))

regressor.add(LSTM(units= 60, activation = 'relu', \
                return_sequences = True))
regressor.add(Dropout(0.3))

regressor.add(LSTM(units= 80, activation = 'relu', \
                return_sequences = True))
```

```
regressor.add(Dropout(0.4))

regressor.add(LSTM(units= 120, activation = 'relu'))
regressor.add(Dropout(0.5))

regressor.add(Dense(units = 1))
```

3. Check the summary of the model using the **summary()** method:

```
regressor.summary()
```

You should get the following output:

```
Model: "sequential"

Layer (type)               Output Shape              Param #
=================================================================
lstm (LSTM)                (None, 60, 50)            11200

dropout (Dropout)          (None, 60, 50)            0

lstm_1 (LSTM)              (None, 60, 60)            26640

dropout_1 (Dropout)        (None, 60, 60)            0

lstm_2 (LSTM)              (None, 60, 80)            45120

dropout_2 (Dropout)        (None, 60, 80)            0

lstm_3 (LSTM)              (None, 120)               96480

dropout_3 (Dropout)        (None, 120)               0

dense (Dense)              (None, 1)                 121
=================================================================
Total params: 179,561
Trainable params: 179,561
Non-trainable params: 0
```

Figure 9.24: Model summary

As you can see from the preceding figure, the summary provides valuable information about all model layers and parameters. This is a good way to make sure that your layers are in the order you wish and that they have the proper output shapes and parameters.

4. Use the **compile()** method to configure your model for training. Choose Adam as your optimizer and mean squared error to measure your loss function:

```
regressor.compile(optimizer='adam', loss = 'mean_squared_error')
```

5. Fit your model and set it to run on **10** epochs. Set your batch size equal to **32**:

```
regressor.fit(X_train, y_train, epochs=10, batch_size=32)
```

You should get the following output:

```
Epoch 1/10
26/26 [==============================] - 7s 111ms/step - loss: 0.0895
Epoch 2/10
26/26 [==============================] - 3s 111ms/step - loss: 0.0222
Epoch 3/10
26/26 [==============================] - 4s 148ms/step - loss: 0.0144
Epoch 4/10
26/26 [==============================] - 3s 129ms/step - loss: 0.0128
Epoch 5/10
26/26 [==============================] - 3s 116ms/step - loss: 0.0117
Epoch 6/10
26/26 [==============================] - 3s 123ms/step - loss: 0.0098
Epoch 7/10
26/26 [==============================] - 4s 143ms/step - loss: 0.0091
Epoch 8/10
26/26 [==============================] - 3s 132ms/step - loss: 0.0117
Epoch 9/10
26/26 [==============================] - 4s 136ms/step - loss: 0.0095
Epoch 10/10
26/26 [==============================] - 3s 132ms/step - loss: 0.0111
<keras.callbacks.History at 0x2a802858a48>
```

Figure 9.25: Training the model

6. Test and predict the stock price and prepare the dataset. Check your data by calling the **head()** function:

```
data_test.head()
```

You should get the following output:

	Date	Open	High	Low	Close	Adj Close	Volume
868	2019-01-02	130.639999	138.479996	130.050003	136.220001	135.547104	12718800
869	2019-01-03	133.789993	135.160004	127.690002	127.989998	127.357750	17638800
870	2019-01-04	130.940002	137.729996	129.699997	136.190002	135.517258	14640500
871	2019-01-07	138.500000	144.889999	136.429993	143.399994	142.691620	17729000
872	2019-01-08	146.690002	146.779999	136.899994	139.830002	139.139282	19650400

Figure 9.26: First five rows of the DataFrame

7. Call the **tail(60)** method to look at the last 60 days of data. You will use this information in the next step:

```
data_training.tail(60)
```

You should get the following output:

	Date	Open	High	Low	Close	Adj Close	Volume
808	2018-10-04	285.269989	286.250000	276.179993	279.290009	277.632599	9780500
809	2018-10-05	278.290009	280.799988	267.540009	269.859985	268.258514	10665900
810	2018-10-08	266.500000	271.160004	260.079987	265.769989	264.192780	10215300
811	2018-10-09	264.940002	268.760010	262.799988	265.540009	263.964203	6837500
812	2018-10-10	261.260010	263.109985	245.600006	245.690002	244.231964	17123500
813	2018-10-11	242.169998	247.559998	234.259995	235.130005	233.734634	18135900
814	2018-10-12	245.509995	249.539993	239.649994	246.539993	245.076920	15205900
815	2018-10-15	246.000000	246.000000	235.339996	235.380005	233.983154	11244000

Figure 9.27: Last 10 rows of the DataFrame

8. Use the **tail(60)** method to create the **past_60_days** variable:

```
past_60_days = data_training.tail(60)
```

9. Add the **past_60_days** variable to your test data with the **append()** function. Set **True** to **ignore_index**. Drop the **Date** and **Adj Close** columns as you will not need that information:

```
df = past_60_days.append(data_test, ignore_index = True)
df = df.drop(['Date', 'Adj Close'], axis = 1)
```

10. Check the DataFrame to make sure that you successfully dropped **Date** and **Adj Close** by using the **head()** function:

```
df.head()
```

You should get the following output:

	Open	High	Low	Close	Volume
0	285.269989	286.250000	276.179993	279.290009	9780500
1	278.290009	280.799988	267.540009	269.859985	10665900
2	266.500000	271.160004	260.079987	265.769989	10215300
3	264.940002	268.760010	262.799988	265.540009	6837500
4	261.260010	263.109985	245.600006	245.690002	17123500

Figure 9.28: Checking the first five rows of the DataFrame

11. Use **scaler.transform** from **StandardScaler** to perform standardization on inputs:

```
inputs = scaler.transform(df)
inputs
```

You should get the following output:

```
array([[0.98500382, 0.97617388, 0.96472665, 0.9627107 , 0.09305696],
       [0.95915875, 0.95622728, 0.93230523, 0.92779115, 0.10278535],
       [0.91550336, 0.9209457 , 0.9043116 , 0.91264582, 0.09783435],
       ...,
       [1.44321835, 1.42886941, 1.44253078, 1.4395483 , 0.05873841],
       [1.45043874, 1.4702631 , 1.45288757, 1.48535462, 0.06383883],
       [1.4857999 , 1.47447198, 1.47240054, 1.4583597 , 0.06169186]])
```

Figure 9.29: DataFrame standardization

From the preceding results, you can see that after standardization, all values are close to **0** now.

12. Split your data into **X_test** and **y_test** datasets. Create a test dataset that has the previous 60 days' stock prices, so that you can test the closing stock price for the 61st day. Here, **X_test** will have two columns. The first column will store the values from 0 to 59. The second column will store values from 1 to 60. In the first column of **y_test**, store the 61st value at index 60, and in the second column, store the 62nd value at index 61. Use a **for** loop to create data in 60 time steps:

```
X_test = []
y_test = []

for i in range(60, inputs.shape[0]):
    X_test.append(inputs[i-60:i])
    y_test.append(inputs[i, 0])
```

13. Convert **X_test** and **y_test** into NumPy arrays:

```
X_test, y_test = np.array(X_test), np.array(y_test)
X_test.shape, y_test.shape
```

You should get the following output:

```
((391, 60, 5), (391,))
```

The preceding result shows that there are **391** observations and for each of them you have the last **60** days' data for the following five features: **Open, High, Low, Close**, and **Volume**. The target variable, on the other hand, contains **391** values.

14. Test some predictions for stock prices by calling **regressor.predict(X_test)**:

```
y_pred = regressor.predict(X_test)
```

15. Before looking at the results, reverse the scaling you did earlier so that the number you get as output will be at the correct scale using the **StandardScaler** utility class that you imported with **scaler.scale_**:

```
scaler.scale_
```

You should get the following output:

```
array([3.70274364e-03, 3.65992009e-03, 3.75248621e-03, 3.70301815e-03,
       1.09875621e-08])
```

Figure 9.30: Using StandardScaler

16. Use the first value in the preceding array to set your scale in preparation for the multiplication of **y_pred** and **y_test**. Recall that you are converting your data back from the scale you did earlier when converting all values to between zero and one:

```
scale = 1/3.70274364e-03
scale
```

You should get the following output:

```
270.0700067909643
```

17. Multiply **y_pred** and **y_test** by **scale** to convert your data back to the proper values:

```
y_pred = y_pred*scale
y_test = y_test*scale
```

18. Use **y_pred** to view predictions for NVIDIA stock:

```
y_pred
```

You should get the following output:

```
array([[143.43747],
       [142.20844],
       [140.99911],
       [139.82808],
       [138.72331],
       [137.71217],
       [136.81802],
       [136.05975],
       [135.45023],
       [134.99947],
       [134.71558],
       [134.60034],
       [134.6484 ],
       [134.85535],
       [135.20615],
```

Figure 9.31: Checking prediction

The preceding results show the predicted Nvidia stock price for the future dates.

19. Plot the real Nvidia stock price and your predictions:

```
plt.figure(figsize=(14,5))
plt.plot(y_test, color = 'black', label = "Real NVDA Stock Price")
plt.plot(y_pred, color = 'gray',\
        label = 'Predicted NVDA Stock Price')
plt.title('NVDA Stock Price Prediction')
plt.xlabel('time')
plt.ylabel('NVDA Stock Price')
plt.legend()
plt.show()
```

You should get the following output:

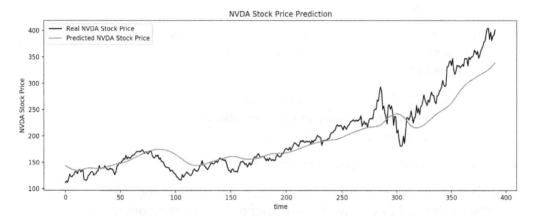

Figure 9.32: NVIDIA stock price visualization

As you can see from the gray line in *Figure 9.32*, your prediction model is pretty accurate, when compared to the actual stock price, which is shown by the black line.

In this exercise, you built an RNN with an LSTM layer for Nvidia stock prediction and completed the training, testing, and prediction steps.

Now, test the knowledge you've gained so far in this chapter in the following activity.

ACTIVITY 9.01: BUILDING AN RNN WITH MULTIPLE LSTM LAYERS TO PREDICT POWER CONSUMPTION

The **household_power_consumption.csv** dataset contains information related to electric power consumption measurements for a household over 4 years with a 1-minute sampling rate. You are required to predict the power consumption of a given minute based on previous measurements.

You are tasked with adapting an RNN model with additional LSTM layers to predict household power consumption at the minute level. You will be building an RNN model with three LSTM layers.

> **NOTE**
>
> You can find the dataset here: https://packt.link/qrloK.

Perform the following steps to complete this activity:

1. Load the data.

2. Prepare the data by combining the **Date** and **Time** columns to form one single **Datetime** column that can be used then to sort the data and fill in missing values.

3. Standardize the data and remove the **Date**, **Time**, **Global_reactive_power**, and **Datetime** columns as they won't be needed for the predictions.

4. Reshape the data for a given minute to include the previous 60 minutes' values.

5. Split the data into training and testing sets with, respectively, data before and after the index **217440**, which corresponds to the last month of data.

6. Define and train an RNN model composed of three different layers of LSTM with **20**, **40**, and **80** units, followed by **50%** dropout and ReLU as the activation function.

7. Make predictions on the testing set with the trained model.

8. Compare the predictions against the actual values on the entire dataset.

   You should get the following output:

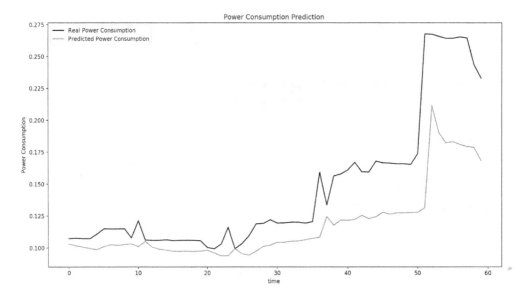

Figure 9.33: Expected output of Activity 9.01

> **NOTE**
>
> The solution to this activity can be found on page 520.

In the next section, you will learn how to apply RNNs to text.

NATURAL LANGUAGE PROCESSING

Natural Language Processing (**NLP**) is a quickly growing field that is both challenging and rewarding. NLP takes valuable data that has traditionally been very difficult for machines to make sense of and turns it into information that can be used. This data can take the form of sentences, words, characters, text, and audio, to name a few. Why is this such a difficult task for machines? To answer that question, consider the following examples.

Recall the two sentences: *it is what it is* and *is it what it is*. These two sentences, though they have completely opposite semantic meanings, would have the exact same representations in this bag of words format. This is because they have the exact same words, just in a different order. So, you know that you need to use a sequential model to process this, but what else? There are several tools and techniques that have been developed to solve these problems. But before you get to that, you need to learn how to preprocess sequential data.

DATA PREPROCESSING

As a quick review, preprocessing generally entails all the steps needed to train your model. Some common steps include data cleaning, data transformation, and data reduction. For natural language processing, more specifically, the steps could be all, some, or none of the following:

- Tokenization
- Padding
- Lowercase conversion
- Removing stop words
- Removing punctuation
- Stemming

The following sections provide a more in-depth description of the steps that you will be using. For now, here's an overview of each step:

- **Dataset cleaning** encompasses the conversion of case to lowercase, the removal of punctuation marks, and so on.

- **Tokenization** is breaking up a character sequence into specified units called tokens.

- **Padding** is a way to make input sentences of different sizes the same by padding them. Padding the sequences means ensuring that the sequences have a uniform length.

- **Stemming** is truncating words down to their stem. For example, the words "rainy" and "raining" both have the stem "rain".

DATASET CLEANING

Here, you create the **clean_text** function, which returns a list containing words once it has been cleaned. You will save all text as lowercase with **lower()** and encode it with **utf8** for character standardization:

```
def clean_text(txt):
    txt = "".join(v for v in txt if v not in string.punctuation)\
            .lower()
    txt = txt.encode("utf8").decode("ascii",'ignore')
    return txt


corpus = [clean_text(x) for x in all_headlines]
```

GENERATING A SEQUENCE AND TOKENIZATION

TensorFlow provides a dedicated class for generating a sequence of N-gram tokens – **Tokenizer** from **keras.preprocessing.text**:

```
from keras.preprocessing.text import Tokenizer

tokenizer = Tokenizer()
```

Once you have instantiated a **Tokenizer()**, you can use the **fit_on_texts()** method to extract tokens from a corpus. This step will attribute an integer index to each unique word from the corpus:

```
tokenizer.fit_on_texts(corpus)
```

After the tokenizer has been trained on a corpus, you can access the indexes allocated to each word from your corpus with the **word_index** attribute:

```
tokenizer.word_index
```

You can convert a sentence into a tokenized version using the **texts_to_sequences()** method:

```
tokenizer.texts_to_sequences([sentence])
```

You can create a function that will generate an N-gram sequence of tokenized sentences from an input corpus with the following snippet:

```
def get_seq_of_tokens(corpus):
    tokenizer.fit_on_texts(corpus)
    all_words = len(tokenizer.word_index) + 1

    input_sequences = []
    for line in corpus:
        token_list = tokenizer.texts_to_sequences([line])[0]
        for i in range(1, len(token_list)):
            n_gram_sequence = token_list[:i+1]
            input_sequences.append(n_gram_sequence)
    return input_sequences, all_words

inp_sequences, all_words = get_seq_of_tokens(corpus)
inp_sequences[:10]
```

The **get_seq_of_tokens()** function trains a **Tokenizer()** on the given corpus. Then you need to iterate through each line of the corpus and convert them into their tokenized equivalents. Finally, for each tokenized sentence, you create the different sequences of N-gram from it.

Next, you will see how you can deal with variable sentence length with padding.

PADDING SEQUENCES

As discussed previously, deep learning models expect fixed-length input. But with text, the length of a sentence can vary. One way to overcome this is to transform all sentences to have the same length. You will need to set the maximum length of sentences. Then, for sentences that are shorter than this threshold, you can add padding, which will add a specific token value to fill the gap. On the other hand, longer sentences will be truncated to fit this constraint. You can use **pad_sequences()** to achieve this:

```
from keras.preprocessing.sequence import pad_sequences
```

You can create the **generate_padded_sequences** function, which will take **input_sequences** and generate the padded version of it:

```
def generate_padded_sequences(input_sequences):
    max_sequence_len = max([len(x) for x in input_sequences])
    input_sequences = np.array(pad_sequences\
                               (input_sequences, \
                                maxlen=max_sequence_len, \
                                padding='pre'))

    predictors, label = input_sequences[:,:-1], \
                        input_sequences[:,-1]
    label = ku.to_categorical(label, num_classes=all_words)
    return predictors, label, max_sequence_len

predictors, label, max_sequence_len = generate_padded_sequences\
                                      (inp_sequences)
```

Now that you know how to process raw text, have a look at the modeling step in the next section.

BACK PROPAGATION THROUGH TIME (BPTT)

There are many types of sequential models. You've already used simple RNNs, deep RNNs, and LSTMs. Let's take a look at a couple of additional models used for NLP.

Remember that you trained feed-forward models by first making a forward pass through the network that goes from input to output. This is the standard feed-forward model where the layers are densely connected. To train this kind of model, you can backpropagate the gradients through the network, taking the derivative of the loss of each weight parameter in the network. Then, you can adjust the parameters to minimize the loss.

But in RNNs, as discussed earlier, your forward pass through the network also consists of going forward in time, updating the cell state based on the input and the previous state, and generating an output, $\mathbf{Y}$. At that time step, computing a loss and then finally summing these losses from the individual time steps gets your total loss.

This means that instead of backpropagating errors through a single feed-forward network at a single time step, errors are backpropagated at each individual time step, and then, finally, across all time steps—all the way from where you are currently, to the beginning of the sequence.

This is why it's called backpropagation through time. As you can see, all errors are flowing back in time to the beginning of your data sequence.

A great example of machine translation and one of the most powerful and widely used applications of RNNs in industry is Google Translate. In machine translation, you input a sequence in one language and the task is to train the RNN to output that sequence in a new language. This is done by employing a dual structure with an encoder that encodes the sentence in its original language into a state vector and a decoder. This then takes that encoded representation as input and decodes it into a new language.

There's a key problem though in this approach: all content that is fed into the encoder structure must be encoded into a single vector. This can become a huge information bottleneck in practice because you may have a large body of text that you want to translate. To get around this problem the researchers at Google developed an extension of RNN called **attention**.

Now, instead of the decoder only having access to the final encoded state, it can access the states of all the time steps in the original sentence. The weights of these vectors that connect the encoder states to the decoder are learned by the network during training. This is called attention because when the network learns, it places its attention on different parts of the input sentence.

In this way, it effectively captures a sort of memory access to the important information in that original sentence. So, with building blocks such as attention and gated cells, like LSTMs, RNNs have really taken off in recent years and are being used in the real world quite successfully.

You should have by now gotten a sense of how RNNs work and why they are so powerful for processing sequential data. You've seen why and how you can use RNNs to perform sequence modeling tasks by defining this recurrence relation. You also learned how you can train RNNs and looked at how gated cells such as LSTMs can help us model long-term dependencies.

In the following exercise, you will see how to use an LSTM model for predicting the next word of a text.

EXERCISE 9.03: BUILDING AN RNN WITH AN LSTM LAYER FOR NATURAL LANGUAGE PROCESSING

In this exercise, you will use an RNN with an LSTM layer to predict the final word of a news headline.

The **Articles.csv** dataset contains raw text that consists of news titles. You will be training an LTSM model that will predict the next word of a given sentence.

> **NOTE**
>
> You can find the dataset here: https://packt.link/RQVoB.

Perform the following steps to complete this exercise:

1. Import the libraries needed:

```
from keras.preprocessing.sequence import pad_sequences
from keras.layers import Embedding, LSTM, Dense, Dropout
from keras.preprocessing.text import Tokenizer
from keras.callbacks import EarlyStopping
from keras.models import Sequential
import keras.utils as ku
import pandas as pd
import numpy as np
import string, os
import warnings
warnings.filterwarnings("ignore")
warnings.simplefilter(action='ignore', category=FutureWarning)
```

You should get the following output:

```
Using TensorFlow backend.
```

2. Load the dataset locally by setting **curr_dir** to **content**. Create the **all_headlines** variable. Use a **for** loop to iterate over the files contained in the folder, and extract the headlines. Remove all headlines with the **Unknown** value. Print the length of **all_headlines**:

```
curr_dir = '/content/'
all_headlines = []
for filename in os.listdir(curr_dir):
    if 'Articles' in filename:
        article_df = pd.read_csv(curr_dir + filename)
        all_headlines.extend(list(article_df.headline.values))
        break

all_headlines = [h for h in all_headlines if h != "Unknown"]
len(all_headlines)
```

The output will be as follows:

```
831
```

3. Create the **clean_text** method to return a list containing words once it has been cleaned. Save all text as lowercase with the **lower()** method and encode it with **utf8** for character standardization. Finally, output 10 headlines from your corpus:

```
def clean_text(txt):
    txt = "".join(v for v in txt \
                  if v not in string.punctuation).lower()
    txt = txt.encode("utf8").decode("ascii",'ignore')
    return txt

corpus = [clean_text(x) for x in all_headlines]
corpus[:10]
```

You should get the following output:

```
['finding an expansive view  of a forgotten people in niger',
 'and now  the dreaded trump curse',
 'venezuelas descent into dictatorship',
 'stain permeates basketball blue blood',
 'taking things for granted',
 'the caged beast awakens',
 'an everunfolding story',
 'oreilly thrives as settlements add up',
 'mouse infestation',
 'divide in gop now threatens trump tax plan']
```

Figure 9.34: Corpus

4. Use **tokenizer.fit** to extract tokens from the corpus. Each integer output corresponds with a specific word. With **input_sequences**, train features that will be a **list []**. With **token_list = tokenizer. texts_to_sequences**, convert each sentence into its tokenized equivalent. With **n_gram_sequence = token_list**, generate the N-gram sequences. Using **input_sequences.append(n_gram_sequence)**, append each N-gram sequence to the list of your features:

```
tokenizer = Tokenizer()

def get_seq_of_tokens(corpus):
    tokenizer.fit_on_texts(corpus)
    all_words = len(tokenizer.word_index) + 1

    input_sequences = []
    for line in corpus:
        token_list = tokenizer.texts_to_sequences([line])[0]
        for i in range(1, len(token_list)):
            n_gram_sequence = token_list[:i+1]
            input_sequences.append(n_gram_sequence)
    return input_sequences, all_words

inp_sequences, all_words = get_seq_of_tokens(corpus)
inp_sequences[:10]
```

You should get the following output:

```
[[169, 17],
 [169, 17, 665],
 [169, 17, 665, 367],
 [169, 17, 665, 367, 4],
 [169, 17, 665, 367, 4, 2],
 [169, 17, 665, 367, 4, 2, 666],
 [169, 17, 665, 367, 4, 2, 666, 170],
 [169, 17, 665, 367, 4, 2, 666, 170, 5],
 [169, 17, 665, 367, 4, 2, 666, 170, 5, 667],
 [6, 80]]
```

Figure 9.35: N-gram tokens

5. Pad the sequences and obtain the **predictors** and **target** variables. Use **pad_sequence** to pad the sequences and make their lengths equal:

```
def generate_padded_sequences(input_sequences):
    max_sequence_len = max([len(x) for x in input_sequences])
    input_sequences = np.array\
                    (pad_sequences(input_sequences, \
                                   maxlen=max_sequence_len, \
                                   padding='pre'))

    predictors, label = input_sequences[:,:-1], \
                        input_sequences[:,-1]
    label = ku.to_categorical(label, num_classes=all_words)
    return predictors, label, max_sequence_len

predictors, label, max_sequence_len = generate_padded_sequences\
                    (inp_sequences)
```

6. Prepare your model for training. Add an input embedding layer with **model.add(Embedding)**. Add a hidden LSTM layer with **100** units and add a dropout of 10%. Then, add a dense layer with a softmax activation function. With the **compile** method, configure your model for training, setting your loss function to **categorical_crossentropy**, and use the Adam optimizer:

```
def create_model(max_sequence_len, all_words):
    input_len = max_sequence_len - 1
    model = Sequential()
```

```
    model.add(Embedding(all_words, 10, input_length=input_len))

    model.add(LSTM(100))
    model.add(Dropout(0.1))

    model.add(Dense(all_words, activation='softmax'))

    model.compile(loss='categorical_crossentropy', \
                  optimizer='adam')

    return model

model = create_model(max_sequence_len, all_words)
model.summary()
```

You should get the following output:

```
Model: "sequential"
```

Layer (type)	Output Shape	Param #
embedding (Embedding)	(None, 18, 10)	24220
lstm (LSTM)	(None, 100)	44400
dropout (Dropout)	(None, 100)	0
dense (Dense)	(None, 2422)	244622

```
Total params: 313,242
Trainable params: 313,242
Non-trainable params: 0
```

Figure 9.36: Model summary

7. Fit your model with **model.fit** and set it to run on **100** epochs. Set **verbose** equal to **5**:

```
model.fit(predictors, label, epochs=100, verbose=5)
```

You should get the following output:

```
Epoch 1/100
Epoch 2/100
Epoch 3/100
Epoch 4/100
Epoch 5/100
Epoch 6/100
Epoch 7/100
Epoch 8/100
Epoch 9/100
Epoch 10/100
Epoch 11/100
Epoch 12/100
Epoch 13/100
Epoch 14/100
Epoch 15/100
Epoch 16/100
```

Figure 9.37: Training the model

8. Write a function that will receive an input text, a model, and the number of next words to be predicted. This function will prepare the input text to be fed into the model that will predict the next word:

```
def generate_text(seed_text, next_words, \
                  model, max_sequence_len):
    for _ in range(next_words):
        token_list = tokenizer.texts_to_sequences\
                     ([seed_text])[0]
        token_list = pad_sequences([token_list], \
                                   maxlen=max_sequence_len-1,\
                                   padding='pre')
        predicted = model.predict_classes(token_list, verbose=0)

        output_word = ""
        for word,index in tokenizer.word_index.items():
            if index == predicted:
                output_word = word
                break
        seed_text += " "+output_word
    return seed_text.title()
```

9. Output some of your generated text with the **print** function. Add your own words for the model to use and generate from. For example, in **the hottest new**, the integer **5** is the number of words output by the model:

```
print (generate_text("the hottest new", 5, model,\
                     max_sequence_len))
print (generate_text("the stock market", 4, model,\
                     max_sequence_len))
print (generate_text("russia wants to", 3, model,\
                     max_sequence_len))
print (generate_text("french citizen", 4, model,\
                     max_sequence_len))
print (generate_text("the one thing", 15, model,\
                     max_sequence_len))
print (generate_text("the coronavirus", 5, model,\
                     max_sequence_len))
```

You should get the following output:

```
The Hottest New The The The The The
The Stock Market The The The The
Russia Wants To The The The
French Citizen The The The The
The One Thing The The The The The The The The The The The The The The The
The Coronavirus The The The The The
```

Figure 9.38: Generated text

In this result, you can see the text generated by your model for each sentence.

In this exercise, you have successfully predicted some news headlines. Not surprisingly, some of them may not be very impressive, but some are not too bad.

Now that you have all the essential knowledge about RNNs, try to test yourself by performing the next activity.

ACTIVITY 9.02: BUILDING AN RNN FOR PREDICTING TWEETS' SENTIMENT

The **tweets.csv** dataset contains a list of tweets related to an airline company. Each of the tweets has been classified as having positive, negative, or neutral sentiment.

You have been tasked to analyze a sample of tweets for the company. Your goal is to build an RNN model that will be able to predict the sentiment of each tweet: either positive or negative.

> **NOTE**
>
> You can find **tweets.csv** here: https://packt.link/dVUd2.

Perform the following steps to complete this activity.

1. Import the necessary packages.

2. Prepare the data (combine the **Date** and **Time** columns, name it **datetime**, sort the data, and fill in missing values).

3. Prepare the text data (tokenize words and add padding).

4. Split the dataset into training and testing sets with, respectively, the first 10,000 tweets and the remaining tweets.

5. Define and train an RNN model composed of two different layers of LSTM with, respectively, **50** and **100** units followed by 20% dropout and ReLU as the activation function.

6. Make predictions on the testing set with the trained model.

   You should get the following output:

```
Epoch 1/2
313/313 [==============================] - 116s 365ms/step - loss: 0.5219 - accuracy: 0.7931
Epoch 2/2
313/313 [==============================] - 116s 370ms/step - loss: 0.5112 - accuracy: 0.7947
<keras.callbacks.History at 0x1ea121ff8c8>
```

Figure 9.39: Expected output of Activity 9.02

> **NOTE**
>
> The solution to this activity can be found on page 526.

SUMMARY

In this chapter, you explored different recurrent models for sequential data. You learned that each sequential data point is dependent on the prior sequence of data points, such as natural language text. You also learned why you must use models that allow for the sequence of data to be used by the model, and sequentially generate the next output.

This chapter introduced RNN models that can make predictions for sequential data. You observed the way RNNs can loop back on themselves, which allows the output of the model to feed back into the input. You reviewed the types of challenges that you face with these models, such as vanishing and exploding gradients, and how to address them.

In the next chapter, you will learn how to utilize custom TensorFlow components to use within your models, including loss functions and layers.

10

CUSTOM TENSORFLOW COMPONENTS

OVERVIEW

In this chapter, you will dive a level deeper into the TensorFlow framework and build custom modules. By the end of it, you will know how to create custom TensorFlow components to use within your models, such as loss functions and layers.

INTRODUCTION

In the previous chapters, you learned how to build CNN or RNN models from predefined TensorFlow modules. You have been using one of the APIs offered by TensorFlow called the sequential API. This API is a great way to start building "simple" deep learning architecture with few lines of code. But if you want to achieve higher performance, you may want to build your own custom architecture. In this case, you will need to use another API called the functional API. Researchers use functional APIs while defining their model architecture. By learning how to use it, you will be able to create custom loss functions or modules, such as a residual block from the ResNet architecture.

TENSORFLOW APIS

When using TensorFlow, you can choose from the sequential, functional, or subclassing APIs to define your models. For most, the sequential API will be the go-to option. However, as time goes by and you are exposed to more complexity, your needs will expand as well.

The **sequential API** is the simplest API used for creating TensorFlow models. It works by stacking different layers one after the other. For example, you will create a sequential model with a first layer that's a convolution layer, followed by a dropout layer, and then a fully connected layer. This model is sequential as the input data will be passed to each defined layer sequentially.

The **functional API** provides more flexibility. You can define models with different layers that interact with each other not in a sequential manner. For instance, you can create two different layers both of which will feed into a third one. This can be easily achieved with the functional API.

Model subclassing allows the user a very low level of control over the entire model. It works by inheriting attributes and methods from TensorFlow classes such as `Layer` or `Model`. You can define your own custom layers or models, but this means you will need to comply with all the requirements of the inherited TensorFlow classes, such as coding mandatory methods.

The following diagram provides a quick overview of the three different APIs offered by TensorFlow:

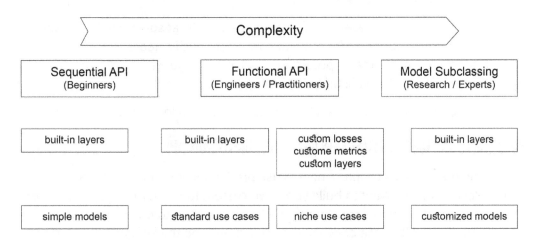

Figure 10.1: Diagram showing a comparison of all three APIs

In the section ahead, you will learn how to define a custom loss function.

IMPLEMENTING CUSTOM LOSS FUNCTIONS

There are several types of loss functions that are commonly used for machine learning. In *Chapter 5, Classification*, you studied different types of loss functions and used them with different classification models. TensorFlow has quite a few built-in loss functions to choose from. The following are just a few of the more common loss functions:

- Mean Absolute Error (MAE)
- Mean Squared Error (MSE)
- Binary cross-entropy
- Categorical cross-entropy
- Hinge
- Huber
- Mean Squared Logarithmic Error (MSLE)

As a quick reminder, you can think of loss functions as a kind of compass that allows you to clearly see what is working in an algorithm and what isn't. The higher the loss, the less accurate the model, and so on.

Although TensorFlow has several loss functions available, at some point, you will most likely need to create your own loss function for your specific needs. For instance, if you are building a model that is predicting stock prices, you want to define a loss function that will penalize substantially incorrect values.

The following section will show you how to build a custom loss function.

BUILDING A CUSTOM LOSS FUNCTION WITH THE FUNCTIONAL API

You saw in the previous chapters how to use predefined loss functions from TensorFlow. But if you want to build your own custom functions, you can use either the functional API or model subclassing. Let's say you want to build a loss function that will raise the difference between the predictions and the actual values to the power of 4:

$$Loss = (pred\text{-}actuals)^4$$

Figure 10.2: Formula for custom loss

While creating a custom loss function, you will always need two arguments: **y_true** (actual values) and **y_pred** (predictions). A loss function will calculate the difference between these two values and return an error value that represents how far the predictions of your model are from the actual values. In the case of MAE, this loss function will return the absolute value of this error. On the other hand, MSE will square the difference between the actual value and the predicted value. But in the preceding example, the error should be raised to the power of **4**.

Let's see how you can implement this using the functional API. Firstly, you will need to import the TensorFlow library using the following command:

```
import tensorflow as tf
```

Then, you will have to create a function called **custom_loss** that takes as input the **y_true** and **y_pred** arguments. You will then use the **pow** function to raise the calculated error to the power of **4**. Finally, you will return the calculated error:

```
def custom_loss(y_true, y_pred):
    custom_loss=tf.math.pow(y_true - y_pred, 4)
    return custom_loss
```

You have created your own custom loss function using the functional API. You can now pass it to the **compile** method, instead of the predefined loss functions, before training your model:

```
model.compile(loss=custom_loss,optimizer=optimizer)
```

After this, you can train your model exactly the same way as you did in previous chapters. TensorFlow will use your custom loss function to optimize the learning process of your model.

BUILDING A CUSTOM LOSS FUNCTION WITH THE SUBCLASSING API

There is another way to define a custom loss function: using the subclassing API. In this case, rather than building a function, you will define a custom class for it. This is quite useful if you want to extend it with additional custom attributes or methods. With subclassing, you can create a custom class that will inherit attributes and methods from the **Loss** class of the **keras.losses** module. You will then need to define the __**init**__**()** and **call()** methods, which are required in the **Loss** class. The __**init**__ method is where you will define all the attributes of your custom class, and the **call()** method is where you will specify the logic for calculating the loss.

The following is a brief example of how you can implement your custom loss, using the subclassing API, where the error should be raised to the power of **4**:

```
class MyCustomLoss(keras.losses.Loss):
    def __init__(self, threshold=1.0, **kwargs):
        super().__init__(**kwargs)

    def call(self, y_true, y_pred):
        return tf.math.pow(y_true - y_pred, 4)
```

In the preceding example, you have reimplemented the same loss function as previously (power of 4) but used subclassing from **keras.losses.Loss**. You started by initializing the attributes of your class in the __**init**__**()** method using the **self** parameter, which refers to the object itself.

Then, in the **call()** method, you defined the logic of your loss function, which calculated the error and raised it to the power of 4.

Now that you're up to speed with loss functions, it's time for you to build one in the next exercise.

EXERCISE 10.01: BUILDING A CUSTOM LOSS FUNCTION

In this exercise, you will create your own custom loss function to train a CNN model to distinguish between images of apples and tomatoes.

You will use the **Apple-or-Tomato** dataset for this exercise. The dataset is a subset of the **Fruits 360** dataset on GitHub. The **Fruits 360** dataset consists of 1,948 total color images with dimensions of 100 by 100 pixels. The **Apple-or-Tomato** dataset has 992 apple images with 662 in the training set and 330 in the test dataset. There are a total of 956 tomato images, with 638 in the training dataset and 318 in the test dataset.

> **Note**
>
> You can get the **Apple-or-Tomato** dataset at the following link: https://packt.link/28kZY.
>
> You can find the **Fruits 360** dataset here: https://github.com/Horea94/Fruit-Images-Dataset/archive/master.zip.

To get started, open a new Colab or Jupyter Notebook. If you are using Google Colab, you will need to download the dataset into your Google Drive first:

1. Open a new Jupyter notebook or Google Colab notebook.

2. If you are using Google Colab, upload your dataset locally with the following code. Otherwise, go to *step 4*. Click on **Choose Files** to navigate to the CSV file and click **Open**. Save the file as **uploaded**. Then, go to the folder where you have saved the dataset:

```
from google.colab import files
uploaded = files.upload()
```

3. Unzip the dataset in the current folder:

```
!unzip \*.zip
```

4. Create a variable, **directory**, that contains the path to the dataset:

```
directory = "/content/gdrive/My Drive/Datasets/apple-or-tomato/"
```

5. Import the **pathlib** library:

```
import pathlib
```

6. Create a variable, **path**, that contains the full path to the dataset using
   pathlib.Path:

```
path = pathlib.Path(directory)
```

7. Create two variables, **train_dir** and **validation_dir**, that take the full
   paths to the train and validation folders, respectively:

```
train_dir = path / 'training_set'
validation_dir = path / 'test_set'
```

8. Create four variables, **train_apple_dir**, **train_tomato_dir**,
   validation_apple_dir, and **validation_tomato_dir**, that take
   the full paths to the **apple** and **tomato** folders for the train and validation
   sets, respectively:

```
train_apple_dir = train_dir / 'apple'
train_tomato_dir = train_dir /'tomato'
validation_apple_dir = validation_dir / 'apple'
validation_tomato_dir = validation_dir / 'tomato'
```

9. Import the **os** package:

```
import os
```

10. Create two variables, called **total_train** and **total_val**, that will get the
    number of images for the training and validation sets, respectively:

```
total_train = len(os.listdir(train_apple_dir)) + \
              len(os.listdir(train_tomato_dir))
total_val = len(os.listdir(validation_apple_dir)) + \
            len(os.listdir(validation_tomato_dir))
```

11. Import **ImageDataGenerator** from the
    tensorflow.keras.preprocessing module:

```
from tensorflow.keras.preprocessing.image import ImageDataGenerator
```

12. Instantiate two **ImageDataGenerator** classes, **train_image_generator**
    and **validation_image_generator**, that will rescale the images by dividing
    by 255:

```
train_image_generator = ImageDataGenerator(rescale=1./255)
validation_image_generator = ImageDataGenerator(rescale=1./255)
```

13. Create three variables, called **batch_size**, **img_height**, and **img_width**, that take the values **32**, **224**, and **224**, respectively:

```
batch_size = 32
img_height = 224
img_width = 224
```

14. Create a data generator called **train_data_gen**, using **flow_from_directory()**, and specify the batch size, the path to the training folder, the value of the **shuffle** parameter, the size of the target, and the class mode:

```
train_data_gen = train_image_generator.flow_from_directory\
                (batch_size=batch_size, directory=train_dir, \
                 shuffle=True, \
                 target_size=(img_height, img_width), \
                 class_mode='binary')
```

15. Create a data generator called **val_data_gen** using **flow_from_directory()** and specify the batch size, the path to the validation folder, the size of the target, and the class mode:

```
val_data_gen = validation_image_generator.flow_from_directory\
                (batch_size=batch_size, directory=validation_dir, \
                 target_size=(img_height, img_width), \
                 class_mode='binary')
```

16. Import **matplotlib** and create a **for** loop that will iterate through five images from **train_data_gen** and plot them:

```
import matplotlib.pyplot as plt

for _ in range(5):
    img, label = train_data_gen.next()
    plt.imshow(img[0])
    plt.show()
```

You should get the following output:

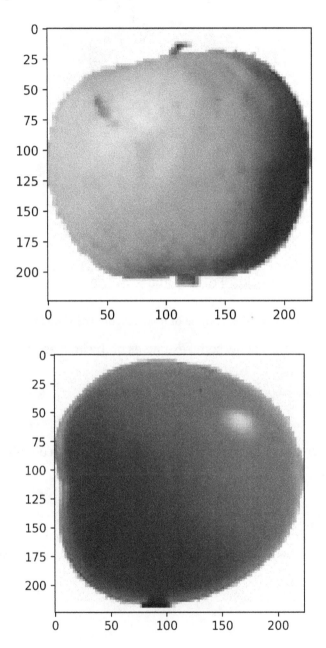

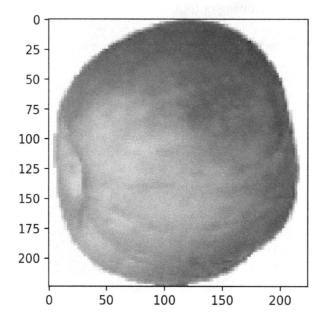

Figure 10.3: Sample of images from the dataset

The preceding results show some examples of the images contained in this dataset.

17. Import the TensorFlow library:

```
import tensorflow as tf
```

18. Create your custom loss function that will square the calculated error:

```
def custom_loss_function(y_true, y_pred):
    print("y_pred ",y_pred)
    print("y_true ", y_true)
    squared_difference = tf.square(float(y_true)-float(y_pred))
    return tf.reduce_mean(squared_difference, axis=-1)
```

19. Import the **NASNetMobile** model from the **tensorflow.keras.applications** module:

```
from tensorflow.keras.applications import NASNetMobile
```

20. Instantiate this model with the ImageNet weights, remove the top layer, and specify the right input dimensions:

```
base_model = NASNetMobile(include_top=False,\
                          input_shape=(100, 100, 3), \
                          weights='imagenet')
```

21. Freeze all the layers of this model so that you are not going to update the model weights of **NASNetMobile**:

```
base_model.trainable = False
```

22. Import the **Flatten** and **Dense** layers from the **tensorflow.keras.layers** module:

```
from tensorflow.keras.layers import Flatten, Dense
```

23. Create a new model that combines the **NASNetMobile** model with two new top layers (with 500 and 1 units, respectively) and ReLu and sigmoid as activation functions:

```
model = tf.keras.Sequential([
    base_model,
    layers.Flatten(),
    layers.Dense(500, activation='relu'),
    layers.Dense(1, activation='sigmoid')
])
```

24. Print the summary of your model:

```
model.summary()
```

You will get the following output:

```
Model: "sequential"
```

Layer (type)	Output Shape	Param #
NASNet (Functional)	(None, 7, 7, 1056)	4269716
flatten (Flatten)	(None, 51744)	0
dense (Dense)	(None, 500)	25872500
dense_1 (Dense)	(None, 1)	501

```
Total params: 30,142,717
Trainable params: 25,873,001
Non-trainable params: 4,269,716
```

Figure 10.4: Model summary

Here, you can see the layers on the left-hand side. You have **Output Shape** shown—for example, **(None, 224, 224, 3)**. Then, the number of parameters is shown under **Param #**. At the bottom, you will find the summary, including trainable and non-trainable parameters.

25. Compile this model by providing your custom loss function, with Adam as the optimizer and accuracy as the metric to be displayed:

```
model.compile(
        optimizer='adam',
        loss=custom_loss_function,
        metrics=['accuracy'])
```

26. Fit the model and provide the train and validation data generators, the number of steps per epoch, and the number of validation steps:

```
history = model.fit(
    Train_data_gen,
    steps_per_epoch=total_train // batch_size,
    epochs=5,
    validation_data=val_data_gen,
    validation_steps=total_val // batch_size)
```

You should get the following output:

```
Epoch 1/5
y_pred  Tensor("sequential/dense_1/Sigmoid:0", shape=(None, 1), dtype=float32)
y_true  Tensor("ExpandDims:0", shape=(None, 1), dtype=float32)
y_pred  Tensor("sequential/dense_1/Sigmoid:0", shape=(None, 1), dtype=float32)
y_true  Tensor("ExpandDims:0", shape=(None, 1), dtype=float32)
40/40 [==============================] - ETA: 0s - loss: 0.1553 - accuracy: 0.8415y_pred  Tensor("sequential/dense_1/Sigmoid:0", shape=(None, 1), dtype=
float32)
y_true  Tensor("ExpandDims:0", shape=(None, 1), dtype=float32)
40/40 [==============================] - 125s 3s/step - loss: 0.1553 - accuracy: 0.8415 - val_loss: 0.1140 - val_accuracy: 0.8859
Epoch 2/5
40/40 [==============================] - 71s 2s/step - loss: 0.0488 - accuracy: 0.9511 - val_loss: 0.0469 - val_accuracy: 0.9531
Epoch 3/5
40/40 [==============================] - 71s 2s/step - loss: 0.0584 - accuracy: 0.9416 - val_loss: 0.0569 - val_accuracy: 0.9422
Epoch 4/5
40/40 [==============================] - 82s 2s/step - loss: 0.0299 - accuracy: 0.9700 - val_loss: 0.0338 - val_accuracy: 0.9656
Epoch 5/5
40/40 [==============================] - 78s 2s/step - loss: 0.0357 - accuracy: 0.9637 - val_loss: 0.0311 - val_accuracy: 0.9688
```

Figure 10.5: Screenshot of training progress

The preceding screenshot shows the information displayed by TensorFlow during the training of your model. You can see the accuracy achieved on the training and validation sets for each epoch. On the fifth epoch, the model is **96%** accurate on both the training set and the validation set.

In this exercise, you have successfully built your own loss function and trained a binary classifier with it to recognize images of apples or tomatoes. In the following section, you will take it a step further and build your own custom layers.

IMPLEMENTING CUSTOM LAYERS

Previously, you looked at implementing your own custom loss function with either the TensorFlow functional API or the subclassing approach. These concepts can also be applied to creating custom layers for a deep learning model. In this section, you will build a ResNet module from scratch.

INTRODUCTION TO RESNET BLOCKS

Residual neural network, or **ResNet**, was first proposed by *Kaiming He* in his paper *Deep Residual Learning for Image Recognition* in 2015. He introduced a new concept called a residual block that tackles the problem of vanishing gradients, which limits the ability of training very deep networks (with a lot of layers).

A residual block is composed of multiple layers. But instead of having a single path where each layer is stacked and executed sequentially, a residual block contains two different paths. The first path has two different convolution layers. The second path, called the **skip connection**, takes the input and forwards it to the last layer of the first path. So, the input of a residual block will go through the first path with the sequence of convolution layers, and its result will be combined with the original input coming from the second path (skip connection), as shown in *Figure 10.6*. Without going too much into the mathematical details, this extra path allows the architecture to pass through the gradients in a deeper layer without impacting the overall performance.

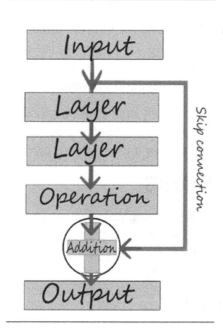

Figure 10.6: Skip connection

As you can see, if you want to build an architecture for the preceding residual block, it will be quite hard with the TensorFlow sequential API. Here, you need to build a very customized layer. This is the reason why you need to use either the functional API or model subclassing instead.

BUILDING CUSTOM LAYERS WITH THE FUNCTIONAL API

In this section, you will see how to use the TensorFlow functional API to build a custom layer.

To start, you will build a function that takes your input as a tensor and adds ReLU and batch normalization to it. For example, in the following code snippet, the **relu_batchnorm_layer** function takes input and then returns a tensor. This makes a composite layer with ReLU activation and batch normalization in succession:

```
def relu_batchnorm_layer(input):
    return BatchNormalization()(ReLU()(input))
```

Now, create a function for your residual block. You'll need to take a tensor as input and pass it to two Conv2D layers. Then, you will add the output of the second Conv2D layer to the original input, which represents the skip connection. The output of this addition will then be passed to the **relu_batchnorm_layer()** function that you defined in the preceding code snippet. The output will be given to another Conv2D layer:

```
def simple_residual_block(input, filters: int, kernel_size: int = 3):
    int_output = Conv2D(filters=filters, kernel_size=kernel_size,
                        padding="same")(input)

    int_output = Conv2D(filters=filters, kernel_size=1, strides=2,
                        padding="same")(int_output)

    output = Add()([int_output,input])
    output = relu_batchnorm_layer(output)
    return output
```

Now, you can use this custom layer in your model. In the following code snippet, you will define a simple model with a Conv2D layer followed by a residual block:

```
inputs = Input(shape=(100, 100, 3))
num_filters = 32

t = BatchNormalization()(inputs)
t = Conv2D(kernel_size=3,
           strides=1,
           filters=32,
           padding="same")(t)
t = relu_batchnorm_layer(t)

t = residual_block(t, filters=num_filters)

t = AveragePooling2D(4)(t)
t = Flatten()(t)
outputs = Dense(1, activation='sigmoid')(t)

model = Model(inputs, outputs)
```

Let's now build custom layers using subclassing in the following section.

BUILDING CUSTOM LAYERS WITH SUBCLASSING

Previously, you looked at how to create a simplified version of a residual block using the functional API. Now, you will see how to use model subclassing to create a custom layer.

To begin, you need to import the **Model** class together with a few layers:

```
from tensorflow.keras.models import Model
from tensorflow.keras.layers import Dense, Dropout, Softmax, concatenate
```

Then, you use model subclassing to create a model with two dense layers. Firstly, define a model subclass denoted as **MyModel**. The objects that you will generate from this class are models with two dense layers.

Define the two dense layers within the **init** method. For instance, the first one can have **64** units and the ReLU activation function, while the second one can have **10** units without an activation function (in this case, the default activation function used is the linear one). After this, in the **call** method, you set up the forward pass by calling the previously defined dense layers. Firstly, you can place the **dense_1** layer to take the inputs and after it, the **dense_2** layer that returns the outputs of the layer:

```
class MyModel (Model) :

  def __init__ (self) :
    super (MyModel, self). __init__ ()
    self.dense_1 = Dense (64, activation='relu')
    self.dense_2 = Dense (10)

  def call(self, inputs) :,
    X = self.dense_1 (inputs)
    return self.dense_2 (X)
```

The next step is to instantiate the model. For this, just call the class with no argument inside the brackets. Next, call the model on a random input to create the weights. For the input, this example uses a one-dimensional vector with **10** elements, but feel free to use a different input. You can then print the summary of the model where you can see the dense layers that you defined before.

Consider the following model summary:

```
model = MyModel ()
model (tf.random.uniform([1,10]))
model.summary ()
```

The resulting output should be like the following:

```
Model: "my_model_3"
```

Layer (type)	Output Shape	Param #
dense_6 (Dense)	multiple	704
dense_7 (Dense)	multiple	650

```
Total params: 1,354
Trainable params: 1,354
Non-trainable params: 0
```

Figure 10.7: Model summary

Now, you can modify the **call** method by including a keyword argument called **training**. This is useful if you want to have different behaviors in training and inference. For example, you can create a dropout layer that will be activated only if **training** is **true**. Firstly, you need to define a dropout layer within the **init** method, given your learning rate of **0.4**. Then, in the **call** method, write an **if** clause with the **training** keyword is set to **true** by default. Inside it, just call the dropout layer:

```
class MyModel(Model):

  def __init__(self):
    super(MyModel, self).__init__()
    self.dense_1 = Dense(64, activation='relu')
    self.dense_2 = Dense(10)
    self.dropout = Dropout(0.4)

  def call(self, inputs, training=True):
    X = self.dense_1(inputs)
    if training:
      X = self.dropout(X)
    return self.dense_2(X)
```

Now, consider the model summary:

```
model = MyModel()
model(tf.random.uniform([1,10]))
model.summary()
```

The summary is displayed as follows, upon running the preceding command:

```
Model: "my_model_4"
```

Layer (type)	Output Shape	Param #
dense_8 (Dense)	multiple	704
dense_9 (Dense)	multiple	650
dropout (Dropout)	multiple	0

```
Total params: 1,354
Trainable params: 1,354
Non-trainable params: 0
```

Figure 10.8: Model summary

In the following exercise, you will build a custom layer.

EXERCISE 10.02: BUILDING A CUSTOM LAYER

The **Healthy-Pneumonia** dataset is a subset of the **National Institute for Health NIH** dataset. The dataset consists of 9,930 total color images with dimensions of 100 by 100 pixels. The **pneumonia-or-healthy** dataset has 1,965 total healthy images with 1,375 images in the training dataset and 590 images in the test dataset.

You will create a custom ResNet block that consists of a Conv2D layer, a batch normalization layer, and a ReLU activation function. You will perform binary classification on the images to distinguish between healthy and pneumonic images.

> **NOTE**
>
> You can get the **pneumonia-or-healthy** dataset here:
> https://packt.link/IOpUX.

To get started, open a new Colab or Jupyter Notebook. If you are using Google Colab, you will need to download the dataset into your Google Drive first:

1. Open a new Jupyter notebook or Google Colab.

2. If you are using Google Colab, you can upload your dataset locally with the following code. Otherwise, go to *step 4*. Click on **Choose Files** to navigate to the CSV file and click **Open**. Save the file as **uploaded**. Then, go to the folder where you saved the dataset:

```
from google.colab import files
uploaded = files.upload()
```

3. Unzip the dataset in the current folder:

```
!unzip \*.zip
```

4. Create a variable, **directory**, that contains the path to the dataset:

```
directory = "/content/gdrive/My Drive/Datasets/pneumonia-or-healthy/"
```

5. Import the **pathlib** library:

```
import pathlib
```

6. Create a variable, **path**, that contains the full path to the data using **pathlib.Path**:

```
path = pathlib.Path(directory)
```

7. Create two variables, called **train_dir** and **validation_dir**, that take the full paths to the train and validation folders, respectively:

```
train_dir = path / 'training_set'
validation_dir = path / 'test_set'
```

8. Create four variables, called **train_healthy_dir**, **train_pneumonia_dir**, **validation_healthy_dir**, and **validation_pneumonia_dir**, that take the full paths to the healthy and pneumonia folders for the train and validation sets, respectively:

```
train_healthy_dir = train_dir / 'healthy'
train_pneumonia_dir = train_dir /'pneumonia'
validation_healthy_dir = validation_dir / 'healthy'
validation_pneumonia_dir = validation_dir / 'pneumonia'
```

9. Import the **os** package:

```
import os
```

10. Create two variables, called **total_train** and **total_val**, to get the number of images for the training and validation sets, respectively:

```
total_train = len(os.listdir(train_healthy_dir)) + \
              len(os.listdir(train_pneumonia_dir))
total_val = len(os.listdir(validation_healthy_dir)) + \
            len(os.listdir(validation_pneumonia_dir))
```

11. Import **ImageDataGenerator** from **tensorflow.keras.preprocessing**:

```
from tensorflow.keras.preprocessing.image import ImageDataGenerator
```

12. Instantiate two **ImageDataGenerator** classes and call them **train_image_generator** and **validation_image_generator**, which will rescale the images by dividing by 255:

```
train_image_generator = ImageDataGenerator(rescale=1./255)
validation_image_generator = ImageDataGenerator(rescale=1./255)
```

13. Create three variables, called **batch_size**, **img_height**, and **img_width**, that take the values **32**, **100**, and **100**, respectively:

```
batch_size = 32
img_height = 100
img_width = 100
```

14. Create a data generator called **train_data_gen** using **flow_from_directory()** and specify the batch size, the path to the training folder, the value of the **shuffle** parameter, the size of the target, and the class mode:

```
train_data_gen = train_image_generator.flow_from_directory\
                 (batch_size=batch_size, directory=train_dir, \
                  shuffle=True, \
                  target_size=(img_height, img_width), \
                  class_mode='binary')
```

15. Create a data generator called **val_data_gen** using
    flow_from_directory() and specify the batch size, the path to the
    validation folder, the size of the target, and the class mode:

```
val_data_gen = validation_image_generator.flow_from_directory\
              (batch_size=batch_size, directory=validation_dir, \
               target_size=(img_height, img_width), \
               class_mode='binary')
```

16. Import **matplotlib** and create a **for** loop that will iterate through five images
    from **train_data_gen** and plot them:

```
import matplotlib.pyplot as plt

for _ in range(5):
    img, label = train_data_gen.next()
    plt.imshow(img[0])
    plt.show()
```

You should see the following output:

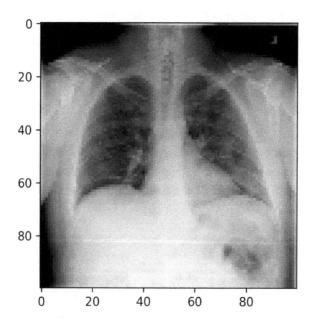

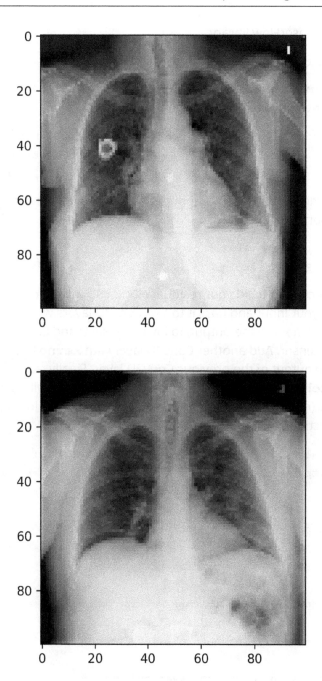

Figure 10.9: Sample of images from the dataset

The preceding results show some examples of the images contained in this dataset.

17. Import the TensorFlow library:

```
import tensorflow as tf
```

18. Import **Input, Conv2D, ReLU, BatchNormalization, Add, AveragePooling2D, Flatten**, and **Dense**:

```
from tensorflow.keras.layers import Input, Conv2D, ReLU, \
                            BatchNormalization, Add, \
                            AveragePooling2D, Flatten, Dense
```

19. Build a function that takes your input as a tensor and adds ReLU and batch normalization to it:

```
def relu_batchnorm_layer(input):
    return BatchNormalization()(ReLU()(input))
```

20. Create a function to build your residual block. You will need to take a tensor (**input**) as your input and pass it to two Conv2D layers with a stride of **2**. Next, add the input to the output, followed by ReLU and batch normalization, returning a tensor. Add another Conv2D layer with **kernel_size=1**. Add its result to the output of the previous Conv2D layer. Finally, apply **relu_batchnorm_layer()** and return its value. You will apply the exact same filters (numbers and dimensions are defined by two input parameters of the construction function) to all Conv2D layers:

```
def residual_block(input, filters: int, kernel_size: int = 3):
    int_output = Conv2D(filters=filters, kernel_size=kernel_size,
                        strides=(2),
                        padding="same")(input)
    int_output = relu_batchnorm_layer(int_output)
    int_output = Conv2D(filters=filters, kernel_size=kernel_size,
                        padding="same")(int_output)

    int_output2 = Conv2D(filters=filters, kernel_size=1, strides=2,
                         padding="same")(input)
    output = Add()([int_output2, int_output])

    output = relu_batchnorm_layer(output)
    return output
```

21. Import the **Model** module:

```
from tensorflow.keras.models import Model
```

22. Use **keras.layers.Input()** to define the input layer to the model. Here, your shape is 100 pixels by 100 pixels and has three colors (RGB):

```
inputs = Input(shape=(100, 100, 3))
```

23. Apply batch normalization to the input, followed by a Conv2D layer with **32** filters of size **3*3**, stride **1**, and **same** padding. Finally, apply the **relu_batchnorm_layer()** function to its output:

```
t = BatchNormalization()(inputs)
t = Conv2D(kernel_size=3,
           strides=1,
           filters=32,
           padding="same")(t)
t = relu_batchnorm_layer(t)
```

24. Provide the output of the previous layer to the **residual_block()** function with **32** filters. Then, pass its output an average pooling layer with four units and then flatten its results before feeding it to a fully connected layer of **1** unit with sigmoid as the activation function:

```
t = residual_block(t, filters=32)

t = AveragePooling2D(4)(t)
t = Flatten()(t)
outputs = Dense(1, activation='sigmoid')(t)
```

25. Instantiate a **Model()** class with the original input and the output of the fully connected layer:

```
model = Model(inputs, outputs)
```

26. Get the summary of your model:

```
model.summary()
```

You will see a summary, including trainable and non-trainable parameters, as follows:

```
Model: "model_2"

Layer (type)                    Output Shape            Param #    Connected to
==================================================================================================
input_4 (InputLayer)            [(None, 100, 100, 3)    0

batch_normalization_70 (BatchNo (None, 100, 100, 3)     12         input_4[0][0]

conv2d_75 (Conv2D)              (None, 100, 100, 32)    896        batch_normalization_70[0][0]

re_lu_67 (ReLU)                 (None, 100, 100, 32)    0          conv2d_75[0][0]

batch_normalization_71 (BatchNo (None, 100, 100, 32)    128        re_lu_67[0][0]

conv2d_76 (Conv2D)              (None, 50, 50, 32)      9248       batch_normalization_71[0][0]

re_lu_68 (ReLU)                 (None, 50, 50, 32)      0          conv2d_76[0][0]

batch_normalization_72 (BatchNo (None, 50, 50, 32)      128        re_lu_68[0][0]

conv2d_78 (Conv2D)              (None, 50, 50, 32)      1056       batch_normalization_71[0][0]

conv2d_77 (Conv2D)              (None, 50, 50, 32)      9248       batch_normalization_72[0][0]

add_32 (Add)                    (None, 50, 50, 32)      0          conv2d_78[0][0]
                                                                   conv2d_77[0][0]

re_lu_69 (ReLU)                 (None, 50, 50, 32)      0          add_32[0][0]

batch_normalization_73 (BatchNo (None, 50, 50, 32)      128        re_lu_69[0][0]

average_pooling2d_2 (AveragePoo (None, 12, 12, 32)      0          batch_normalization_73[0][0]

flatten_2 (Flatten)             (None, 4608)            0          average_pooling2d_2[0][0]

dense_2 (Dense)                 (None, 1)               4609       flatten_2[0][0]
==================================================================================================
Total params: 25,453
Trainable params: 25,255
Non-trainable params: 198
```

Figure 10.10: Model summary

27. Compile the model by providing binary cross-entropy as the loss function, Adam as the optimizer, and accuracy as the metric to be displayed:

```
model.compile(
        optimizer='adam',
        loss=binary_crossentropy,
        metrics=['accuracy'])
```

28. Fit the model and provide the train and validation data generators, the number of epochs, the steps per epoch, and the validation steps:

```
history = model.fit(
    Train_data_gen,
    steps_per_epoch=total_train // batch_size,
    epochs=5,
    validation_data=val_data_gen,
    validation_steps=total_val // batch_size
)
```

You should get output like the following:

```
Epoch 1/5
85/85 [==============================] - 41s 480ms/step - loss: 0.4527 - accuracy: 0.7837 - val_loss: 0.5725 - val_accuracy: 0.6884
Epoch 2/5
85/85 [==============================] - 41s 478ms/step - loss: 0.4363 - accuracy: 0.7914 - val_loss: 0.5215 - val_accuracy: 0.7405
Epoch 3/5
85/85 [==============================] - 41s 478ms/step - loss: 0.4441 - accuracy: 0.7877 - val_loss: 0.5386 - val_accuracy: 0.7396
Epoch 4/5
85/85 [==============================] - 41s 480ms/step - loss: 0.4281 - accuracy: 0.8032 - val_loss: 0.5843 - val_accuracy: 0.7231
Epoch 5/5
85/85 [==============================] - 41s 481ms/step - loss: 0.3967 - accuracy: 0.8227 - val_loss: 0.6919 - val_accuracy: 0.7109
```

Figure 10.11: Screenshot of training progress

The preceding screenshot shows the information displayed by TensorFlow during the training of your model. You can see the accuracy achieved on the training and validation sets for each epoch.

In this exercise, you created your own custom layer for the network. Now, let's test the knowledge you have gained so far in the following activity.

ACTIVITY 10.01: BUILDING A MODEL WITH CUSTOM LAYERS AND A CUSTOM LOSS FUNCTION

The **table-or-glass** dataset is a subset of images taken from the **Open Images V6** dataset. The **Open Images V6** dataset has around 9 million images. The **table-or-glass** dataset consists of 7,484 total color images with dimensions of 100 by 100 pixels. The **table-or-glass** dataset has 3,741 total glass images with 2,618 in the training and 1,123 in the test dataset. There are a total of 3,743 table images with 2,618 in the training and 1,125 in the test dataset. You are required to train a more complex model that can distinguish images of glasses and tables using custom ResNet blocks and a custom loss function.

> **NOTE**
>
> You can find the dataset here: https://packt.link/bE5F6.

The following steps will help you to complete this activity:

1. Import the dataset and unzip the file into a local folder.

2. Create the list of images for both the training and testing sets.

3. Analyze the distribution of the target variable.

4. Preprocess the images (standardization and reshaping).

5. Create a custom loss function that will calculate the average squared error.

6. Create a custom residual block constructor function.

7. Train your model.

8. Print the learning curves for accuracy and loss.

> **NOTE**
>
> The solution to this activity can be found on page 532.

SUMMARY

This chapter demonstrated how to build and utilize custom TensorFlow components. You learned how to design and implement custom loss functions, layers, and residual blocks. Using the TensorFlow functional API or model subclassing allows you to build more complex deep learning models that may be a better fit for your projects.

In the next and final chapter, you will explore and build generative models that can learn patterns and relationships within data, and use those relationships to generate new, unique data.

11

GENERATIVE MODELS

OVERVIEW

This chapter introduces you to generative models—their components, how they function, and what they can do. You will start with generative **long short-term memory** (**LSTM**) networks and how to use them to generate new text. You will then learn about **generative adversarial networks** (**GANs**) and how to create new data, before moving on to **deep convolutional generative adversarial networks** (**DCGANs**) and creating your own images.

By the end of the chapter, you will know how to effectively use different types of GANs and generate various types of new data.

INTRODUCTION

In this chapter, you will explore generative models, which are types of unsupervised learning algorithms that generate completely new artificial data. Generative models differ from predictive models in that they aim to generate new samples from the same distribution of training data. While the purpose of these models may be very different from those covered in other chapters, you can and will use many of the concepts learned in prior chapters, including loading and preprocessing various data files, hyperparameter tuning, and building convolutional and **recurrent neural networks** (**RNNs**). In this chapter, you will learn about one way to generate new samples from a training dataset, which is to use LSTM models to complete sequences of data based on initial seed data.

Another way that you will learn about is the concept of two neural networks competing against one another in an adversarial way, that is, a generator generating samples and a discriminator trying to distinguish between the generated and real samples. As both models train simultaneously, the generator generates more realistic samples as the discriminator can more accurately distinguish between the "real" and "fake" data over time. These networks working together are called GANs. Generative models can be used to generate new text data, audio samples, and images.

In this chapter, you will focus primarily on three areas of generative models – text generation or language modeling, GANs, and DCGANs.

TEXT GENERATION

In *Chapter 9, Recurrent Neural Networks*, you were introduced to **natural language processing** (**NLP**) and text generation (also known as language modeling), as you worked with some sequential data problems. In this section, you will be extending your sequence model for text generation using the same dataset to generate extended headlines.

Previously in this book, you saw that sequential data is data in which each point in the dataset is dependent on the point prior and the order of the data is important. Recall the example with the bag of words from *Chapter 9, Recurrent Neural Networks*. With the *bag-of-words* approach, you simply used a set of word counts to derive meaning from their use. As you can see in *Figure 11.1*, these two sentences have completely opposite semantic meanings, but would be identical in a bag-of-words format. While this may be an effective strategy for some problems, it's not an ideal approach for predicting the next word or words.

it is what it is
is it what it is

Figure 11.1: An example of identical words with differing semantics

Consider the following example of a language model. You are given a sentence or a phrase, **yesterday I took my car out for a**, and are asked to predict the word that comes next in the sequence. Here, an appropriate word to complete the sequence would be **drive**.

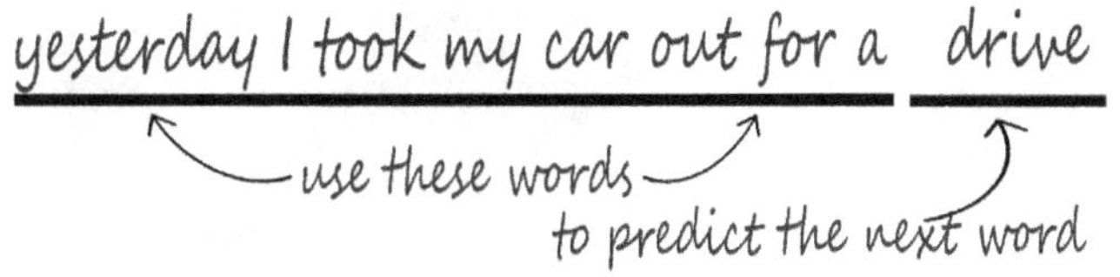

yesterday I took my car out for a   drive
— use these words —
to predict the next word

Figure 11.2: Sentence example

To be successful in working with sequential data, you need a neural network capable of storing the value of the sequence. For this, you can use RNNs and LSTMs. LSTMs that are used for generating new sequences, such as text generation or language modeling, are known as generative LSTMs.

Let's do a simple review of RNNs and LSTMs.

Essentially, RNNs loop back on themselves, storing information and repeating the process, in a continuous cycle. Information is first transformed into vectors so that it can be processed by machines. The RNN then processes the vector sequence one at a time. As the RNN processes each vector, the vector gets passed through the previous hidden state. In this way, the hidden state retains information from the previous step, acting as a type of memory. It does this by combining the input and the previous hidden state with a tanh function that compresses the values between **−1** and **1**.

Essentially, this is how the RNN functions. RNNs don't need a lot of computation and work well with short sequences. Simply put, RNNs are networks that have loops that allow information to persist over time.

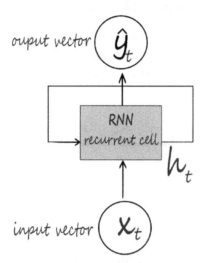

Figure 11.3: RNN data flow

RNNs do come with a couple of challenges—most notably, the exploding and vanishing gradient problems.

The **exploding gradient problem** is what happens when gradients become too large for optimization. The opposite problem may occur where your gradients are too small. This is what is known as the **vanishing gradient problem**. This happens when gradients become increasingly smaller as you make repeated multiplications. Since the size of the gradient determines the size of the weight updates, exploding or vanishing gradients mean that the network can no longer be trained. This is a very real problem when it comes to training RNNs since the output of the networks feeds back into the input. The vanishing and exploding gradient issues were covered in *Chapter 9, Recurrent Neural Networks*, and more details of how these issues are solved can be found there.

LSTMs can selectively control the flow of information within each LSTM node. With added control, you can more easily adjust the model to prevent potential problems with gradients.

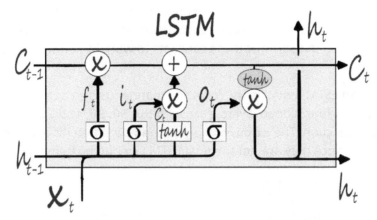

Figure 11.4: LSTM architecture

So, what enables LSTMs to track and store information throughout many time steps? You'll recall from *Chapter 9, Recurrent Neural Networks*, that the key building block behind the LSTM is the structure called a *gate*, which allows the LSTM to selectively add or remove information to its cell state.

Gates consist of a bounding function such as sigmoid or tanh. For example, if the function were sigmoid, it would force its input to be between zero and one. Intuitively, you can think of this as capturing how much of the information passed through the gate should be retained. This should be between zero and one, effectively *gating* the flow of information.

LSTMs process information through four simple steps.

They first forget their irrelevant history. Second, they perform a computation to store relevant parts of new information, and thirdly, they use these two steps together to selectively update their internal state. Finally, they generate an output.

Figure 11.5: LSTM processing steps

This was a bit of a refresher on LSTMs and how they can selectively control and regulate the flow of information. Now that you've reviewed LSTMs and their architecture, you can put some of these concepts to work by reviewing your code and LSTM model.

You can create an LSTM model in the following manner using a sequential model. This LSTM contains four hidden layers, each with **50**, **60**, **80**, and **120** units and a ReLU activation function. The **return_sequences** parameter is set to **True** for all but the last layer since they are not the final LSTM layer in the network:

```
regressor = Sequential()

regressor.add(LSTM(units= 50, activation = 'relu', \
                   return_sequences = True, \
                   input_shape = (X_train.shape[1], 5)))
regressor.add(Dropout(0.2))

regressor.add(LSTM(units= 60, activation = 'relu', \
                   return_sequences = True))
regressor.add(Dropout(0.3))

regressor.add(LSTM(units= 80, activation = 'relu', \
              return_sequences = True))
regressor.add(Dropout(0.4))

regressor.add(LSTM(units= 120, activation = 'relu'))
regressor.add(Dropout(0.5))

regressor.add(Dense(units = 1))
```

Now that you've recalled how to create RNNs with LSTM layers, you'll next learn how to apply them to natural language text and generate new text in a sequence.

EXTENDING NLP SEQUENCE MODELS TO GENERATE TEXT

NLP takes data in the form of natural language that has traditionally been very difficult for machines to make sense of and turns it into data that can be useful for machine learning applications. This data can take the form of characters, words, sentences, or paragraphs. You will be focusing on text generation in this section.

As a quick review, *preprocessing* generally entails all the steps needed to train your model. Some common steps include *data cleaning*, *transformation*, and *data reduction*. For NLP, more specifically, the steps could be all or some of the following:

- **Dataset cleaning** encompasses the conversion of the case to lowercase, removing punctuation.

- **Tokenization** is breaking up a character sequence into specified units called tokens.

- **Padding** is a way to make input sentences of different sizes the same by padding them.

- **Padding the sequences** refers to making sure that the sequences have a uniform length.

- **Stemming** is truncating words down to their stem. For example, the words `rainy` and `raining` both have the stem `rain`.

Let's take a closer look at what the process looks like.

DATASET CLEANING

Here, you create a function, **clean_text**, that returns a list of words after cleaning. Now, save all text as lowercase with **lower()** method, encoded with **utf8** for character standardization. Finally, output 10 headlines from your corpus:

```
def clean_text(txt):
    txt = "".join(v for v in txt \
                    if v not in string.punctuation).lower()
    txt = txt.encode("utf8").decode("ascii",'ignore')
    return txt


corpus = [clean_text(x) for x in all_headlines]
corpus[:10]
```

Cleaning the text in this manner is a great way to standardize text to input into a model. Converting all words to lowercase in the same encoding ensures consistency of the text. It also ensures that capitalization or different encodings of the same words are not treated as different words by any model that is created.

GENERATING A SEQUENCE AND TOKENIZATION

Neural networks expect input data in a consistent, numerical format. Much like how images are processed for image classification models, where each image is represented as a three-dimensional array, and are often resized to meet the expectations of the model, text must be processed similarly. Luckily, Keras has a number of utility classes and functions to aid with processing text data for neural networks. One such class is **Tokenizer**, which vectorizes a text corpus by converting the corpus into a sequence of integers. The following code imports the **Tokenizer** class from Keras:

```
from keras.preprocessing.text import Tokenizer
```

GENERATING A SEQUENCE OF N-GRAM TOKENS

Here, you create a function named **get_seq_of_tokens**. With **tokenizer.fit_on_texts**, you extract tokens from the corpus. Each integer output corresponds with a specific word. The **input_seq** parameter is initialized as an empty list, **[]**. With **token_list = tokenizer.texts_to_sequences**, you convert text to the tokenized equivalent. With **n_gram_sequence = token_list**, you generate the n-gram sequences. Using **input_seq.append(n_gram_sequence)**, you append each sequence to the list of your features:

```
tokenizer = Tokenizer()

def get_seq_of_tokens(corpus):
    tokenizer.fit_on_texts(corpus)
    all_words = len(tokenizer.word_index) + 1

    input_seq = []
    for line in corpus:
        token_list = tokenizer.texts_to_sequences([line])[0]
        for i in range(1, len(token_list)):
            n_gram_sequence = token_list[:i+1]
```

```
        input_seq.append(n_gram_sequence)
    return input_seq, all_words

your_sequences, all_words = get_seq_of_tokens(corpus)
your_sequences[:10]
```

get_seq_of_tokens ensures that a corpus is broken up into sequences of equal length. If a corpus is too short for the network's expected input, the resultant sequence will have to be padded.

PADDING SEQUENCES

Here, you create a **generate_padded_sequences** function that takes **input_seq** as input. The **pad_sequences** function is used to pad the sequences to make their lengths equal. In the function, first, the maximum sequence length is determined by calculating the length of each input sequence. Once the maximum sequence length is determined, all other sequences are padded to match. Next, the **predictors** and **label** parameters are created. The **label** parameter is the last word of the sequence, and the **predictors** parameter is all the preceding words. Finally, the **label** parameter is converted to a categorical array:

```
def generate_padded_sequences(input_seq):
    max_sequence_len = max([len(x) for x in input_seq])
    input_seq = np.array(pad_sequences\
                        (input_seq, maxlen=max_sequence_len, \
                        padding='pre'))

    predictors, label = input_seq[:,:-1],input_seq[:,-1]
    label = keras.utils.to_categorical(label, num_classes=all_words)
    return predictors, label, max_sequence_len

predictors, label, max_sequence_len = generate_padded_sequences\
                                (your_sequences)
```

Now that you have learned some preprocessing and cleaning steps for working with natural language, including cleaning, generating n-gram sequences, and padding sequences for consistent lengths, you are ready for your first exercise of the chapter, that is, text generation.

EXERCISE 11.01: GENERATING TEXT

In this exercise, you will use the LSTM model from *Exercise 9.02, Building an RNN with LSTM Layer Nvidia Stock Prediction*, to extend your prediction sequence and generate new text. In that exercise, you created an LSTM model to predict the stock price of Nvidia by feeding the historical stock prices to the model. The model was able to use LSTM layers to understand patterns in the historical stock prices for future predictions.

In this exercise, you will use the same principle applied to text, by feeding the historical headlines to the model. You will use the **articles.csv** dataset for this exercise. The dataset contains 831 news headlines from the New York Times in CSV format. Along with the headlines, the dataset also contains several attributes about the news article, including the publication date, print page, and keywords. You are required to generate new news headlines using the given dataset.

> **NOTE**
>
> You can find **articles.csv** here: http://packt.link/RQVoB.

Perform the following steps to complete this exercise:

1. Open a new Jupyter or Colab notebook.

2. Import the following libraries:

```
from keras.preprocessing.sequence import pad_sequences
from keras.models import Sequential
from keras.layers import Embedding, LSTM, Dense, Dropout
import tensorflow.keras.utils as ku
from keras.preprocessing.text import Tokenizer
import pandas as pd
import numpy as np
from keras.callbacks import EarlyStopping
import string, os
import warnings
warnings.filterwarnings("ignore")
warnings.simplefilter(action='ignore', category=FutureWarning)
```

You should get the following output:

```
Using TensorFlow backend.
```

3. Load the dataset locally by setting **your_dir** to **content/**. Create a **your_headlines** parameter as an empty list and use a **for** loop to iterate over:

```
your_dir = 'content/'
your_headlines = []
for filename in os.listdir(your_dir):
    if 'Articles' in filename:
        article_df = pd.read_csv(your_dir + filename)
        your_headlines.extend(list(article_df.headline.values))
        break


your_headlines = [h for h in your_headlines if h != "Unknown"]
len(our_headlines)
```

The output will represent the number of headlines in your dataset:

```
831
```

4. Now, create a **clean_text** function to return a list of cleaned words. Convert the text to lowercase with **lower()** method and encode it with **utf8** for character standardization. Finally, output 20 headlines from your corpus:

```
def clean_text(txt):
    txt = "".join(v for v in txt \
                   if v not in string.punctuation).lower()
    txt = txt.encode("utf8").decode("ascii",'ignore')
    return txt


corpus = [clean_text(x) for x in all_headlines]
corpus[60:80]
```

You should get the following output:

```
['lets go for a win on opioids',
 'floridas vengeful governor',
 'how to end the politicization of the courts',
 'when dr king came out against vietnam',
 'britains trains dont run on time blame capitalism',
 'questions for no license plates here using art to transcend prison walls',
 'dry spell',
 'are there subjects that should be offlimits to artists or to certain artists in particular',
 'that is great television',
 'thinking in code',
 'how gorsuchs influence could be greater than his vote',
 'new york today how to ease a hangover',
 'trumps gifts to china',
 'at penn station rail mishap spurs large and lasting headache',
 'chemical attack on syrians ignites worlds outrage',
 'adventure is still on babbos menu',
 'swimming in the fast lane',
 'a national civics exam',
 'obama adviser is back in the political cross hairs',
 'the hippies have won']
```

Figure 11.6: Corpus

5. With **tokenizer.fit**, extract tokens from the corpus. Each integer output corresponds to a specific word. The **input_seq** parameter is initialized as an empty list, **[]**. With **token_list = tokenizer.texts_to_sequences**, you convert each sentence into its tokenized equivalent. With **n_gram_sequence = token_list**, you generate the n-gram sequences. Using **input_seq.append(n_gram_sequence)**, you append each sequence to a list of features:

```
tokenizer = Tokenizer()

def get_seq_of_tokens(corpus):
    tokenizer.fit_on_texts(corpus)
    all_words = len(tokenizer.word_index) + 1

    input_seq = []
    for line in corpus:
        token_list = tokenizer.texts_to_sequences([line])[0]
        for i in range(1, len(token_list)):
            n_gram_sequence = token_list[:i+1]
            input_seq.append(n_gram_sequence)
```

```
    return input_seq, all_words

your_sequences, all_words = get_seq_of_tokens(corpus)
your_sequences[:20]
```

You should get the following output:

```
[[169, 17],
 [169, 17, 665],
 [169, 17, 665, 367],
 [169, 17, 665, 367, 4],
 [169, 17, 665, 367, 4, 2],
 [169, 17, 665, 367, 4, 2, 666],
 [169, 17, 665, 367, 4, 2, 666, 170],
 [169, 17, 665, 367, 4, 2, 666, 170, 5],
 [169, 17, 665, 367, 4, 2, 666, 170, 5, 667],
 [6, 80],
 [6, 80, 1],
 [6, 80, 1, 668],
 [6, 80, 1, 668, 10],
 [6, 80, 1, 668, 10, 669],
 [670, 671],
 [670, 671, 129],
 [670, 671, 129, 672],
 [673, 674],
 [673, 674, 368],
 [673, 674, 368, 675]]
```

Figure 11.7: n-gram tokens

The output shows the n-gram tokens of the headlines. For each headline, the number of n-grams is determined by the length of the headline.

6. Pad the sequences and obtain the variables, **predictors** and **target**:

```
def generate_padded_sequences(input_seq):
    max_sequence_len = max([len(x) for x in input_seq])
    input_seq = np.array(pad_sequences\
                        (input_seq, maxlen=max_sequence_len, \
                         padding='pre'))

    predictors, label = input_seq[:,:-1],input_seq[:,-1]
    label = ku.to_categorical(label, num_classes=all_words)
    return predictors, label, max_sequence_len

predictors, label, \
max_sequence_len = generate_padded_sequences(inp_seq)
```

7. Prepare your model for training. Add an input embedding layer with **model.add(Embedding)**, a hidden LSTM layer with **model.add(LSTM(100))**, and a dropout of 10%. Then, add the output layer with **model.add(Dense)** using the softmax activation function. With **compile()** method, configure your model for training, setting your loss function to **categorical_crossentropy**. Use the Adam optimizer:

```
def create_model(max_sequence_len, all_words):
    input_len = max_sequence_len - 1
    model = Sequential()

    model.add(Embedding(all_words, 10, input_length=input_len))

    model.add(LSTM(100))
    model.add(Dropout(0.1))

    model.add(Dense(all_words, activation='softmax'))

    model.compile(loss='categorical_crossentropy', \
                  optimizer='adam')

    return model

model = create_model(max_sequence_len, all_words)
model.summary()
```

You should get the following output:

```
Model: "sequential"

_____
Layer (type)                 Output Shape              Param #
=================================================================
embedding (Embedding)        (None, 18, 10)            24220

_____
lstm (LSTM)                  (None, 100)               44400

_____
dropout (Dropout)            (None, 100)               0

_____
dense (Dense)                (None, 2422)              244622

=================================================================
Total params: 313,242
Trainable params: 313,242
Non-trainable params: 0
```

Figure 11.8: Model summary

8. Fit the model and set **epochs** to **200** and **verbose** to **5**:

```
model.fit(predictors, label, epochs=200, verbose=5)
```

You should get the following output:

```
Epoch 1/200
Epoch 2/200
Epoch 3/200
Epoch 4/200
Epoch 5/200
Epoch 6/200
Epoch 7/200
Epoch 8/200
Epoch 9/200
Epoch 10/200
Epoch 11/200
Epoch 12/200
Epoch 13/200
Epoch 14/200
Epoch 15/200
Epoch 16/200
```

Figure 11.9: Training the model

9. Create a function that will generate a headline given a starting seed text, the number of words to generate, the model, and the maximum sequence length. The function will include a **for** loop to iterate over the number of words to generate. In each iteration, the tokenizer will tokenize the text, and then pad the sequence before predicting the next word in the sequence. Next, the iteration will convert the token back into a word and add it to the sentence. Once the **for** loop completes, the generated headline will be returned:

```
def generate_text(seed_text, next_words, model, max_sequence_len):
    for _ in range(next_words):
        token_list = tokenizer.texts_to_sequences([seed_text])[0]
        token_list = pad_sequences([token_list], \
                                   maxlen = max_sequence_len-1, \
                                   padding='pre')
        predicted = model.predict\
                    (token_list, verbose=0)

        output_word = ""
        for word,index in tokenizer.word_index.items():
            if index == predicted.any():
                output_word = word
                break
        seed_text += " "+output_word
    return seed_text.title()
```

10. Finally, output some of your generated text with the **print** function by printing the output of the function you created in *Step 9*. Use the **10 ways**, **europe looks to**, **best way**, **homeless in**, **unexpected results**, and **critics warn** seed words with the corresponding number of words to generate; that is, **11**, **8**, **10**, **10**, **10**, and **10**, respectively:

```
print (generate_text("10 ways", 11, model, max_sequence_len))
print (generate_text("europe looks to", 8, model, \
                     max_sequence_len))
print (generate_text("best way", 10, model, max_sequence_len))
print (generate_text("homeless in", 10, model, max_sequence_len))
print (generate_text("unexpected results", 10, model,\
                     max_sequence_len))
print (generate_text("critics warn", 10, model, \
                     max_sequence_len))
```

You should get the following output:

```
10 Ways To Teach And Learn About Poetry With The New York Times
Europe Looks To State Or Florence Hold Part On Its Chemical
Best Way To Stand A Small Equal Large And Lasting Headache Russian
Homeless In The Dreaded Trump Curse Partisan Tool Not The Old Days
Unexpected Results Trumps Fat History Lesson In Play Catch Thats Tricky Say
Critics Warn Leader Moves To Hold Early Elections Shakier Flowers 5 Home
```

Figure 11.10: Generated text

The output shows the generated headlines with the seed text provided. The words generated are limited to what was included in the training dataset, which itself was fairly limited in size, leading to some nonsensical results.

Now that you've generated text with an LSTM in your first exercise, let's move on to working with images by using GANs to generate new images based on a given dataset.

GENERATIVE ADVERSARIAL NETWORKS

GANs are networks that generate new, synthetic data by learning patterns and underlying representations from a training dataset. The GAN does this by using two networks that compete with one another in an adversarial fashion. These networks are called the **generator** and **discriminator**.

To see how these networks compete with one another, consider the following example. The example will skip over a few details that will make more sense as you get to them later in the chapter.

Imagine two entities: a money counterfeiter and a business owner. The counterfeiter attempts to make a currency that looks authentic to fool the business owner into thinking the currency is legitimate. By contrast, the business owner tries to identify any fake bills, so that they don't end up with just a piece of worthless paper instead of real currency.

This is essentially what GANs do. The counterfeiter in this example is the generator, and the business owner is the discriminator. The generator creates an image and passes it to the discriminator. The discriminator checks whether the image is real or not, and both networks compete against each other, driving improvements within one another.

The generator's mission is to create a synthetic sample of data that can fool the discriminator. The generator will try to trick the discriminator into thinking that the sample is real. The discriminator's mission is to be able to correctly classify a synthetic sample created by the generator.

Figure 11.11: GAN-generated images

The next sections will look a bit closer at the generator and discriminator and how they function individually, before considering both in combination in the *The Adversarial Network* section.

THE GENERATOR NETWORK

As discussed, GANs are utilized for unsupervised learning tasks in machine learning. GANs consist of two models (a generator and a discriminator) that automatically discover and learn the patterns in input data. The two models compete with one another to analyze, capture, and create variations within data. GANs can be used to generate new data that looks like it could have come from the original data.

First up is the generator model. How does the generator create synthetic data?

The generator receives input as a *fixed-length random vector* called the **latent vector**, which goes into the generator network. This is sometimes referred to as the **random noise seed**. A new sample is generated from it. The generated instance is then sent to the discriminator for classification. Through random noise, the generator learns which outputs were more convincing and continues to improve in that direction.

Figure 11.12: Input and output model in the generator network

In the following figure, you can see that the discriminator takes input from both real data and the generator. The generator neural network attempts to generate data that looks real to the discriminator.

The generator doesn't get to see what the real data is. The main goal of the generator is to convince the discriminator to classify its output as real.

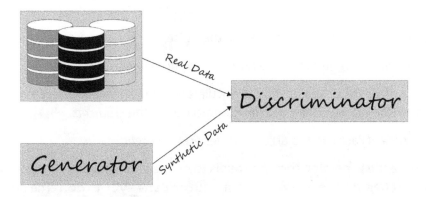

Figure 11.13: Two sources of data for the discriminator model

The GAN includes the following components:

- Noisy input vector
- Discriminator network
- Generator loss

Backpropagation is used to adjust the weights in the optimal direction by calculating a weight's impact on the output. The backpropagation method is used to obtain gradients and these gradients can help change the generator weights.

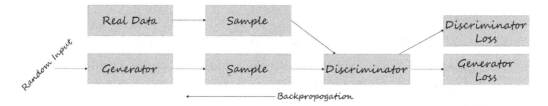

Figure 11.14: Backpropagation in GAN

The basic procedure of a single generator iteration looks something like this:

1. Based on real data from a dataset, *sample random noise* is used.

2. The *generator* produces *output* from the noise.

3. The *discriminator* classifies the output as "*real*" or "*fake*."

4. The *loss* from this classification is calculated, followed by *backpropagation through the generator* and *discriminator* to obtain the *gradients*.

5. The *gradients* are used to adjust the generator *weights*.

Now, to code the generator, the first step is to define your generator model. You begin by creating your generator function with **define_your_gen**. The number of outputs of your generator should match the size of the data you are trying to synthesize. Therefore, the final layer of your generator should be a dense layer with the number of units equal to the expected size of the output:

```
model.add(Dense(n_outputs, activation='linear'))
```

The model will not compile because it does not directly fit the generator model.

The code block will look something like the following:

```
def define_your_gen(latent_dim, n_outputs=2):
    model = Sequential()
    model.add(Dense(5, activation='relu', \
                    kernel_initializer='he_uniform', \
                    input_dim=latent_dim))
    model.add(Dense(n_outputs, activation='linear'))
    return model
```

The generator composes one half of the GAN; the other half is the discriminator.

THE DISCRIMINATOR NETWORK

A **discriminator** is a neural network model that learns to identify real data from the fake data that the generator sends as input. The two sources of training data are the authentic data samples and the fake generator samples:

- Real data instances are used by the discriminator as positive samples during the training.

- Synthetic data instances created by the generator are used as fake examples during the training process.

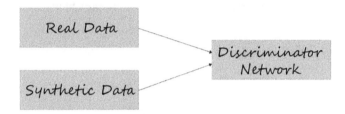

Figure 11.15: Inputs for the discriminator network

During the discriminator training process, the discriminator is connected to the generator and discriminator loss. It requires both real data and synthetic data from the generator, but only uses the discriminator loss for weight updates.

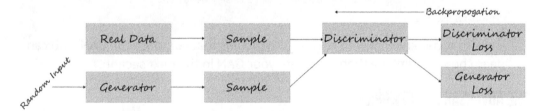

Figure 11.16: Backpropagation with discriminator loss

Now let's take a look at how the discriminator works with some code.

Your first step is to define your discriminator model with **define_disc()**.

The model takes a vector from your generator and makes a prediction as to whether the sample is real or fake. Therefore, you use binary classification.

You're creating a simple GAN, so you will only need one hidden layer. Use **model.add(Dense(25)** to create the hidden layer.

Again, your activation function will be ReLU with **activation='relu'** and the **he_uniform** weight initialization with **kernel_initializer='he_uniform'**.

Your output layer will only need a single node for binary classification. To ensure your output is zero or one, you will use the sigmoid activation function:

```
model.add(Dense(1, activation='sigmoid'))
```

The model will attempt to minimize your loss function. Use Adam for your stochastic gradient descent:

```
model.compile(loss='binary_crossentropy', \
              optimizer='adam', metrics=['accuracy'])
```

Here's a look at your discriminator model code:

```
def define_disc(n_inputs=2):
    model = Sequential()
    model.add(Dense(25, activation='relu', \
                    kernel_initializer='he_uniform', \
                    input_dim=n_inputs))
    model.add(Dense(1, activation='sigmoid'))
    model.compile(loss='binary_crossentropy', \
                  optimizer='adam', metrics=['accuracy'])
    return model
```

Now that you know how to create both models that compose the GAN, you can learn how to combine them to create your GAN in the next section.

THE ADVERSARIAL NETWORK

GANs consist of two networks, a generator, which is represented as $G(x)$, and a discriminator, represented as $D(x)$. Both networks play an adversarial game. The generator network tries to learn the underlying distribution of the training data and generates similar samples, while the discriminator network tries to catch the fake samples generated by the generator.

The generator network takes a sample and generates a fake sample of data. The generator is trained to increase the probability of the discriminator network making mistakes. The discriminator network decides whether the data is generated or taken from the real sample using binary classification with the help of a sigmoid function. The sigmoid function ensures that the output is zero or one.

The following list represents an overview of a typical GAN at work:

1. First, a *noise vector* or the *input vector* is fed to the generator network.

2. The generator creates synthetic data samples.

3. Authentic data is passed to the discriminator along with the synthetic data.

4. The discriminator then identifies the data and classifies it as real or fake.

5. The model is trained and the loss backpropagated into both the discriminator and generator networks.

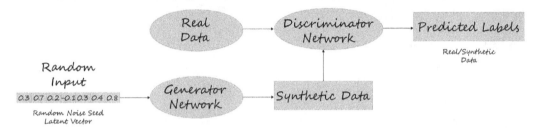

Figure 11.17: GAN model with input and output

To code an adversarial network, the following steps are necessary. Each of these is described in detail in the following sections:

1. Combine the generator and discriminator models in your GAN.

2. Generate real samples with class labels.

3. Create points in latent space to use as input for the generator.

4. Use the generator to create fake samples.

5. Evaluate the discriminator performance.

6. Train the generator and discriminator.

7. Create the latent space, generator, discriminator, and GAN, and train the GAN on the training data.

Now that you've explored the inner workings of the generator and discriminator, take a look at how you can combine the models to compete with one another.

COMBINING THE GENERATIVE AND DISCRIMINATIVE MODELS

The **define_your_gan()** function creates your combined model.

While creating the combined GAN model, freeze the weights of the discriminator model by specifying **discriminator.trainable = False**. This prevents the discriminator weights from getting updated while you update the generator weights.

Now, you can add both models with **model.add(generator)** and **model.add(discriminator)**.

Then, specify **binary_crossentropy** as the loss function and Adam as your optimizer while compiling your model:

```
def define_your_gan(generator, discriminator):
    discriminator.trainable = False
    model = Sequential()
    model.add(generator)
    model.add(discriminator)
    model.compile(loss='binary_crossentropy', optimizer='adam')
    return model
```

GENERATING REAL SAMPLES WITH CLASS LABELS

Now extract real samples from the dataset to inspect fake samples against them. You can use the **generate_real()** function defined previously. In the first line of the function, **rand(n) - 0.5**, create random numbers on **n** in the range of **-0.5** to **0.5**. Use **hstack** to stack your array. Now you can generate class labels with **y = ones((n, 1))**:

```
def generate_real(n):
    X1 = rand(n) - 0.5
    X2 = X1 * X1
    X1 = X1.reshape(n, 1)
    X2 = X2.reshape(n, 1)
    X = hstack((X1, X2))
    y = ones((n, 1))
    return X, y
```

CREATING LATENT POINTS FOR THE GENERATOR

Next, use the generator model to create fake samples. You need to generate the same number of points in the latent space with your **gen_latent_points()** function. These latent points will be passed to the generator to create samples. This function generates uniformly random samples from NumPy's **randn** function. The number will correspond to the latent dimension multiplied by the number of samples to generate. This array of random numbers will then be reshaped to match the expected input of the generator:

```
def gen_latent_points(latent_dim, n):
    x_input = randn(latent_dim * n)
    x_input = x_input.reshape(n, latent_dim)
    return x_input
```

USING THE GENERATOR TO GENERATE FAKE SAMPLES AND CLASS LABELS

The **gen_fake()** function generates fake samples with a class label of zero. This function generates the latent points using the function created in the previous step. Then, the generator will generate samples based on the latent points. Finally, the class label, **y**, is generated as an array of zeros representing the fact that this is synthetic data:

```
def gen_fake(generator, latent_dim, n):
    x_input = gen_latent_points(latent_dim, n)
    X = generator.predict(x_input)
    y = zeros((n, 1))
    return X, y
```

EVALUATING THE DISCRIMINATOR MODEL

The following **performance_summary()** function is used to plot both real and fake data points. The function generates real values and synthetic data and evaluates the performance of the discriminator via its accuracy in identifying the synthetic images. Then, it finally plots both the real and synthetic images for visual review:

```
def performance_summary(epoch, generator, \
                        discriminator, latent_dim, n=100):
    x_real, y_real = generate_real(n)
    _, acc_real = discriminator.evaluate\
                (x_real, y_real, verbose=0)
    x_fake, y_fake = gen_fake\
                (generator, latent_dim, n)
```

```
    _, acc_fake = discriminator.evaluate\
                  (x_fake, y_fake, verbose=0)
    print(epoch, acc_real, acc_fake)
    plt.scatter(x_real[:, 0], x_real[:, 1], color='green')
    plt.scatter(x_fake[:, 0], x_fake[:, 1], color='red')
    plt.show()
```

TRAINING THE GENERATOR AND DISCRIMINATOR

Now, train your model with the **train()** function. This function contains a **for** loop to iterate through the epochs. At each epoch, real data is sampled with a size equal to half the batch, and then synthetic data is generated. Then, the discriminator trains on the real, followed by the synthetic, data. Then, the GAN model is trained. When the epoch number is a multiple of the input argument, **n_eval**, a performance summary is generated:

```
def train(g_model, d_model, your_gan_model, \
          latent_dim, n_epochs=1000, n_batch=128, n_eval=100):
    half_batch = int(n_batch / 2)
    for i in range(n_epochs):
        x_real, y_real = generate_real(half_batch)
        x_fake, y_fake = gen_fake\
                         (g_model, latent_dim, half_batch)
        d_model.train_on_batch(x_real, y_real)
        d_model.train_on_batch(x_fake, y_fake)
        x_gan = gen_latent_points(latent_dim, n_batch)
        y_gan = ones((n_batch, 1))
        your_gan_model.train_on_batch(x_gan, y_gan)
        if (i+1) % n_eval == 0:
            performance_summary(i, g_model, d_model, latent_dim)
```

CREATING THE LATENT SPACE, GENERATOR, DISCRIMINATOR, GAN, AND TRAINING DATA

You can combine all the steps to build and train the model. Here, **latent_dim** is set to **5**, representing five latent dimensions:

```
latent_dim = 5
generator = define_gen(latent_dim)
discriminator = define_discrim()
your_gan_model = define_your_gan(generator, discriminator)
train(generator, discriminator, your_gan_model, latent_dim)
```

In this section, you learned about GANs, different components, the generator and discriminator, and how you combine them to create an adversarial network. You will now use these concepts to generate sequences with your own GAN.

EXERCISE 11.02: GENERATING SEQUENCES WITH GANS

In this exercise, you will use a GAN to create a model that generates a quadratic function ($y=x^2$) for values of x between -0.5 and 0.5. You will create a generator that will simulate the normal distribution and then square the values to simulate the quadratic function. You will also create a discriminator that will discriminate between a true quadratic function and the output from the generator. Next, you will combine them to create your GAN model. Finally, you will train your GAN model and evaluate your model, comparing the results from the generator against a true quadratic function.

Perform the following steps to complete this exercise:

1. Open a new Jupyter or Colab notebook and import the following libraries:

```
from keras.models import Sequential
from numpy import hstack, zeros, ones
from numpy.random import rand, randn
from keras.layers import Dense
import matplotlib.pyplot as plt
```

2. Define the generator model. Begin by creating your generator function with **define_gen**.

    Use Keras' **linear** activation function for the last layer of the generator network because the output vector should consist of continuous real values as a normal distribution does. The first element of the output vector has a range of **[-0.5,0.5]**. Since you will only consider values of x between these two values, the second element has a range of **[0.0,0.25]**:

```
def define_gen(latent_dim, n_outputs=2):
    model = Sequential()
    model.add(Dense(15, activation='relu', \
            kernel_initializer='he_uniform', \
            input_dim=latent_dim))
    model.add(Dense(n_outputs, activation='linear'))
    return model
```

3. Now, with **define_disc()**, define your discriminator. The discriminator network has a binary output that identifies whether the input is real or fake. For this reason, use sigmoid as the activation function and binary cross-entropy as your loss.

   You're creating a simple GAN, so use one hidden layer with **25** nodes. Use ReLU activation and **he_uniform** weight initialization. Your output layer will only need a single node for binary classification. Use Adam as your optimizer. The model will attempt to minimize your loss function:

```
def define_disc(n_inputs=2):
    model = Sequential()
    model.add(Dense(25, activation='relu', \
                    kernel_initializer='he_uniform', \
                    input_dim=n_inputs))
    model.add(Dense(1, activation='sigmoid'))
    model.compile(loss='binary_crossentropy', \
                  optimizer='adam', metrics=['accuracy'])
    return model
```

4. Now, add both models with **model.add(generator)** and **model.add(discriminator)**. Then, specify binary cross-entropy as your loss function and Adam as your optimizer, while compiling your model:

```
def define_your_gan(generator, discriminator):
    discriminator.trainable = False
    model = Sequential()
    model.add(generator)
    model.add(discriminator)
    model.compile(loss='binary_crossentropy', optimizer='adam')
    return model
```

5. Extract real samples from your dataset to inspect fake samples against them. Use the **generate_real()** function defined previously. **rand(n) - 0.5** creates random numbers on **n** in the range of **-0.5** to **0.5**. Use **hstack** to stack your array. Now, generate class labels with **y = ones((n, 1))**:

```
def generate_real(n):
    X1 = rand(n) - 0.5
    X2 = X1 * X1
    X1 = X1.reshape(n, 1)
    X2 = X2.reshape(n, 1)
    X = hstack((X1, X2))
    y = ones((n, 1))
    return X, y
```

6. Next, set the generator model to create fake samples. Generate the same number of points in the latent space with your **gen_latent_points()** function. Then, pass them to the generator and use them to create samples:

```
def gen_latent_points(latent_dim, n):
    x_input = randn(latent_dim * n)
    x_input = x_input.reshape(n, latent_dim)
    return x_input
```

7. Use the generator to generate fake samples with class labels:

```
def gen_fake(generator, latent_dim, n):
    x_input = gen_latent_points(latent_dim, n)
    X = generator.predict(x_input)
    y = zeros((n, 1))
    return X, y
```

8. Evaluate the discriminator model. The **performance_summary()** function will plot both real and fake data points:

```python
def performance_summary(epoch, generator, \
                        discriminator, latent_dim, n=100):
    x_real, y_real = generate_real(n)
    _, acc_real = discriminator.evaluate\
                (x_real, y_real, verbose=0)
    x_fake, y_fake = gen_fake\
                (generator, latent_dim, n)
    _, acc_fake = discriminator.evaluate\
                (x_fake, y_fake, verbose=0)
    print(epoch, acc_real, acc_fake)
    plt.scatter(x_real[:, 0], x_real[:, 1], color='green')
    plt.scatter(x_fake[:, 0], x_fake[:, 1], color='red')
    plt.show()
```

9. Now, train your model with the **train()** function:

```python
def train(g_model, d_model, your_gan_model, \
        latent_dim, n_epochs=1000, \
        n_batch=128, n_eval=100):
    half_batch = int(n_batch / 2)
    for i in range(n_epochs):
        x_real, y_real = generate_real(half_batch)
        x_fake, y_fake = gen_fake\
                        (g_model, latent_dim, half_batch)
        d_model.train_on_batch(x_real, y_real)
        d_model.train_on_batch(x_fake, y_fake)
        x_gan = gen_latent_points(latent_dim, n_batch)
        y_gan = ones((n_batch, 1))
        your_gan_model.train_on_batch(x_gan, y_gan)
        if (i+1) % n_eval == 0:
            performance_summary(i, g_model, d_model, latent_dim)
```

10. Create a parameter for the latent dimension and set it equal to **5**. Then, create a generator, discriminator, and GAN using the respective functions. Train the generator, discriminator, and GAN models using the **train** function:

```
latent_dim = 5
generator = define_gen(latent_dim)
discriminator = define_disc()
your_gan_model = define_your_gan(generator, discriminator)
train(generator, discriminator, your_gan_model, latent_dim)
```

You will get the following output:

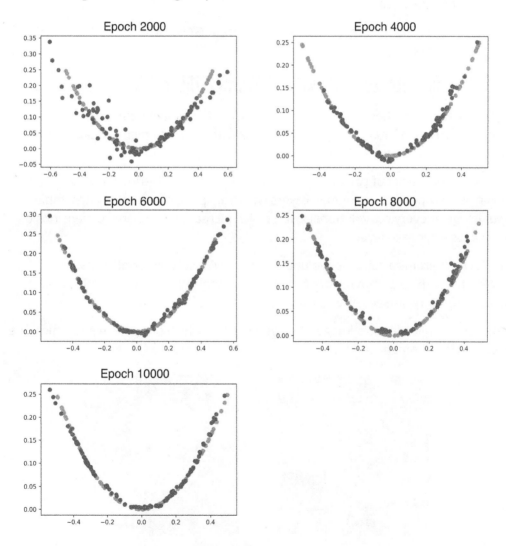

Figure 11.18: Distribution of real and fake data

The output shows the generator progressively improving by generating points that more closely resemble a quadratic function. In early epochs, the points generated by the generator, indicated by the blue dots, show little similarity to the true quadratic function, indicated by the red dots. However, by the final epoch, the points generated by the generator almost lie on top of the true points, demonstrating that the generator has almost captured the true underlying function – the quadratic.

In this exercise, you utilized the different components of a generative model to create data that fits a quadratic function. As you can see in *Figure 11.18*, by the final epoch, the fake data resembles the real data, showing that the generator can capture the quadratic function well.

Now it's time for the final section of the book, on DCGANs, where you'll be creating your own images.

DEEP CONVOLUTIONAL GENERATIVE ADVERSARIAL NETWORKS (DCGANS)

DCGANs use convolutional neural networks instead of simple neural networks for both the discriminator and the generator. They can generate higher-quality images and are commonly used for this purpose.

The generator is a set of convolutional layers with fractional stride convolutions, also known as transpose convolutions. Layers with transpose convolutions upsample the input image at every convolutional layer, which increases the spatial dimensions of the images after each layer.

The discriminator is a set of convolutional layers with stride convolutions, so it downsamples the input image at every convolutional layer, decreasing the spatial dimensions of the images after each layer.

Consider the following two images. Can you identify which one is fake and which one is real? Take a moment and look carefully at each of them.

Figure 11.19: Face example

You may be surprised to find out that neither of the images shown is of real people. These images were created using images of real people, but they are not of real people. They were created by two competing neural networks.

As you know, a GAN is composed of two different neural networks: the discriminator and the generator. What looks different right away is that each of these networks has different inputs and outputs. This is key to understanding how GANs can do what they do.

For the discriminator, the input is an image—a 3D tensor (height, width, color). The output is a single number that is used to make the classification. In *Figure 11.20*, you can see [0.95]. It implies there is a 95% chance that the tomato image is real.

For the generator, the input is a generated random seed vector of numbers. The output is an image.

The generator network learns to generate images similar to the ones in the dataset, while the discriminator learns to discriminate the original images from the generated ones. In this competitive fashion, they learn to generate realistic images like the ones in the training dataset.

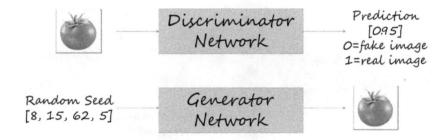

Figure 11.20: Discriminator and generator networks

Let's take a look at how the generator trains. One of the key points to take away from *Figure 11.20* is that the generator network has *weights static*, while the discriminator network shows *weights trained*. This is important because this enables you to differentiate how the GAN loss function changes from updates to the weights on the generator and discriminator independently.

Note that **X** (the random seed) is fed into the model to produce **y**. Your model outputs what you predict.

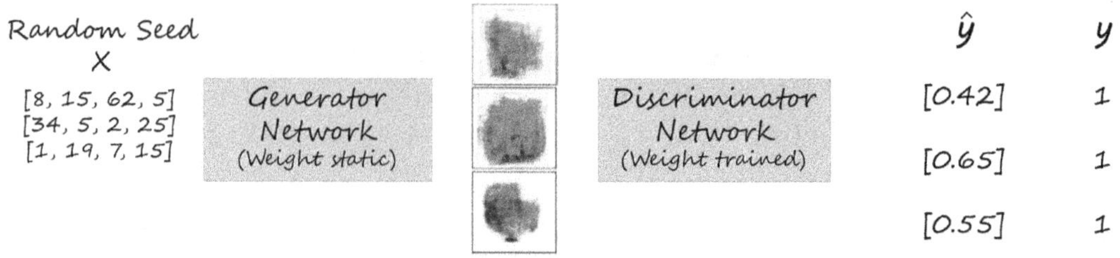

Figure 11.21: How the generator is trained

Another important point to keep in mind is that the generator trains without ever seeing any of the real data. The generator's only goal is to fool the discriminator.

Now, consider the training process of the discriminator network. The discriminator is trained on a training dataset consisting of an equal number of real and fake (generated) images. The real images are sampled randomly from the original dataset and are labeled as one. An equal number of fake images is generated using the generator network and are labeled as zero.

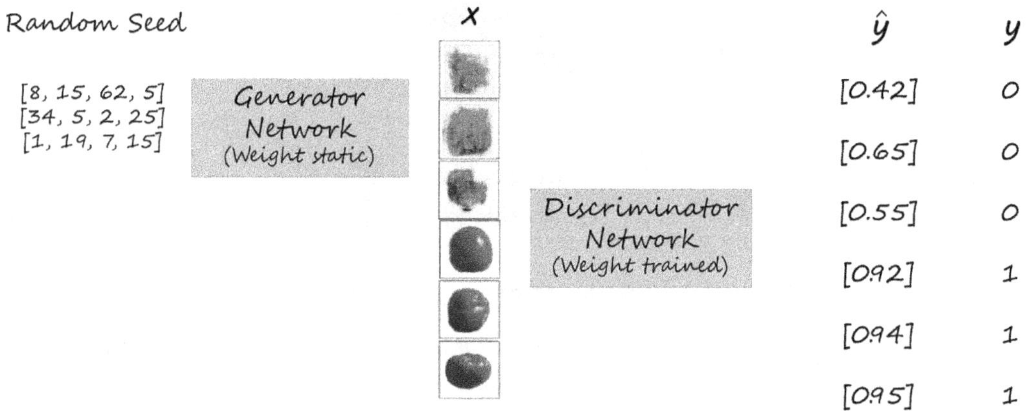

Figure 11.22: How the discriminator is trained

The core differences between the original "vanilla" GAN and DCGAN correspond to the differences in the architecture. Pooling layers of the vanilla GAN are replaced with transposed convolutions in the generator and stride convolutions in the discriminator of the DCGAN. The generator and discriminator of DCGANs both use batch normalization layers, except for the generator output layer and the discriminator input layer. Also, the fully connected hidden layers of DCGANs are removed. Finally, the activation functions in DCGANs are generally different to reflect the use of convolutional layers. In the generator, ReLU is used for all layers except for the output layer, where tanh is used, and for the discriminator, Leaky ReLU is used for all layers.

TRAINING A DCGAN

To start, you're going to set all the constants that will define your DCGAN.

The resolution of the images that you want to generate is specified by the **gen_res** parameter. The final resolution will be **32*gen_res** for the height and width of the image. You will use **gen_res = 3**, which results in an image resolution of **96x96**.

Image channels, **img_chan**, are simply how many numbers per pixel the image has. For color, you need a pixel value for each of the three color channels: **red, green, and blue** (**RGB**). So, your image channel should be set to **3**.

Your preview image rows and columns (**img_rows** and **img_cols**) will be how many images you want to display in a row and a column. For example, if you were to choose a preview image row of **4**, and a preview column value of **4**, you would get a total of 16 images displayed.

data_path is where your data is stored on your computer. This provides the path needed for the code to access and store data.

epoch is the number of passes when training the data.

Batch size, **num_batch**, is the number of training samples per iteration.

Buffer size, **num_buffer**, is the random shuffle that is used. You will simply set this to your dataset size.

Seed vector, **seed_vector**, is the size of the vector of seeds that will be used to generate images.

Consider the following sample to see how to initialize all the constants that define your DCGAN:

```
gen_res = 3
gen_square = 32 * gen_res
img_chan = 3
img_rows = 5
img_cols = 5
img_margin = 16
seed_vector = 200
data_path = '/content/drive/MyDrive/Datasets\
            '/apple-or-tomato/training_set/'
epochs = 1000
num_batch = 32
num_buffer = 1000
```

Now you can build the generator and the discriminator. Start by defining your generator function with **def create_generator**, using **seed_size** and **channels** as arguments:

```
def create_generator(seed_size, channels):
    model = Sequential()
```

Now, you will create the generated image that is going to come from an *input seed*; different seed numbers will generate different images and your seed size will determine how many different images are generated.

Next, add a dense layer with **4*4*256** as the dimensionality of your output space, and use the ReLU activation function. **input_dim** is an input shape, which you will have equal to **seed_size**.

Use the following code to add a layer that reshapes your inputs to match your output space of **4*4*256**:

```
model.add(Reshape((4,4,256)))
```

Your **UpSampling2D** layer is a simple layer that doubles the dimensions of input. It must be followed by a convolutional layer (**Conv2D**):

```
model.add(UpSampling2D())
```

Add your **Conv2D** layer with **256** as your input. You can choose **kernel_size=3** for your **3x3** convolution filter. With **padding="same"**, you can ensure that the layer's outputs will have the same spatial dimensions as its inputs:

```
model.add(Conv2D(256,kernel_size=3,padding="same"))
```

Use batch normalization to normalize your individual layers and help prevent gradient problems. Momentum can be anywhere in the range of **0.0** to **0.99**. Here, use **momentum=0.8**:

```
model.add(BatchNormalization(momentum=0.8))
```

On your final CNN layer, you will use the tanh activation function to ensure that your output images are in the range **−1** to **1**:

```
model.add(Conv2D(channels,kernel_size=3,padding="same"))
model.add(Activation("tanh"))
```

The complete code block should look like this:

```
def create_generator(seed_size, channels):
    model = Sequential()

    model.add(Dense(4*4*256,activation="relu", \
                    input_dim=seed_size))
    model.add(Reshape((4,4,256)))

    model.add(UpSampling2D())
    model.add(Conv2D(256,kernel_size=3,padding="same"))
    model.add(BatchNormalization(momentum=0.8))
    model.add(Activation("relu"))

    model.add(UpSampling2D())
    model.add(Conv2D(256,kernel_size=3,padding="same"))
    model.add(BatchNormalization(momentum=0.8))
    model.add(Activation("relu"))

    model.add(UpSampling2D())
    model.add(Conv2D(128,kernel_size=3,padding="same"))
```

```
      model.add(BatchNormalization(momentum=0.8))
      model.add(Activation("relu"))

      if gen_res>1:
        model.add(UpSampling2D(size=(gen_res,gen_res)))
        model.add(Conv2D(128,kernel_size=3,padding="same"))
        model.add(BatchNormalization(momentum=0.8))
        model.add(Activation("relu"))

      model.add(Conv2D(channels,kernel_size=3,padding="same"))
      model.add(Activation("tanh"))

      return model
```

Now you can define your discriminator:

```
def create_discriminator(image_shape):
    model = Sequential()
```

Here, use a **Conv2D** layer. You can choose **kernel_size=3** for your **3x3** convolution filter. With **strides=2**, you specify how many strides are for your "sliding window." Set **input_shape=image_shape** to ensure they match, and again, with **padding="same"**, you ensure that the layer's outputs will have the same spatial dimensions as its inputs. Add a LeakyReLU activation function after the **Conv2D** layer for all discriminator layers:

```
model.add(Conv2D(32, kernel_size=3, \
                 strides=2, input_shape=image_shape, \
                 padding="same"))
model.add(LeakyReLU(alpha=0.2))
```

The **Flatten** layer converts your data into a single feature vector for input into your last layer:

```
model.add(Flatten())
```

For your activation function, use sigmoid for binary classification output:

```
model.add(Dense(1, activation='sigmoid'))
```

The complete code block should look like this:

```
def create_discriminator(image_shape):
    model = Sequential()

    model.add(Conv2D(32, kernel_size=3, strides=2, \
                     input_shape=image_shape,
                     padding="same"))
    model.add(LeakyReLU(alpha=0.2))

    model.add(Dropout(0.25))
    model.add(Conv2D(64, kernel_size=3, strides=2, \
                     padding="same"))
    model.add(ZeroPadding2D(padding=((0,1),(0,1))))
    model.add(BatchNormalization(momentum=0.8))
    model.add(LeakyReLU(alpha=0.2))

    model.add(Dropout(0.25))
    model.add(Conv2D(128, kernel_size=3, strides=2, \
                     padding="same"))
    model.add(BatchNormalization(momentum=0.8))
    model.add(LeakyReLU(alpha=0.2))

    model.add(Dropout(0.25))
    model.add(Conv2D(256, kernel_size=3, strides=1, \
                     padding="same"))
    model.add(BatchNormalization(momentum=0.8))
    model.add(LeakyReLU(alpha=0.2))

    model.add(Dropout(0.25))
    model.add(Conv2D(512, kernel_size=3, \
                     strides=1, padding="same"))
    model.add(BatchNormalization(momentum=0.8))
    model.add(LeakyReLU(alpha=0.2))

    model.add(Dropout(0.25))
    model.add(Flatten())
    model.add(Dense(1, activation='sigmoid'))

    return model
```

Next, create your loss functions. Since the outputs of the discriminator and generator networks are different, you need to define two separate loss functions for them. Moreover, they need to be trained separately in independent passes through the networks.

You can use **tf.keras.losses.BinaryCrossentropy** for **cross_entropy**. This calculates the loss between true and predicted labels. Then, define the **discrim_loss** function from your **real_output** and **fake_output** parameters using **tf.ones** and **tf.zeros** to calculate **total_loss**:

```
cross_entropy = tf.keras.losses.BinaryCrossentropy()

def discrim_loss(real_output, fake_output):
    real_loss = cross_entropy(tf.ones_like(real_output), \
                                 real_output)
    fake_loss = cross_entropy(tf.zeros_like(fake_output), \
                                 fake_output)
    total_loss = real_loss + fake_loss
    return total_loss

def gen_loss(fake_output):
    return cross_entropy(tf.ones_like(fake_output), \
                             fake_output)
```

The Adam optimizer is used for the generator and discriminator, with the same learning rate and momentum:

```
gen_optimizer = tf.keras.optimizers.Adam(1.5e-4,0.5)
disc_optimizer = tf.keras.optimizers.Adam(1.5e-4,0.5)
```

Here, you have your individual training step. It's very important that you only modify one network's weights at a time. With **tf.GradientTape()**, you can train the discriminator and generator at the same time, but separately from one another. This is how TensorFlow does automatic differentiation. It calculates the derivatives. You'll see that it creates two "tapes" – **gen_tape** and **disc_tape**.

Then, create **real_output** and **fake_output** for the discriminator. Use this for the generator loss (**g_loss**). Now, you can calculate the discriminator loss (**d_loss**), calculate the gradients of both the generator and discriminator with **gradients_of_generator** and **gradients_of_discriminator**, and apply them:

```python
@tf.function
def train_step(images):
    seed = tf.random.normal([num_batch, seed_vector])

    with tf.GradientTape() as gen_tape, \
            tf.GradientTape() as disc_tape:
        gen_imgs = generator(seed, training=True)

        real_output = discriminator(images, training=True)
        fake_output = discriminator(gen_imgs, training=True)

        g_loss = gen_loss(fake_output)
        d_loss = discrim_loss(real_output, fake_output)


    gradients_of_generator = gen_tape.gradient(\
        g_loss, generator.trainable_variables)
    gradients_of_discriminator = disc_tape.gradient(\
        d_loss, discriminator.trainable_variables)

    gen_optimizer.apply_gradients(zip(
        gradients_of_generator, generator.trainable_variables))
    disc_optimizer.apply_gradients(zip(
        gradients_of_discriminator,
        discriminator.trainable_variables))
    return g_loss,d_loss
```

Next, create a number of fixed seeds with **fixed_seeds**, a seed for each image displayed, and for each seed vector. This is done so you can track the same images, observing the changes over time. With **for epoch in range**, you are tracking your time. Loop through each batch with **for image_batch in dataset**. Now, continue to track your loss for both the generator and discriminator with **generator_loss** and **discriminator_loss**. Now you have a nice display of all this information as it trains:

```
def train(dataset, epochs):
    fixed_seed = np.random.normal\
                (0, 1, (img_rows * img_cols, seed_vector))
    start = time.time()

    for epoch in range(epochs):
        epoch_start = time.time()

        g_loss_list = []
        d_loss_list = []

        for image_batch in dataset:
            t = train_step(image_batch)
            g_loss_list.append(t[0])
            d_loss_list.append(t[1])

        generator_loss = sum(g_loss_list) / len(g_loss_list)
        discriminator_loss = sum(d_loss_list) / len(d_loss_list)

        epoch_elapsed = time.time()-epoch_start
        print (f'Epoch {epoch+1}, gen loss={generator_loss}', \
                f'disc loss={discriminator_loss},'\
                f' {time_string(epoch_elapsed)}')
        save_images(epoch,fixed_seed)

    elapsed = time.time()-start
    print (f'Training time: {time_string(elapsed)}')
```

In this last section, you took an additional step in using generative networks. You learned how to train a DCGAN and how to utilize the generator and discriminator together to create your very own images.

In the next exercise, you will implement what you have learned so far in this section.

EXERCISE 11.03: GENERATING IMAGES WITH DCGAN

In this exercise, you will generate your own images from scratch using a DCGAN. You will build your DCGAN with a generator and discriminator that both have convolutional layers. Then, you will train your DCGAN on images of a tomato, and throughout the training process, you will output generated images from the generator to track the performance of the generator.

> **NOTE**
>
> You can find **tomato-or-apple** dataset here:
> https://packt.link/6Z8vW.

For this exercise, it is recommended that you use Google Colab:

1. Load Google Colab and Google Drive:

```
try:
    from google.colab import drive
    drive.mount('/content/drive', force_remount=True)
    COLAB = True
    print("Note: using Google CoLab")
    %tensorflow_version 2.x
except:
    print("Note: not using Google CoLab")
    COLAB = False
```

Your output should look something like this:

```
Mounted at /content/drive
Note: using Google Colab
```

2. Import the relevant libraries:

```
import tensorflow as tf
from tensorflow.keras.layers
import Input, Reshape, Dropout, Dense
from tensorflow.keras.layers
import Flatten, BatchNormalization
from tensorflow.keras.layers
import Activation, ZeroPadding2D
from tensorflow.keras.layers import LeakyReLU
```

```
from tensorflow.keras.layers import UpSampling2D, Conv2D
from tensorflow.keras.models
import Sequential, Model, load_model
from tensorflow.keras.optimizers import Adam
import zipfile
import numpy as np
from PIL import Image
from tqdm import tqdm
import os
import time
import matplotlib.pyplot as plt
from skimage.io import imread
```

3. Format a time string to track your time usage:

```
def time_string(sec_elapsed):
    hour = int(sec_elapsed / (60 * 60))
    minute = int((sec_elapsed % (60 * 60)) / 60)
    second = sec_elapsed % 60
    return "{}:{:>02}:{:>05.2f}".format(hour, minute, second)
```

4. Set the generation resolution to **3**. Also, set **img_rows** and **img_cols** to **5** and **img_margin** to **16** so that your preview images will be a **5x5** array (25 images) with a 16-pixel margin.

Set **seed_vector** equal to **200**. Set **data_path** to where you stored your image dataset. As you can see, you are using Google Drive here. If you don't know your data path, you can simply locate where your files are, right-click, and select **Copy Path**. Set your epochs to **1000**.

Finally, print the parameters:

```
gen_res = 3
gen_square = 32 * gen_res
img_chan = 3
img_rows = 5
img_cols = 5
img_margin = 16
seed_vector = 200
data_path = '/content/drive/MyDrive/Datasets'\
            '/apple-or-tomato/training_set/'
epochs = 5000
num_batch = 32
```

```
num_buffer = 60000


print(f"Will generate a resolution of {gen_res}.")
print(f"Will generate {gen_square}px square images.")
print(f"Will generate {img_chan} image channels.")
print(f"Will generate {img_rows} preview rows.")
print(f"Will generate {img_cols} preview columns.")
print(f"Our preview margin equals {img_margin}.")
print(f"Our data path is: {data_path}.")
print(f"Our number of epochs are: {epochs}.")
print(f"Will generate a batch size of {num_batch}.")
print(f"Will generate a buffer size of {num_buffer}.")
```

Your output should look something like this:

```
Will generate a resolution of 3.
Will generate 96px square images.
Will generate 3 image channels.
Will generate 5 preview rows.
Will generate 5 preview columns.
Our preview margin equals 16.
Our data path is: apple-or-tomato/training_set/.
Our number of epochs are: 1000.
Will generate a batch size of 32.
Will generate a buffer size of 60000.
```

Figure 11.23: Output showing parameters

5. Load and preprocess the images. Here, you will save a NumPy preprocessed file. Load the previous training NumPy file. The name of the binary file of the images has the dimensions of the images encoded in it:

```
training_binary_path = os.path.join(data_path,\
        f'training_data_{gen_square}_{gen_square}.npy')


print(f"Looking for file: {training_binary_path}")


if not os.path.isfile(training_binary_path):
    start = time.time()
    print("Loading training images...")


    train_data = []
    images_path = os.path.join(data_path, 'tomato')
```

```
        for filename in tqdm(os.listdir(images_path)):
            path = os.path.join(images_path,filename)
            images = Image.open(path).resize((gen_square,
                gen_square),Image.ANTIALIAS)
            train_data.append(np.asarray(images))
        train_data = np.reshape(train_data, (-1,gen_square,
                gen_square,img_chan))
        train_data = train_data.astype(np.float32)
        train_data = train_data / 127.5 - 1.

        print("Saving training images...")
        np.save(training_binary_path,train_data)
        elapsed = time.time()-start
        print (f'Image preprocessing time: {time_string(elapsed)}')
    else:
        print("Loading the training data...")
        train_data = np.load(training_binary_path)
```

6. Batch and shuffle the data. Use the **tensorflow.data.Dataset** object library to use its functions to shuffle the dataset and create batches:

```
train_dataset = tf.data.Dataset.from_tensor_slices(train_data) \
                    .shuffle(num_buffer).batch(num_batch)
```

7. Build the generator:

```
def create_generator(seed_size, channels):
    model = Sequential()

    model.add(Dense(4*4*256,activation="relu", \
                    input_dim=seed_size))
    model.add(Reshape((4,4,256)))

    model.add(UpSampling2D())
    model.add(Conv2D(256,kernel_size=3,padding="same"))
    model.add(BatchNormalization(momentum=0.8))
    model.add(Activation("relu"))

    model.add(UpSampling2D())
    model.add(Conv2D(256,kernel_size=3,padding="same"))
    model.add(BatchNormalization(momentum=0.8))
    model.add(Activation("relu"))
```

```
model.add(UpSampling2D())
model.add(Conv2D(128,kernel_size=3,padding="same"))
model.add(BatchNormalization(momentum=0.8))
model.add(Activation("relu"))

if gen_res>1:
    model.add(UpSampling2D(size=(gen_res,gen_res)))
    model.add(Conv2D(128,kernel_size=3,padding="same"))
    model.add(BatchNormalization(momentum=0.8))
    model.add(Activation("relu"))

model.add(Conv2D(channels,kernel_size=3,padding="same"))
model.add(Activation("tanh"))

return model
```

8. Build the discriminator:

```
def create_discriminator(image_shape):
    model = Sequential()

    model.add(Conv2D(32, kernel_size=3, strides=2, \
                    input_shape=image_shape,
                    padding="same"))
    model.add(LeakyReLU(alpha=0.2))

    model.add(Dropout(0.25))
    model.add(Conv2D(64, kernel_size=3, \
                    strides=2, padding="same"))
    model.add(ZeroPadding2D(padding=((0,1),(0,1))))
    model.add(BatchNormalization(momentum=0.8))
    model.add(LeakyReLU(alpha=0.2))

    model.add(Dropout(0.25))
    model.add(Conv2D(128, kernel_size=3, strides=2, \
                    padding="same"))
    model.add(BatchNormalization(momentum=0.8))
    model.add(LeakyReLU(alpha=0.2))

    model.add(Dropout(0.25))
```

```
model.add(Conv2D(256, kernel_size=3, strides=1, \
                 padding="same"))
model.add(BatchNormalization(momentum=0.8))
model.add(LeakyReLU(alpha=0.2))

model.add(Dropout(0.25))
model.add(Conv2D(512, kernel_size=3, strides=1, \
                 padding="same"))
model.add(BatchNormalization(momentum=0.8))
model.add(LeakyReLU(alpha=0.2))

model.add(Dropout(0.25))
model.add(Flatten())
model.add(Dense(1, activation='sigmoid'))

return model
```

9. During the training process, display generated images to get some insight into
    the progress that's been made. Save the images. At regular intervals of 100
    epochs, save a grid of images to evaluate the progress:

```
def save_images(cnt,noise):
    img_array = np.full((
      img_margin + (img_rows * (gen_square+img_margin)),
      img_margin + (img_cols * (gen_square+img_margin)), 3),
      255, dtype=np.uint8)

    gen_imgs = generator.predict(noise)

    gen_imgs = 0.5 * gen_imgs + 0.5

    img_count = 0
    for row in range(img_rows):
    for col in range(img_cols):
        r = row * (gen_square+16) + img_margin
        c = col * (gen_square+16) + img_margin
        img_array[r:r+gen_square,c:c+gen_square] \
            = gen_imgs[img_count] * 255
        img_count += 1
```

```
output_path = os.path.join(data_path, 'output')
if not os.path.exists(output_path):
os.makedirs(output_path)

filename = os.path.join(output_path, f"train-{cnt}.png")
im = Image.fromarray(img_array)
im.save(filename)
```

10. Now, create a generator that generates noise:

```
generator = create_generator(seed_vector, img_chan)

noise = tf.random.normal([1, seed_vector])
gen_img = generator(noise, training=False)

plt.imshow(gen_img[0, :, :, 0])
```

Your output should look something like this:

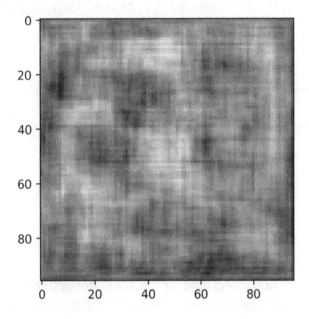

Figure 11.24: Output showing noise

11. View one of the images generated by typing in the following commands:

```
img_shape = (gen_square,gen_square,img_chan)

discriminator = create_discriminator(img_shape)
decision = discriminator(gen_img)
print(decision)
```

Your output should look something like this:

```
tf.Tensor([[0.4994658]], shape=(1,1), dtype=float32)
```

12. Create your loss functions. Since the outputs of the discriminator and generator networks are different, you need to define two separate loss functions for them. Moreover, they need to be trained separately in independent passes through the networks. Use **tf.keras.losses.BinaryCrossentropy** for **cross_entropy**. This calculates the loss between true and predicted labels. Then, define the **discrim_loss** function from **real_output** and **fake_output** using **tf.ones** and **tf.zeros** to calculate **total_loss**:

```
cross_entropy = tf.keras.losses.BinaryCrossentropy()

def discrim_loss(real_output, fake_output):
    real_loss = cross_entropy(tf.ones_like(real_output), \
                        real_output)
    fake_loss = cross_entropy(tf.zeros_like(fake_output), \
                        fake_output)
    total_loss = real_loss + fake_loss
    return total_loss

def gen_loss(fake_output):
    return cross_entropy(tf.ones_like(fake_output), \
                    fake_output)
```

13. Create two Adam optimizers (one for the generator and one for the discriminator), using the same learning rate and momentum for each:

```
gen_optimizer = tf.keras.optimizers.Adam(1.5e-4,0.5)
disc_optimizer = tf.keras.optimizers.Adam(1.5e-4,0.5)
```

14. Create a function to implement an individual training step. With **tf.GradientTape()**, train the discriminator and generator at the same time, but separately from one another.

Then, create **real_output** and **fake_output** for the discriminator. Use this for the generator loss (**g_loss**). Then, calculate the discriminator loss (**d_loss**) and calculate the gradients of both the generator and discriminator with **gradients_of_generator** and **gradients_of_discriminator**, and apply them:

```
@tf.function
def train_step(images):
    seed = tf.random.normal([num_batch, seed_vector])

    with tf.GradientTape() as gen_tape, \
        tf.GradientTape() as disc_tape:
    gen_imgs = generator(seed, training=True)

        real_output = discriminator(images, training=True)
        fake_output = discriminator(gen_imgs, training=True)

        g_loss = gen_loss(fake_output)
        d_loss = discrim_loss(real_output, fake_output)


    gradients_of_generator = gen_tape.gradient(\
        g_loss, generator.trainable_variables)
    gradients_of_discriminator = disc_tape.gradient(\
        d_loss, discriminator.trainable_variables)

    gen_optimizer.apply_gradients(zip(
        gradients_of_generator, generator.trainable_variables))
    disc_optimizer.apply_gradients(zip(
        gradients_of_discriminator,
        discriminator.trainable_variables))
    return g_loss,d_loss
```

15. Create an array number of fixed seeds with **fixed_seeds** equal to the number of images displayed along one dimension and the seed vector along the other dimension so that you can track the same images. This allows you to see how individual seeds evolve over time. Loop through each batch with **for image_batch in dataset**. Continue to track your loss for both the generator and discriminator with **generator_loss** and **discriminator_loss**. You get a nice display of all this information as it trains:

```
def train(dataset, epochs):
    fixed_seed = np.random.normal(0, 1, (img_rows * img_cols,
                                          seed_vector))

    start = time.time()

    for epoch in range(epochs):
    epoch_start = time.time()

    g_loss_list = []
    d_loss_list = []

    for image_batch in dataset:
        t = train_step(image_batch)
        g_loss_list.append(t[0])
        d_loss_list.append(t[1])

    generator_loss = sum(g_loss_list) / len(g_loss_list)
    discriminator_loss = sum(d_loss_list) / len(d_loss_list)

    epoch_elapsed = time.time()-epoch_start
    print (f'Epoch {epoch+1}, gen loss={generator_loss}', \
            f'disc loss={discriminator_loss},'\
            f' {time_string(epoch_elapsed)}')
    save_images(epoch,fixed_seed)

    elapsed = time.time()-start
    print (f'Training time: {time_string(elapsed)}')
```

16. Train on your training dataset:

```
train(train_dataset, epochs)
```

Your output should look something like this:

```
Epoch 1, gen loss=2.6256332397460938,disc loss=0.44953447580337524, 0:01:11.57
Epoch 2, gen loss=3.0611958503723145,disc loss=0.5719443559646606, 0:01:02.51
Epoch 3, gen loss=3.2427496910095215,disc loss=0.6098432540893555, 0:00:52.03
Epoch 4, gen loss=2.688871145248413,disc loss=1.0773324966430664, 0:00:55.30
Epoch 5, gen loss=2.158496618270874,disc loss=1.3300279378890991, 0:01:02.03
Epoch 6, gen loss=2.2589030265808105,disc loss=1.1785333156585693, 0:00:52.75
Epoch 7, gen loss=2.331599712371826,disc loss=1.199823021888733, 0:00:59.40
Epoch 8, gen loss=2.2243056297302246,disc loss=1.176113486289978, 0:00:56.94
Epoch 9, gen loss=2.140230178833008,disc loss=1.3953707218170166, 0:00:57.23
Epoch 10, gen loss=2.0387909412384033,disc loss=1.342348575592041, 0:00:57.40
Epoch 11, gen loss=2.004645824432373,disc loss=1.1653015613555908, 0:00:59.84
Epoch 12, gen loss=2.2482171058654785,disc loss=1.3021776676177979, 0:01:00.65
Epoch 13, gen loss=2.155640125274658,disc loss=1.3053537607192993, 0:00:56.82
Epoch 14, gen loss=2.0571579933166504,disc loss=1.4974734783172607, 0:00:56.89
Epoch 15, gen loss=1.9703991413116455,disc loss=1.1519067287445068, 0:00:55.92
Epoch 16, gen loss=2.1521060466766357,disc loss=1.2173162698745728, 0:01:02.76
Epoch 17, gen loss=2.1796844005584717,disc loss=0.9764562845230103, 0:00:58.10
Epoch 18, gen loss=2.1656653881073,disc loss=1.0634845495224, 0:00:58.65
Epoch 19, gen loss=2.1950347423553467,disc loss=1.0944265127182007, 0:01:01.07
Epoch 20, gen loss=2.051949977874756,disc loss=0.9224456548690796, 0:00:57.41
```

Figure 11.25: Training output

17. Take a closer look at the generated images, **train-0**, **train-100**, **train-250**, **train-500**, and **train-999**. These images were automatically saved during the training process, as specified in the **train** function:

```
a = imread('/content/drive/MyDrive/Datasets'\
            '/apple-or-tomato/training_set/output/train-0.png')
plt.imshow(a)
```

You will get output like the following:

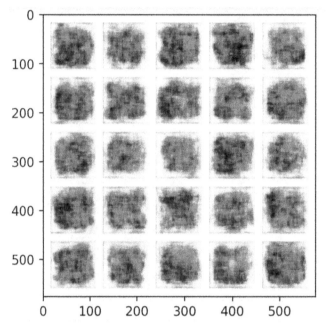

Figure 11.26: Output images after first epoch completed

Now, run the following commands:

```
a = imread('/content/drive/MyDrive/Datasets'\
           '/apple-or-tomato/training_set/output/train-100.png')
plt.imshow(a)
```

You will get output like the following:

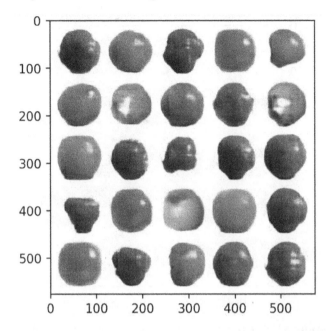

Figure 11.27: Output images after 101st epoch completed

Also, run the following commands:

```
a = imread('/content/drive/MyDrive/Datasets'\
          '/apple-or-tomato/training_set/output/train-500.png')
plt.imshow(a)
```

You will get output like the following:

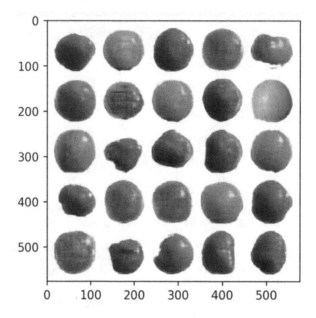

Figure 11.28: Output images after 501st epoch completed

Now, run the following commands:

```
a = imread('/content/drive/MyDrive/Datasets'\
           '/apple-or-tomato/training_set/output/train-999.png')
plt.imshow(a)
```

You will get output like the following:

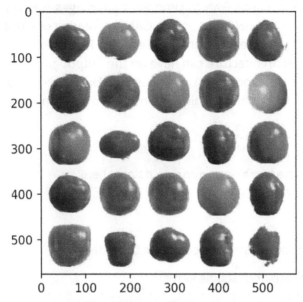

Figure 11.29: Output images after 1,000th epoch completed

The output shows that after 1,000 epochs, the images of the synthetic tomatoes generated by the generator look very similar to real tomatoes and the images improve during the training process.

In this exercise, you created your own images with a DCGAN. As you can see from *Figure 11.29*, the results are impressive. While some of the images are easy to determine as fake, others look very real.

In the next section, you will complete a final activity to put all that you've learned in this chapter to work and generate your own images with a GAN.

ACTIVITY 11.01: GENERATING IMAGES USING GANS

In this activity, you will build a GAN to generate new images. You will then compare the results between a DCGAN and a vanilla GAN by creating one of each and training them on the same dataset for the same 500 epochs. This activity will demonstrate the difference that model architecture can have on the output and show why having an appropriate model is so important. You will use the **banana-or-orange** dataset. You'll only be using the banana training set images to train and generate new images.

> **NOTE**
>
> You can find **banana-or-orange** dataset here:
> https://packt.link/z6TCy.

Perform the following steps to complete the activity:

1. Load Google Colab and Google Drive.

   Import the relevant libraries, including **tensorflow, numpy, zipfile, tqdm, zipfile, skimage, time**, and **os**.

2. Create a function to format a time string to track your time usage.

3. Set the generation resolution to **3**. Also, set **img_rows** and **img_cols** to **5** and **img_margin** to **16** so that your preview images will be a **5x5** array (25 images) with a 16-pixel margin. Set **seed_vector** equal to **200, data_path** to where you stored your image dataset, and **epochs** to **500**. Finally, print the parameters.

4. If a NumPy preprocessed file exists from prior execution, then load it into memory; otherwise, preprocess the data and save the image binary.

5. Batch and shuffle the data. Use the **tensorflow.data.Dataset** object library to use its functions to shuffle the dataset and create batches.

6. Build the generator for the DCGAN.

7. Build the discriminator for the DCGAN.

8. Build the generator for the vanilla GAN.

9. Build the discriminator for the vanilla GAN.

10. Create a function to generate and save images that can be used to view progress during the model's training.

11. Next, initialize the generator for the DCGAN and view the output.

12. Initialize the generator for the vanilla GAN and view the output.

13. Print the decision of the DCGAN discriminator evaluated on the seed image.

14. Print the decision of the vanilla GAN discriminator evaluated on the seed image.

15. Create your loss functions. Since the output of both the discriminator and generator networks is different, you can define two separate loss functions for them. Moreover, they need to be trained separately in independent passes through the networks. Both GANs can utilize the same loss functions for their discriminators and generators. You can use **tf.keras.losses.BinaryCrossentropy** for **cross_entropy**. This calculates the loss between true and predicted labels. Then, define the **discrim_loss** function from **real_output** and **fake_output** using **tf.ones** and **tf.zeros** to calculate **total_loss**.

16. Create two Adam optimizers, one for the generator and one for the discriminator. Use the same learning rate and momentum for each.

17. Create **real_output** and **fake_output** for the discriminator. Use this for the generator loss (**g_loss**). Then, calculate the discriminator loss (**d_loss**) and the gradients of both the generator and discriminator with **gradients_of_generator** and **gradients_of_discriminator** and apply them. Encapsulate these steps within a function, passing in the generator, discriminator, and images and returning the generator loss (**g_loss**) and discriminator loss (**d_loss**).

18. Next, create a number of fixed seeds with **fixed_seeds** equal to the number of images to display so that you can track the same images. This allows you to see how individual seeds evolve over time, tracking your time with **for epoch in range**. Now, loop through each batch with **for image_batch in dataset**. Continue to track your loss for both the generator and discriminator with **generator_loss** and **discriminator_loss**. Now, you have a nice display of all this information as it trains.

19. Train the DCGAN model on your training dataset.

20. Train the vanilla model on your training dataset.

21. View your images generated by the DCGAN model after the 100th epoch.

22. View your images generated by the DCGAN model after the 500th epoch.

23. View your images generated by the vanilla GAN model after the 100th epoch.

24. View your images generated by the vanilla GAN model after the 500th epoch.

> **NOTE**
>
> The solution to this activity can be found on page 541.

SUMMARY

In this chapter, you learned about a very exciting class of machine learning models called generative models. You discovered the amazing potential of this new and continually developing field in machine learning by using a generative LSTM on a language modeling challenge to generate textual output.

Then, you learned about generative adversarial models. You implemented a GAN to generate data for a normal distribution of points. You also went even further into deep convolutional neural networks (DCGANS), discovering how to use one of the most powerful applications of GANs while creating new images of tomatoes and bananas that exhibited human-recognizable characteristics of the fruits on which they were trained.

We hope you enjoyed the final chapter of *The TensorFlow Workshop* and the book as a whole.

Let's take a look back at the amazing journey that you have completed. First, you started by learning the basics of TensorFlow and how to perform operations on the building blocks of ANNs—tensors. Then, you learned how to load and preprocess a variety of data types in TensorFlow, including tabular data, images, audio files, and text.

Next, you learned about a variety of resources that can be used in conjunction with TensorFlow to aid in your development, including TensorBoard for visualizing important components of your model, TensorFlow Hub for accessing pre-trained models, and Google Colab for building and training models in a managed environment. Then, you dived into building sequential models to solve regression and classification.

To improve model performance, you then learned about regularization and hyperparameter tuning, which are used to ensure that your models perform well not only on the data they are trained upon, but also on new, unseen data. From there, you explored convolutional neural networks, which are an excellent choice when working with image data. After that, you learned in-depth how to utilize pre-trained networks to solve your own problems and fine-tune them to your own data. Then, you learned how to build and train RNNs, which are best used when working with sequential data, such as stock prices or even natural language. In the later part of the book, you explored more advanced TensorFlow capabilities using the Functional API and how to develop anything you might need in TensorFlow, before finally learning how to use TensorFlow for more creative endeavors via generative models.

With this book, you have not only taken your first steps in TensorFlow, but also now learned how to create models and provide solutions to complex problems. It's been an exciting journey from beginning to end, and we wish you luck in your continuing progress.

APPENDIX

CHAPTER 1: INTRODUCTION TO MACHINE LEARNING WITH TENSORFLOW

ACTIVITY 1.01: PERFORMING TENSOR ADDITION IN TENSORFLOW

Solution:

1. Import the TensorFlow library:

```
import tensorflow as tf
```

2. Create two tensors with a rank **0** using TensorFlow's **Variable** class:

```
var1 = tf.Variable(2706, tf.int32)
var2 = tf.Variable(2386, tf.int32)
```

3. Create a new variable to add the two scalars created and print the result:

```
var_sum = var1 + var2
var_sum.numpy()
```

This will result in the following output:

```
5092
```

This output shows the total revenue for **Product A** at **Location X**.

4. Create two tensors, a scalar of rank **0** and a vector of rank **1**, using TensorFlow's **Variable** class:

```
scalar1 = tf.Variable(95, tf.int32)
vector1 = tf.Variable([2706, 2799, 5102], \
                      tf.int32)
```

5. Create a new variable as the sum of the scalar and vector created and print the result:

```
vector_scalar_sum = scalar1 + vector1
vector_scalar_sum.numpy()
```

This will result in the following output:

```
array([2801, 2894, 5197])
```

The result is the new sales goal for **Salesperson 1** at **Location X**.

6. Now create three tensors with a rank of 2, representing the revenue for each product, salesperson, and location, using TensorFlow's **Variable** class:

```
matrix1 = tf.Variable([[2706, 2799, 5102], \
                        [2386, 4089, 5932]], tf.int32)
matrix2 = tf.Variable([[5901, 1208, 645], \
                        [6235, 1098, 948]], tf.int32)
matrix3 = tf.Variable([[3908, 2339, 5520], \
                        [4544, 1978, 4729]], tf.int32)
```

7. Create a new variable as the sum of the three tensors created and print the result:

```
matrix_sum = matrix1 + matrix2 + matrix3
matrix_sum.numpy()
```

This will result in the following output:

$$array([[12515, \quad 6346, \ 11267],$$
$$[13165, \quad 7165, \ 11609]])$$

Figure 1.42: The output of the matrix summation as a NumPy variable

The result represents the total revenue for each product at each location.

In this activity, you performed addition on tensors with ranks **0**, **1**, and **2**, and showed that scalars (tensors of rank 0) can be added to tensors of other ranks, known as scalar addition.

ACTIVITY 1.02: PERFORMING TENSOR RESHAPING AND TRANSPOSITION IN TENSORFLOW

Solution:

1. Import the TensorFlow library:

```
import tensorflow as tf
```

2. Create a one-dimensional array with 24 elements using TensorFlow's **Variable** class. Verify the shape of the matrix:

```
array1 = tf.Variable([*range(24)])
array1.shape.as_list()
```

This will result in the following output:

```
[24]
```

3. Reshape the matrix so that it has 12 rows and 2 columns using TensorFlow's **reshape** function. Verify the shape of the new matrix:

```
reshape1 = tf.reshape(array1, shape=[12, 2])
reshape1.shape.as_list()
```

This will result in the following output:

```
[12, 2]
```

4. Reshape the matrix so that it has a shape of **3x4x2** using TensorFlow's **reshape** function. Verify the shape of the new matrix:

```
reshape2 = tf.reshape(array1, shape=[3, 4, 2])
reshape2.shape.as_list()
```

This will result in the following output:

```
[3, 4, 2]
```

5. Verify that the rank of this new tensor is of rank **3** by using TensorFlow's **rank** function:

```
tf.rank(reshape2).numpy()
```

This will result in the following output:

```
3
```

6. Transpose the tensor created in *step 3*. Verify the shape of the new tensor:

```
transpose1 = tf.transpose(reshape1)
transpose1.shape.as_list()
```

This will result in the following output:

```
[2, 12]
```

In this activity, you have practiced performing tensor reshaping and transposition on tensors of various ranks and learned how to change the rank of a tensor by reshaping it. You simulated the grouping of 24 school children into class projects of varying sizes using TensorFlow's **reshape** and **transpose** functions.

ACTIVITY 1.03: APPLYING ACTIVATION FUNCTIONS

Solution:

1. Import the TensorFlow library:

```
import tensorflow as tf
```

2. Create a **3x4** tensor as an input in which the rows represent the sales from various sales representatives, the columns represent various vehicles available at the dealership, and values represent the average percentage difference from the MSRP. The values can be positive or negative depending on whether the salesperson was able to sell for more or less than the MSRP:

```
input1 = tf.Variable([[-0.013, 0.024, 0.06, 0.022], \
                      [0.001, -0.047, 0.039, 0.016], \
                      [0.018, 0.030, -0.021, -0.028]], \
                      tf.float32)
```

3. Create a **4x1 weights** tensor with a shape of **4x1** representing the MSRP of the cars:

```
weights = tf.Variable([[19995.95], [24995.50], \
                       [36745.50], [29995.95]], \
                       tf.float32)
```

4. Create a bias tensor of size **3x1** representing the fixed costs associated with each salesperson:

```
bias = tf.Variable([[-2500.0],[-2500.0],[-2500.0]], \
                    tf.float32)
```

5. Matrix multiply the input by the weight to show the average deviation from the MSRP on all cars and add the bias to subtract the fixed costs of the salesperson:

```
output = tf.matmul(input1,weights) + bias
output
```

The following is the output:

```
<tf.Tensor: shape=(3, 1), dtype=float32, numpy=
array([[  704.58545],
       [-1741.7827 ],
       [-3001.75   ]], dtype=float32)>
```

Figure 1.43: The output of the matrix multiplication

6. Apply a ReLU activation function to highlight the net-positive salespeople:

```
output = tf.keras.activations.relu(output)
output
```

This will result in the following output:

```
<tf.Tensor: shape=(3, 1), dtype=float32, numpy=
array([[704.58545],
       [  0.     ],
       [  0.     ]], dtype=float32)>
```

Figure 1.44: The output after applying the activation function

This result shows the result of salespeople that had net-positive sales; those with net-negative sales are zeroed.

In this activity, you performed tensor multiplication on tensors of various sizes, tensor addition, and also applied an activation function. You began by defining the tensors, followed by matrix multiplying two of them, then adding a bias tensor, and finally applying an activation function to the result.

CHAPTER 2: LOADING AND PROCESSING DATA

ACTIVITY 2.01: LOADING TABULAR DATA AND RESCALING NUMERICAL FIELDS WITH A MINMAX SCALER

Solution:

1. Open a new Jupyter notebook to implement this activity. Save the file as **Activity2-01.ipnyb**.

2. In a new Jupyter Notebook cell, import the pandas library, as follows:

```
import pandas as pd
```

3. Create a new pandas DataFrame named **df** and read the **Bias_correction_ucl.csv** file into it. Examine whether your data is properly loaded by printing the resultant DataFrame:

```
df = pd.read_csv('Bias_correction_ucl.csv')
```

> **NOTE**
>
> Make sure you change the path (highlighted) to the CSV file based on its location on your system. If you're running the Jupyter notebook from the same directory where the CSV file is stored, you can run the preceding code without any modification.

4. Drop the **date** column using the **drop** method. Since you're dropping the columns, pass **1** to the **axis** argument and **True** to the **inplace** argument:

```
df.drop('Date', inplace=True, axis=1)
```

5. Plot a histogram of the **Present_Tmax** column that represents the maximum temperature across dates and weather stations across the dataset:

```
ax = df['Present_Tmax'].hist(color='gray')
ax.set_xlabel("Normalized Temperature")
ax.set_ylabel("Frequency")
```

The output will be as follows:

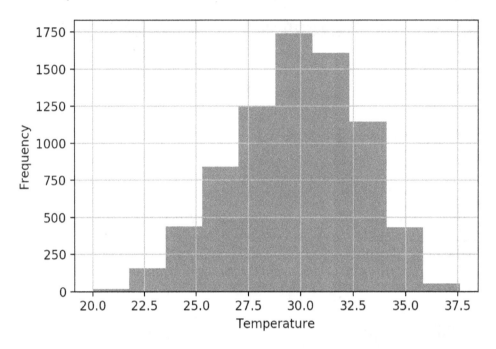

Figure 2.20: A Temperature versus Frequency histogram of the Present_Tmax column

The resultant histogram shows the distribution of values for the
Present_Tmax column.

6. Import **MinMaxScaler** and use it to fit and transform the feature DataFrame:

```
from sklearn.preprocessing import MinMaxScaler
scaler = MinMaxScaler()
df2 = scaler.fit_transform(df)
df2 = pd.DataFrame(df2, columns=df.columns)
```

7. Plot a histogram of the transformed **Present_Tmax** column:

```
ax = df2['Present_Tmax'].hist(color='gray')
ax.set_xlabel("Normalized Temperature")
ax.set_ylabel("Frequency")
```

The output will be as follows:

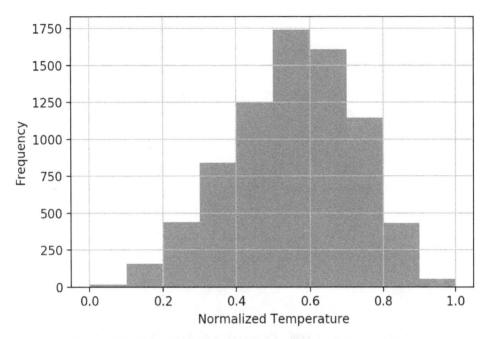

Figure 2.21: A histogram of the rescaled Present_Tmax column

The resultant histogram shows that the temperature values range from **0** to **1**, as evidenced by the range on the *x* axis of the histogram. By using **MinMaxScaler**, the values will always have a minimum value of **0** and a maximum value of **1**.

In this activity, you have performed some further preprocessing of the numerical fields. Here, you scaled the numerical fields so that they have a minimum value of **0** and a maximum value of **1**. This could be beneficial over the standard scaler if the numerical fields are not normally distributed. It also ensures the resulting fields are bound between a minimum and maximum value.

ACTIVITY 2.02: LOADING IMAGE DATA FOR BATCH PROCESSING

Solution:

1. Open a new Jupyter notebook to implement this activity. Save the file as **Activity2-02.ipnyb**.

2. In a new Jupyter Notebook cell, import the **ImageDataGenerator** class from Keras' preprocessing package:

```
from tensorflow.keras.preprocessing.image \
    import ImageDataGenerator
```

3. Instantiate the **ImageDataGenerator** class and pass the **rescale** argument with a value of **1/255** to convert image values so that they're between **0** and **1**:

```
train_datagen = ImageDataGenerator(rescale = 1./255,\
                                   shear_range = 0.2,\
                                   rotation_range= 180,\
                                   zoom_range = 0.2,\
                                   horizontal_flip = True)
```

4. Use the data generator's **flow_from_directory** method to direct the data generator to the image data. Pass in the arguments of the target size, the batch size, and the class mode:

```
training_set = train_datagen.flow_from_directory\
               ('image_data',\
                target_size = (64, 64),\
                batch_size = 25,\
                class_mode = 'binary')
```

5. Create a function to display the images in the batch:

```
import matplotlib.pyplot as plt

def show_batch(image_batch, label_batch):\
    lookup = {v: k for k, v in
        training_set.class_indices.items()}
    label_batch = [lookup[label] for label in \
                   label_batch]
    plt.figure(figsize=(10,10))
    for n in range(25):
        ax = plt.subplot(5,5,n+1)
        plt.imshow(image_batch[n])
```

```
        plt.title(label_batch[n].title())
        plt.axis('off')
```

6. Take a batch from the data generator and pass it to the function to display the images and their labels:

```
image_batch, label_batch = next(training_set)
show_batch(image_batch, label_batch)
```

The output will be as follows:

Figure 2.22: Augmented images from a batch

he output shows a batch of 25 images and their respective labels that have been augmented by rotation, zooming, and shearing. The augmented images show the same objects but with different pixel values, which helps create more robust models.

ACTIVITY 2.03: LOADING AUDIO DATA FOR BATCH PROCESSING

Solution:

1. Open a new Jupyter notebook to implement this activity. Save the file as **Activity2-03.ipnyb**.

2. In a new Jupyter Notebook cell, import the TensorFlow and **os** libraries:

```
import tensorflow as tf
import os
```

3. Create a function that will load and then return an audio file using TensorFlow's **read_file** function followed by the **decode_wav** function, respectively. Return the transpose of the resultant tensor:

```
def load_audio(file_path, sample_rate=44100):
    # Load audio at 44.1kHz sample-rate
    audio = tf.io.read_file(file_path)
    audio, sample_rate = tf.audio.decode_wav\
                         (audio,\
                          desired_channels=-1,\
                          desired_samples=sample_rate)
    return tf.transpose(audio)
```

4. Load in the paths to the audio data as a list using **os.list_dir**:

```
prefix = " ../Datasets/data_speech_commands_v0.02"\
         "/zero/"
paths = [os.path.join(prefix, path) for path in \
         os.listdir(prefix)]
```

5. Create a function that will take a dataset object, shuffle it, and load the audio using the function you created in *Step 2*. Then, apply the absolute value and the **log1p** function to the dataset. This function adds **1** to each value then takes the logarithm. Next, repeat the dataset object, batch it, and prefetch it with a buffer size equal to the batch size:

```
def prep_ds(ds, shuffle_buffer_size=1024, \
            batch_size=16):
    # Randomly shuffle (file_path, label) dataset
    ds = ds.shuffle(buffer_size=shuffle_buffer_size)
    # Load and decode audio from file paths
    ds = ds.map(load_audio)
    # Take the absolute value
    ds = ds.map(tf.abs)
    # Apply log1p function
    ds = ds.map(tf.math.log1p)
    # Repeat dataset forever
    ds = ds.repeat()
    # Prepare batches
    ds = ds.batch(batch_size)
    # Prefetch
    ds = ds.prefetch(buffer_size=batch_size)

    return ds
```

6. Create a dataset object using TensorFlow's **from_tensor_slices** function and pass in the paths to the audio files. Then, apply the function you created in *Step 5* to the dataset object:

```
ds = tf.data.Dataset.from_tensor_slices(paths)
train_ds = prep_ds(ds)
```

7. Take the first batch of the dataset and print it out:

```
for x in train_ds.take(1):\
    print(x)
```

The output will look as follows:

```
tf.Tensor(
[[[9.1548543e-05 1.2206286e-04 2.7462048e-04 ... 0.0000000e+00
    0.0000000e+00 0.0000000e+00]]

 [[9.1548543e-05 9.1548543e-05 1.2206286e-04 ... 0.0000000e+00
    0.0000000e+00 0.0000000e+00]]

 [[5.1866431e-04 2.1948551e-03 2.3470970e-03 ... 0.0000000e+00
    0.0000000e+00 0.0000000e+00]]

 ...

 [[3.3817268e-03 3.8378409e-03 4.4760513e-03 ... 0.0000000e+00
    0.0000000e+00 0.0000000e+00]]

 [[9.3252808e-03 2.2361631e-02 2.5847148e-02 ... 0.0000000e+00
    0.0000000e+00 0.0000000e+00]]

 [[2.1339520e-03 4.2633601e-03 4.2025824e-03 ... 0.0000000e+00
    0.0000000e+00 0.0000000e+00]]], shape=(16, 1, 44100), dtype=float32)
```

Figure 2.23: A batch of the audio data

The output shows the first batch of MFCC spectrum values in tensor form.

8. Plot the first audio file from the batch:

```
import matplotlib.pyplot as plt
plt.plot(x[0,:,:].numpy().T, color = 'gray')
plt.xlabel('Sample')
plt.ylabel('Value'))
```

The output will look as follows:

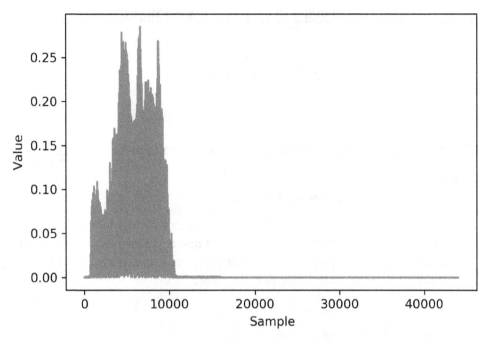

Figure 2.24: A visual representation of the batch of the preprocessed audio data

The preceding plot shows the preprocessed audio data. You can see that the values are non-negative, with a minimum value of **0**, and that the data is logarithmically scaled.

CHAPTER 3: TENSORFLOW DEVELOPMENT

ACTIVITY 3.01: USING TENSORBOARD TO VISUALIZE TENSOR TRANSFORMATIONS

Solution:

1. Import the TensorFlow library and set a seed:

```
import tensorflow as tf
tf.random.set_seed(42)
```

2. Set the log directory and initialize a file writer object to write the trace:

```
logdir = 'logs/'
writer = tf.summary.create_file_writer(logdir)
```

3. Create a TensorFlow function to multiply two tensors and add a value of **1** to all elements in the resulting tensor using the **ones_like** function to create a tensor of the same shape as the result of the matrix multiplication. Then, apply a sigmoid function to each value of the tensor:

```
@tf.function
def my_func(x, y):
    r1 = tf.matmul(x, y)
    r2 = r1 + tf.ones_like(r1)
    r3 = tf.keras.activations.sigmoid(r2)
    return r3
```

4. Create two tensors with the shape **5x5x5**:

```
x = tf.random.uniform((5, 5, 5))
y = tf.random.uniform((5, 5, 5))
```

5. Turn on graph tracing:

```
tf.summary.trace_on(graph=True, profiler=True)
```

6. Apply the function to the two tensors and export the trace to the log directory:

```
z = my_func(x, y)
with writer.as_default():
    tf.summary.trace_export(name="my_func_trace",\
                            step=0,\
                            profiler_outdir=logdir)
```

7. Launch TensorBoard in the command line and view the graph in a browser:

```
tensorboard --logdir=./logs
```

You should get something like the following image:

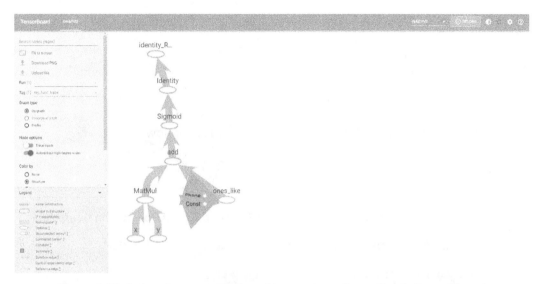

Figure 3.19: A visual representation of tensor transformation in TensorBoard

The result represents the graph created for the tensor transformation. You can see in the bottom left at the beginning of the graph that a matrix multiplication is performed on the tensors named **x** and **y** on the node named `MatMul`. In the bottom right is the creation of the tensor using the `ones_like` function. The input nodes represent the shape of the tensor and the value, which is a constant value. Upon the creation of the two tensors, they are input into a node representing the addition function, after which the output is input to a node representing the application of the sigmoid function. The final nodes represent the creation of the output tensor.

In this activity, you created functions for tensor transformation, and then presented a visual representation of the transformation in TensorBoard.

ACTIVITY 3.02: PERFORMING WORD EMBEDDING FROM A PRE-TRAINED MODEL FROM TENSORFLOW HUB

Solution:

1. Import TensorFlow and TensorFlow Hub and print the version of the library:

```
import tensorflow as tf
import tensorflow_hub as hub
print('TF version: ', tf.__version__)
print('HUB version: ', hub.__version__)
```

You should get the versions of TensorFlow and TensorFlow Hub.

```
 TF version: 2.2.0
 Hub version: 0.8.0
```

Figure 3.20: The output of the versions of TensorFlow and TensorFlow Hub in Google Colab

2. Set the handle for the module for the universal sentence encoder:

```
module_handle ="https://tfhub.dev/google"\
            "/universal-sentence-encoder/4"
```

3. Use the TensorFlow Hub **KerasLayer** class to create a hub layer, passing in the following arguments: **module_handle**, **input_shape**, and **dtype**:

```
hub_layer = hub.KerasLayer(module_handle, input_shape=[],\
                          dtype=tf.string)
```

4. Create a list containing a string to encode with the encoder:

```
text = ['The TensorFlow Workshop']
```

5. Apply **hub_layer** to the text to embed the sentence as a vector:

```
hub_layer(text)
```

You should get the following output:

```
<tf.Tensor: shape=(1, 512), dtype=float32, numpy=
array([[-0.01592658, -0.01910833, -0.00460122, -0.04786165, -0.0090545 ,
        -0.05658781, -0.04260132,  0.06827556,  0.03513585,  0.01448399,
        -0.00549438,  0.04602941,  0.02016041,  0.0008662 , -0.01191864,
         0.07414375, -0.03241738, -0.04448074,  0.00137551, -0.06724778,
        -0.02604278,  0.01092253, -0.01246114,  0.03847544,  0.00819034,
         0.06088841, -0.02359939, -0.05117927, -0.01725158, -0.02764523,
         0.04102336, -0.03135261,  0.06100909, -0.02693282, -0.07294274,
        -0.00857984, -0.04100463, -0.01803453,  0.04117068, -0.01969654,
        -0.04563987,  0.0257121 , -0.03328102, -0.05113809, -0.03377022,
         0.07439086, -0.02235463, -0.00438892, -0.00755636,  0.07249703,
        -0.07135288, -0.05469208,  0.01436193,  0.0396053 , -0.01475235,
        -0.03984744,  0.05067959,  0.07571234,  0.03281045, -0.00155282,
        -0.07548428,  0.01494772, -0.04175217,  0.03947704, -0.0147364 ,
        -0.01756434, -0.00077199,  0.00788859, -0.07518636,  0.04074219,
        -0.02049077, -0.03601787, -0.01753781,  0.03299529,  0.05840027,
        -0.03444539,  0.0186691 ,  0.03436609,  0.05346094,  0.02573053,
        -0.05013486, -0.05430874, -0.04835197,  0.03301562, -0.03129521,
         0.04714367, -0.07143752, -0.02783648, -0.02234376, -0.0619083 ,
        -0.05527468,  0.02779463,  0.04658304, -0.02259884, -0.05570157,
         0.06667245, -0.02903359, -0.05355389,  0.06542732, -0.05243086,
        -0.03966407,  0.01379365, -0.03453102,  0.07174195, -0.00385802,
        -0.03642376, -0.01343285,  0.00164682, -0.05571308,  0.01775301,
         0.03831774,  0.00128905, -0.0665922 , -0.01266254,  0.00407203,
        -0.07047658,  0.04188056, -0.01210087, -0.04976766,  0.03678571,
```

Figure 3.21: The output of the embedding vector

Here, you can see that the text has been converted to a 512-dimensional embedding vector. The embedding vector is a one-dimensional tensor that maps the text into a vector of continuous variables as shown in the preceding figure.

In this activity, you used the Google Colab environment to download a model from TensorFlow Hub. You used a universal sentence encoder to embed a sentence into a 512-dimensional vector. This activity has shown that with a few short lines of code on powerful remote servers, you can access state-of-the-art machine learning models for any application.

CHAPTER 4: REGRESSION AND CLASSIFICATION MODELS

ACTIVITY 4.01: CREATING A MULTI-LAYER ANN WITH TENSORFLOW

Solution:

1. Open a new Jupyter notebook to implement this activity.

2. Import the TensorFlow and pandas libraries:

```
import tensorflow as tf
import pandas as pd
```

3. Load in the dataset using the pandas **read_csv** function:

```
df = pd.read_csv('superconductivity.csv')
```

> **NOTE**
>
> Make sure you change the path (highlighted) to the CSV file based on its location on your system. If you're running the Jupyter notebook from the same directory where the CSV file is stored, you can run the preceding code without any modification.

4. Drop the **date** column and drop any rows that have null values:

```
df.dropna(inplace=True)
```

5. Create target and feature datasets:

```
target = df['critical_temp']
features = df.drop('critical_temp', axis=1)
```

6. Rescale the feature dataset:

```
from sklearn.preprocessing import StandardScaler
scaler = StandardScaler()
feature_array = scaler.fit_transform(features)
features = pd.DataFrame(feature_array, columns=features.columns)
```

7. Initialize a Keras model of the **Sequential** class:

```
model = tf.keras.Sequential()
```

8. Add an input layer to the model using the model's **add** method, and set **input_shape** to be the number of columns in the feature dataset. Add four hidden layers of sizes **64**, **32**, **16**, and **8** to the model with the first having a ReLU activation function, then add an output layer with one unit:

```
model.add(tf.keras.layers.InputLayer\
        (input_shape=features.shape[1],), \
        name='Input_layer'))
model.add(tf.keras.layers.Dense(64, activation='relu', \
                                name='Dense_layer_1'))
model.add(tf.keras.layers.Dense(32, name='Dense_layer_2'))
model.add(tf.keras.layers.Dense(16, name='Dense_layer_3'))
model.add(tf.keras.layers.Dense(8, name='Dense_layer_4'))
model.add(tf.keras.layers.Dense(1, name='Output_layer'))
```

9. Compile the model with an RMSprop optimizer with a learning rate equal to **0.001** and the mean squared error for the loss:

```
model.compile(tf.optimizers.RMSprop(0.001), loss='mse')
```

10. Create a TensorBoard callback:

```
tensorboard_callback = tf.keras.callbacks\
                        .TensorBoard(log_dir="./logs")
```

11. Fit the model to the training data for **100** epochs, with a batch size equal to **32** and a validation split equal to 20%:

```
model.fit(x=features.to_numpy(), y=target.to_numpy(), \
        epochs=100, callbacks=[tensorboard_callback], \
        batch_size=32, validation_split=0.2)
```

You should get the following output:

```
Train on 17010 samples, validate on 4253 samples
Epoch 1/100
17010/17010 [==============================] - 2s 104us/sample - loss: 2285.8694 - val_loss: 652.0398
Epoch 2/100
17010/17010 [==============================] - 1s 46us/sample - loss: 1992.0314 - val_loss: 594.5726
Epoch 3/100
17010/17010 [==============================] - 1s 50us/sample - loss: 1730.1853 - val_loss: 522.3722
Epoch 4/100
17010/17010 [==============================] - 1s 50us/sample - loss: 1497.6520 - val_loss: 500.8754
Epoch 5/100
17010/17010 [==============================] - 1s 50us/sample - loss: 1292.5218 - val_loss: 463.0061
Epoch 6/100
17010/17010 [==============================] - 1s 48us/sample - loss: 1113.1502 - val_loss: 421.8601
Epoch 7/100
17010/17010 [==============================] - 1s 55us/sample - loss: 961.1294 - val_loss: 385.5281
Epoch 8/100
17010/17010 [==============================] - 1s 53us/sample - loss: 806.4936 - val_loss: 377.2940
Epoch 9/100
17010/17010 [==============================] - 1s 54us/sample - loss: 683.4299 - val_loss: 357.6940
Epoch 10/100
17010/17010 [==============================] - 1s 54us/sample - loss: 580.5784 - val_loss: 335.1988
Epoch 11/100
17010/17010 [==============================] - 1s 52us/sample - loss: 498.0246 - val_loss: 328.5940
Epoch 12/100
17010/17010 [==============================] - 1s 62us/sample - loss: 430.3305 - val_loss: 339.9765
Epoch 13/100
17010/17010 [==============================] - 1s 56us/sample - loss: 379.7621 - val_loss: 304.2136
Epoch 14/100
17010/17010 [==============================] - 1s 50us/sample - loss: 341.5881 - val_loss: 303.6295
Epoch 15/100
17010/17010 [==============================] - 1s 57us/sample - loss: 313.5454 - val_loss: 293.1223
Epoch 16/100
17010/17010 [==============================] - 1s 59us/sample - loss: 290.2986 - val_loss: 281.2344
Epoch 17/100
17010/17010 [==============================] - 1s 50us/sample - loss: 275.2153 - val_loss: 276.7121
Epoch 18/100
17010/17010 [==============================] - 1s 51us/sample - loss: 262.7841 - val_loss: 276.8066
Epoch 19/100
17010/17010 [==============================] - 1s 50us/sample - loss: 254.1752 - val_loss: 261.5233
Epoch 20/100
17010/17010 [==============================] - 1s 53us/sample - loss: 247.0332 - val_loss: 286.3439
```

Figure 4.16: The output of the fitting process showing the epoch,
training time per sample, and loss after each epoch

12. Evaluate the model on the training data:

```
loss = model.evaluate(features.to_numpy(), target.to_numpy())
print('loss:', loss)
```

This will result in the following output:

```
loss: 165.735601268987
```

13. Visualize the model architecture and model-fitting process in TensorBoard by calling the following on the command line:

```
tensorboard --logdir=logs/
```

The model architecture should look like the following:

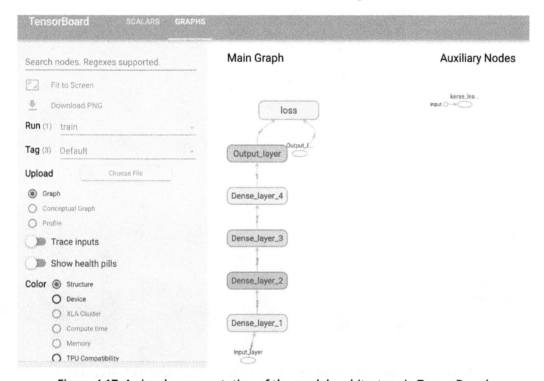

Figure 4.17: A visual representation of the model architecture in TensorBoard

14. Visualize the model-fitting process in TensorBoard. You should get the following output:

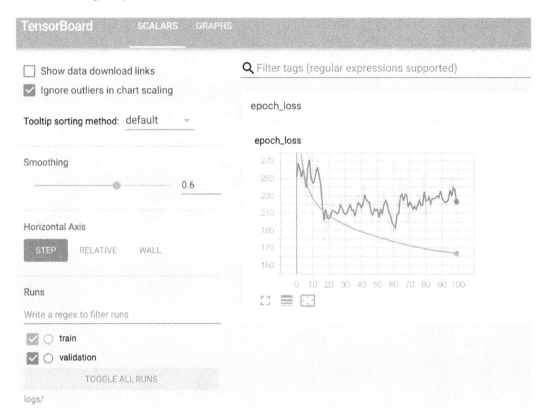

Figure 4.18: A visual representation of the loss as a function of an epoch on the training and validation split in TensorBoard

During the model-fitting process, the loss on the training and validation sets is calculated after each epoch and displayed in TensorBoard in the **SCALARS** tab. From TensorBoard, you can see that the mean squared error reduces after each epoch consistently on the training set but plateaus on the validation set.

In this activity, you have further practiced building models in TensorFlow and viewing its architecture and training process in TensorBoard. During this section, you have learned how to build, train, and evaluate ANNs using TensorFlow for regression tasks. You used Keras layers of the **Dense** class as an easy way to create fully connected layers that include activation functions on the output of the layers. The layers can be created simply by passing in the number of units desired in the layer. Keras configures the initialization of the weights and biases, as well as any other additional parameters that are common in a machine learning workflow.

ACTIVITY 4.02: CREATING A MULTI-LAYER CLASSIFICATION ANN WITH TENSORFLOW

Solution:

1. Open a new Jupyter notebook to implement this activity.

2. Import the TensorFlow and pandas libraries:

```
import tensorflow as tf
import pandas as pd
```

3. Load in the dataset using the pandas **read_csv** function:

```
df = pd.read_csv('superconductivity.csv')
```

> **NOTE**
>
> Make sure you change the path (highlighted) to the CSV file based on its location on your system. If you're running the Jupyter notebook from the same directory where the CSV file is stored, you can run the preceding code without any modification.

4. Drop any rows that have null values:

```
df.dropna(inplace=True)
```

5. Set the target values to **true** when values of the **critical_temp** column are above **77.36** and **false** when below. The feature dataset is the remaining columns in the dataset:

```
target = df['critical_temp'].apply(lambda x: 1 if x>77.36 else 0)
features = df.drop('critical_temp', axis=1)
```

6. Rescale the feature dataset:

```
from sklearn.preprocessing import StandardScaler
scaler = StandardScaler()
feature_array = scaler.fit_transform(features)
features = pd.DataFrame(feature_array, columns=features.columns)
```

7. Initialize a Keras model of the **Sequential** class:

```
model = tf.keras.Sequential()
```

8. Add an input layer to the model using the model's **add** method and set **input_shape** to the number of columns in the feature dataset. Add three hidden layers of sizes **32**, **16**, and **8** to the model, then add an output layer with **1** unit and a sigmoid activation function:

```
model.add(tf.keras.layers.InputLayer\
          (input_shape=features.shape[1], \
           name='Input_layer'))
model.add(tf.keras.layers.Dense(32, name='Hidden_layer_1'))
model.add(tf.keras.layers.Dense(16, name='Hidden_layer_2'))
model.add(tf.keras.layers.Dense(8, name='Hidden_layer_3'))
model.add(tf.keras.layers.Dense(1, name='Output_layer', \
                                activation='sigmoid'))
```

9. Compile the model with an RMSprop optimizer with a learning rate equal to **0.0001** and binary cross-entropy for the loss and compute the accuracy metric:

```
model.compile(tf.optimizers.RMSprop(0.0001), \
              loss= 'binary_crossentropy', metrics=['accuracy'])
```

10. Create a TensorBoard callback:

```
tensorboard_callback = tf.keras.callbacks.TensorBoard\
                       (log_dir="./logs")
```

11. Fit the model to the training data for **50** epochs and a validation split equal to 20%:

```
model.fit(x=features.to_numpy(), y=target.to_numpy(),\
          epochs=50, callbacks=[tensorboard_callback],\
          validation_split=0.2)
```

You should get the following output:

```
Train on 17010 samples, validate on 4253 samples
Epoch 1/50
17010/17010 [==============================] - 1s 87us/sample - loss: 0.5612 - accuracy: 0.6971 - val_loss: 0.4220 -
val_accuracy: 0.8451
Epoch 2/50
17010/17010 [==============================] - 1s 48us/sample - loss: 0.3891 - accuracy: 0.8028 - val_loss: 0.1949 -
val_accuracy: 0.9744
Epoch 3/50
17010/17010 [==============================] - 1s 52us/sample - loss: 0.3345 - accuracy: 0.8253 - val_loss: 0.1034 -
val_accuracy: 0.9718
Epoch 4/50
17010/17010 [==============================] - 1s 50us/sample - loss: 0.3161 - accuracy: 0.8285 - val_loss: 0.0780 -
val_accuracy: 0.9732
Epoch 5/50
17010/17010 [==============================] - 1s 48us/sample - loss: 0.3066 - accuracy: 0.8320 - val_loss: 0.0678 -
val_accuracy: 0.9767
Epoch 6/50
17010/17010 [==============================] - 1s 48us/sample - loss: 0.3008 - accuracy: 0.8378 - val_loss: 0.0628 -
val_accuracy: 0.9746
Epoch 7/50
17010/17010 [==============================] - 1s 48us/sample - loss: 0.2964 - accuracy: 0.8413 - val_loss: 0.0594 -
val_accuracy: 0.9737
Epoch 8/50
17010/17010 [==============================] - 1s 53us/sample - loss: 0.2927 - accuracy: 0.8419 - val_loss: 0.0568 -
val_accuracy: 0.9767
Epoch 9/50
17010/17010 [==============================] - 1s 53us/sample - loss: 0.2902 - accuracy: 0.8430 - val_loss: 0.0564 -
val_accuracy: 0.9753
Epoch 10/50
17010/17010 [==============================] - 1s 50us/sample - loss: 0.2881 - accuracy: 0.8468 - val_loss: 0.0544 -
val_accuracy: 0.9751
Epoch 11/50
17010/17010 [==============================] - 1s 50us/sample - loss: 0.2862 - accuracy: 0.8463 - val_loss: 0.0537 -
val_accuracy: 0.9781
Epoch 12/50
17010/17010 [==============================] - 1s 50us/sample - loss: 0.2842 - accuracy: 0.8474 - val_loss: 0.0539 -
val_accuracy: 0.9751
Epoch 13/50
17010/17010 [==============================] - 1s 50us/sample - loss: 0.2830 - accuracy: 0.8481 - val_loss: 0.0527 -
val_accuracy: 0.9779
```

Figure 4.19: The output of the fitting process showing the epoch, training time per sample, loss, and accuracy after each epoch, and evaluated on the validation split

12. Evaluate the model on the training data:

```
loss, accuracy = model.evaluate(features.to_numpy(), \
                                target.to_numpy())
print(f'loss: {loss}, accuracy: {accuracy}')
```

This will display the following output:

```
loss: 0.21984571637242145, accuracy: 0.8893383145332336
```

13. Visualize the model architecture and model-fitting process in TensorBoard by calling the following on the command line:

```
tensorboard --logdir=logs/
```

You should get a screen similar to the following in the browser:

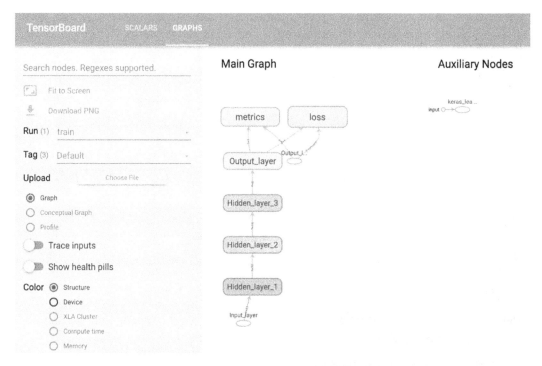

Figure 4.20: A visual representation of the model architecture in TensorBoard

The loss function can be visualized as follows:

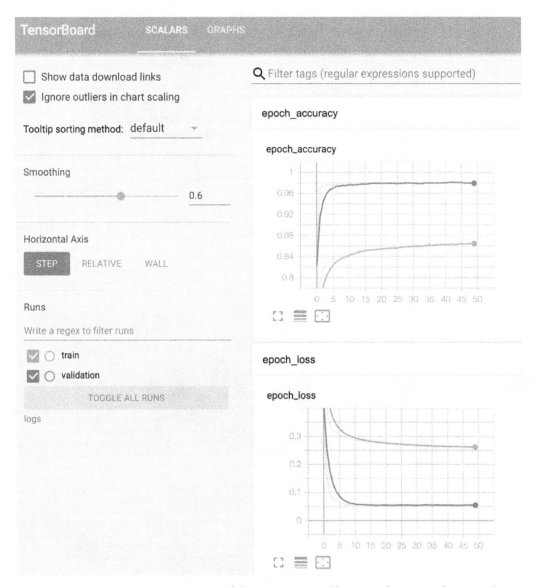

Figure 4.21: A visual representation of the accuracy and loss as a function of an epoch on the training and validation split in TensorBoard

During the model-fitting process, the accuracy and loss on the training and validation sets are calculated after each epoch and displayed in TensorBoard in the **SCALARS** tab. From TensorBoard, you can see that the loss metric (binary cross-entropy) reduces after each epoch consistently on the training set but plateaus on the validation set.

In this activity, you have practiced building classification models in TensorFlow by building a multi-layer ANN to determine whether a material will exhibit superconductivity above or below the boiling point of nitrogen. Moreover, you used TensorBoard to view the models' architecture and monitor key metrics during the training process, including the loss and the accuracy of the models.

CHAPTER 5: CLASSIFICATION MODELS

ACTIVITY 5.01: BUILDING A CHARACTER RECOGNITION MODEL WITH TENSORFLOW

Solution:

1. Open a new Jupyter notebook.

2. Import the pandas library and use **pd** as the alias:

```
import pandas as pd
```

3. Create a variable called **file_url** that contains the URL to the dataset:

```
file_url = 'https://raw.githubusercontent.com/PacktWorkshops'\
           '/The-TensorFlow-Workshop/master/Chapter05'\
           '/dataset/letter-recognition.data'
```

4. Load the dataset into a **DataFrame()** function called **data** using **read_csv()** method, provide the URL to the CSV file, and set **header=None** as the dataset doesn't provide column names. Print the first five rows using **head()** method.

```
data = pd.read_csv(file_url, header=None)
data.head()
```

The expected output will be as follows:

	0	1	2	3	4	5	6	7	8	9	10	11	12	13	14	15	16
0	0	1	2	3	4	5	6	7	8	9	10	11	12	13	14	15	16
1	0	2	8	3	5	1	8	13	0	6	6	10	8	0	8	0	8
2	1	5	12	3	7	2	10	5	5	4	13	3	9	2	8	4	10
3	2	4	11	6	8	6	10	6	2	6	10	3	7	3	7	3	9
4	3	7	11	6	6	3	5	9	4	6	4	4	10	6	10	2	8

Figure 5.42: First five rows of the data

You can see that the dataset contains **17** columns and they are all numeric. Column **0** is the **target** variable, and each value corresponds to a letter of the alphabet.

5. Extract the target variable (column **0**) using the **pop()** method and save it in a variable called **target**:

```
target = data.pop(0)
```

6. Split **data** into a training set by keeping the first 15,000 observations and save it in a variable called **X_train**. Perform the same split on **target** and save the first 15,000 cases in a variable called **y_train**:

```
X_train = data[:15000]
y_train = target[:15000]
```

7. Split **data** into a test set by keeping the last 5,000 observations and save it in a variable called **X_test**. Perform the same split on **target** and save the last 5,000 cases in a variable called **y_test**:

```
X_test = data[15000:]
y_test = target[15000:]
```

8. Import the TensorFlow library and use **tf** as the alias:

```
import tensorflow as tf
```

9. Set the seed as **8** using **tf.random.set_seed()** to get reproducible results:

```
tf.random.set_seed(8)
```

10. Instantiate a sequential model using **tf.keras.Sequential()** and store it in a variable called **model**:

```
model = tf.keras.Sequential()
```

11. Import the **Dense()** class from **tensorflow.keras.layers**:

```
from tensorflow.keras.layers import Dense
```

12. Create a fully connected layer of **512** units with **Dense()** and specify ReLu as the activation function and the input shape as **(16,)**, which corresponds to the number of features from the dataset. Save it in a variable called **fc1**:

```
fc1 = Dense(512, input_shape=(16,), activation='relu')
```

13. Create a fully connected layer of **512** units with **Dense()** and specify ReLu as the activation function. Save it in a variable called **fc2**:

```
fc2 = Dense(512, activation='relu')
```

14. Create a fully connected layer of **128** units with **Dense()** and specify ReLu as the activation function. Save it in a variable called **fc3**:

```
fc3 = Dense(128, activation='relu')
```

15. Create a fully connected layer of **128** units with **Dense()** and specify ReLu as the activation function. Save it in a variable called **fc4**:

```
fc4 = Dense(128, activation='relu')
```

16. Create a fully connected layer of **26** units with **Dense()** and specify softmax as the activation function. Save it in a variable called **fc5**:

```
fc5 = Dense(26, activation='softmax')
```

17. Sequentially add all five fully connected layers to the model using **add()** method.

```
model.add(fc1)
model.add(fc2)
model.add(fc3)
model.add(fc4)
model.add(fc5)
```

18. Print the summary of the model using **summary()** method.

```
model.summary()
```

The expected output will be as follows:

```
Model: "sequential"
```

Layer (type)	Output Shape	Param #
dense (Dense)	(None, 512)	8704
dense_1 (Dense)	(None, 512)	262656
dense_2 (Dense)	(None, 128)	65664
dense_3 (Dense)	(None, 128)	16512
dense_4 (Dense)	(None, 26)	3354

```
Total params: 356,890
Trainable params: 356,890
Non-trainable params: 0
```

Figure 5.43: Summary of the model architecture

The preceding output shows that there are five layers in your model (as expected) and also tells you the number of parameters at each layer.

19. Instantiate **SparseCategoricalCrossentropy()** from **tf.keras.losses** and save it in a variable called **loss**:

```
loss = tf.keras.losses.SparseCategoricalCrossentropy()
```

20. Instantiate **Adam()** from **tf.keras.optimizers** with **0.001** as the learning rate and save it in a variable called **optimizer**:

```
optimizer = tf.keras.optimizers.Adam(0.001)
```

21. Compile the model using **compile()** method, specify the optimizer and loss parameters you just created, and use accuracy as the metric to be reported:

```
model.compile(optimizer=optimizer, loss=loss, \
              metrics=['accuracy'])
```

22. Start the model training process using **fit()** method on the training set for five epochs:

```
model.fit(X_train, y_train, epochs=5)
```

The expected output will be as follows:

```
Epoch 1/5
469/469 [==============================] - 4s 8ms/step - loss: 1.1898 - accuracy: 0.6385
Epoch 2/5
469/469 [==============================] - 3s 7ms/step - loss: 0.5809 - accuracy: 0.8138
Epoch 3/5
469/469 [==============================] - 3s 7ms/step - loss: 0.4236 - accuracy: 0.8623
Epoch 4/5
469/469 [==============================] - 3s 7ms/step - loss: 0.3464 - accuracy: 0.8845
Epoch 5/5
469/469 [==============================] - 4s 8ms/step - loss: 0.2767 - accuracy: 0.9065
<tensorflow.python.keras.callbacks.History at 0x201feb29088>
```

Figure 5.44: Logs of the training process

The preceding output shows the logs of each epoch during the training of the model. Note that it took around 2 seconds to process a single epoch, and the accuracy score increased from **0.6229** (first epoch) to **0.9011** (fifth epoch).

23. Evaluate the performance of the model on the test set using **evaluate()** method.

```
model.evaluate(X_test, y_test)
```

The expected output will be as follows:

```
157/157 [==============================] - 1s 4ms/step - loss: 0.3329 - accuracy: 0.8898
[0.3329078257083893, 0.8898220062255859]
```

Figure 5.45: Performance of the model on the test set

24. Predict the probabilities for each class on the test set using **predict()** method. Save it in a variable called **preds_proba**:

```
preds_proba = model.predict(X_test)
```

25. Convert the class probabilities into a single predicted value using **argmax ()** method with **axis=1**:

```
preds = preds_proba.argmax(axis=1)
```

26. Import **confusion_matrix** from **tensorflow.math**:

```
from tensorflow.math import confusion_matrix
```

27. Print the confusion matrix on the test set:

```
confusion_matrix(y_test, preds)
```

The expected output will be as follows:

```
<tf.Tensor: shape=(26, 26), dtype=int32, numpy=
array([[155,    0,    0,    0,    0,    0,    0,    0,    0,    0,    0,    0,    0,
           1,    0,    0,    0,    1,    0,    0,   12,   15,    0,    0,    0,    0],
       [  0,  184,    2,    0,    0,    2,    0,    0,    1,    0,    0,    0,    0,
          12,    2,    0,    0,    0,    0,    1,    0,    0,    1,    0,    0,    0],
       [  0,    2,  214,    0,    0,    0,    0,    0,    0,    0,    0,    0,    0,
           0,    0,    0,    0,    0,    0,    0,    0,    0,    0,    0,    0,    0],
       [  0,    0,   10,  172,    0,    0,    0,    1,    0,    1,    0,    5,    0,
           0,    0,    2,    0,    1,    1,    0,    0,    3,    0,    2,    0,    0],
       [  0,    0,    2,    0,  152,    0,    1,    0,    0,    1,    0,    5,    1,
           1,    6,    3,    0,    5,    1,    1,    8,    0,   19,    0,    2,    0],
       [  0,    0,    0,    0,    2,  178,    2,    0,    0,    0,    1,    0,    0,
           3,    0,    0,    0,    0,    0,    2,    0,    0,    4,    0,    0,    6],
       [  1,    1,   13,    0,    1,    0,  153,    0,    0,    0,    0,    1,    0,
           0,    0,    2,    0,    0,    0,    0,    1,    0,    0,    0,    0,    0],
       [  0,    0,    2,    0,    0,    0,    0,  198,    0,    0,    0,    0,    0,
           0,    0,    2,    0,    0,    0,    0,    0,    4,    0,    0,    0,    0],
```

Figure 5.46: Confusion matrix of the test set

The preceding output shows the model is correctly predicting the 26 letters of the alphabet most of the time (most of the values are located on the diagonal). It achieved an accuracy score of around 0.89 for both the training and test sets. This activity concludes the section on multi-class classification. In the section ahead, you will look at another type of classification called multi-label.

ACTIVITY 5.02: BUILDING A MOVIE GENRE TAGGING A MODEL WITH TENSORFLOW

Solution:

1. Open a new Jupyter notebook.

2. Import the pandas library and use **pd** as the alias:

    ```
    import pandas as pd
    ```

3. Create a variable called **feature_url** that contains the URL to the dataset:

    ```
    feature_url = 'https://raw.githubusercontent.com'\
                  '/PacktWorkshops'/The-TensorFlow-Workshop'\
                  '/master/Chapter05'/dataset/IMDB-F-features.csv'
    ```

4. Load the dataset into a DataFrame called **feature** using **read_csv()** method and provide the URL to the CSV file. Print the first five rows using the **head()** method:

    ```
    feature = pd.read_csv(feature_url)
    feature.head()
    ```

    The expected output will be as follows:

	experience	deep	star	situation	hand	birthday	queen	space	de	bond	...	bad	share	owner	today	married	doctor
0	0.0	0.0	0.0	0.0	0.0	0.0	0.0	1.0	0.0	0.0	...	0.0	0.0	0.0	0.0	0.0	0.0
1	0.0	0.0	0.0	0.0	0.0	0.0	0.0	0.0	0.0	0.0	...	0.0	0.0	0.0	0.0	0.0	0.0
2	0.0	0.0	0.0	0.0	0.0	0.0	0.0	0.0	0.0	0.0	...	0.0	0.0	0.0	0.0	0.0	0.0
3	0.0	0.0	0.0	0.0	0.0	0.0	0.0	0.0	0.0	0.0	...	0.0	0.0	0.0	0.0	0.0	0.0
4	0.0	0.0	0.0	0.0	0.0	0.0	0.0	0.0	0.0	0.0	...	0.0	0.0	0.0	0.0	0.0	0.0

Figure 5.47: The first five rows of the features

5. Create a variable called **target_url** that contains the URL to the dataset:

    ```
    target_url = 'https://raw.githubusercontent.com'\
                 '/PacktWorkshops/The-TensorFlow-Workshop'\
                 '/master/Chapter05'/dataset/IMDB-F-targets.csv'
    ```

6. Load the dataset into a DataFrame called **target** using **read_csv()** method and provide the URL to the CSV file. Print the first five rows using the **head()** method:

    ```
    target = pd.read_csv(target_url)
    target.head()
    ```

The expected output will be as follows:

	Sci-Fi	Crime	Romance	Animation	Music	Comedy	War	Horror	Film-Noir	Adventure	...	Action	Documentary	Musical
0	0.0	0.0	0.0	0.0	0.0	1.0	0.0	0.0	0.0	0.0	...	0.0	0.0	0.0
1	0.0	0.0	0.0	0.0	0.0	1.0	0.0	0.0	0.0	0.0	...	0.0	0.0	0.0
2	0.0	0.0	0.0	1.0	0.0	1.0	0.0	0.0	0.0	0.0	...	0.0	0.0	0.0
3	0.0	0.0	0.0	0.0	0.0	1.0	0.0	0.0	0.0	0.0	...	0.0	0.0	0.0
4	0.0	0.0	0.0	0.0	0.0	1.0	0.0	0.0	0.0	0.0	...	0.0	0.0	1.0

5 rows × 28 columns

Figure 5.48: The first five rows of the targets

7. Split the data into a training set by keeping the first 15,000 observations and save it in a variable called **X_train**. Perform the same split on **target** and save the first 15,000 cases in a variable called **y_train**:

```
X_train = feature[:15000]
y_train = target[:15000]
```

8. Split the data into a test set by keeping the last 5,000 observations and save it in a variable called **X_test**. Perform the same split on **target** and save the last 5,000 cases in a variable called **y_test**:

```
X_test = feature[15000:]
y_test = target[15000:]
```

9. Import the TensorFlow library and use **tf** as the alias:

```
import tensorflow as tf
```

10. Set the seed for **tensorflow** as 8 using **tf.random.set_seed()**. This will help to get reproducible results:

```
tf.random.set_seed(8)
```

11. Instantiate a sequential model using **tf.keras.Sequential()** and store it in a variable called **model**:

```
model = tf.keras.Sequential()
```

12. Import the **Dense()** class from **tensorflow.keras.layers**:

```
from tensorflow.keras.layers import Dense
```

13. Create a fully connected layer of **512** units with **Dense()** and specify ReLu as the activation function and the input shape as **(1001,)** which corresponds to the number of features from the dataset. Save it in a variable called **fc1**:

```
fc1 = Dense(512, input_shape=(1001,), activation='relu')
```

14. Create a fully connected layer of **512** units with **Dense()** and specify ReLu as the activation function. Save it in a variable called **fc2**:

```
fc2 = Dense(512, activation='relu')
```

15. Create a fully connected layer of **128** units with **Dense()** and specify ReLu as the activation function. Save it in a variable called **fc3**:

```
fc3 = Dense(128, activation='relu')
```

16. Create a fully connected layer of **128** units with **Dense()** and specify ReLu as the activation function. Save it in a variable called **fc4**:

```
fc4 = Dense(128, activation='relu')
```

17. Create a fully connected layer of **28** units with **Dense()** and specify sigmoid as the activation function. Save it in a variable called **fc5**:

```
fc5 = Dense(28, activation='sigmoid')
```

18. Sequentially add all five fully connected layers to the model using **add()** method.

```
model.add(fc1)
model.add(fc2)
model.add(fc3)
model.add(fc4)
model.add(fc5)
```

19. Print the summary of the model using **summary()** method.

```
model.summary()
```

The expected output will be as follows:

```
Model: "sequential"
```

Layer (type)	Output Shape	Param #
dense (Dense)	(None, 512)	513024
dense_1 (Dense)	(None, 512)	262656
dense_2 (Dense)	(None, 128)	65664
dense_3 (Dense)	(None, 128)	16512
dense_4 (Dense)	(None, 28)	3612

```
Total params: 861,468
Trainable params: 861,468
Non-trainable params: 0
```

Figure 5.49: Summary of the model architecture

20. Instantiate **BinaryCrossentropy()** from **tf.keras.losses** and save it in a variable called **loss**:

```
loss = tf.keras.losses.BinaryCrossentropy()
```

21. Instantiate **Adam()** from **tf.keras.optimizers** with **0.001** as the learning rate and save it in a variable called **optimizer**:

```
optimizer = tf.keras.optimizers.Adam(0.001)
```

22. Compile the model using **compile()** method and specify the optimizer and loss parameters that were just created, with accuracy as the metric to be reported:

```
model.compile(optimizer=optimizer, loss=loss, \
              metrics=['accuracy'])
```

23. Start the model training process using the **fit()** method on the training set for **20** epochs:

```
model.fit(X_train, y_train, epochs=20)
```

The expected output will be as follows:

```
Epoch 1/20
469/469 [==============================] - 8s 14ms/step - loss: 0.2252 - accuracy: 0.1390
Epoch 2/20
469/469 [==============================] - 6s 13ms/step - loss: 0.2006 - accuracy: 0.1831
Epoch 3/20
469/469 [==============================] - 6s 12ms/step - loss: 0.1779 - accuracy: 0.2517
Epoch 4/20
469/469 [==============================] - 6s 12ms/step - loss: 0.1336 - accuracy: 0.3787
Epoch 5/20
469/469 [==============================] - 6s 12ms/step - loss: 0.0904 - accuracy: 0.4996
Epoch 6/20
469/469 [==============================] - 6s 12ms/step - loss: 0.0638 - accuracy: 0.5653
Epoch 7/20
469/469 [==============================] - 6s 13ms/step - loss: 0.0459 - accuracy: 0.5978
Epoch 8/20
469/469 [==============================] - 6s 12ms/step - loss: 0.0339 - accuracy: 0.6193
Epoch 9/20
469/469 [==============================] - 6s 12ms/step - loss: 0.0278 - accuracy: 0.6205
Epoch 10/20
469/469 [==============================] - 6s 12ms/step - loss: 0.0226 - accuracy: 0.6238
```

Figure 5.50: Logs of the training process

You can observe that the model is trained for 20 epochs and that accuracy is improving, achieving **61.67%** after the ninth epoch.

24. Evaluate the performance of the model on the test set using the **evaluate()** method:

```
model.evaluate(X_test, y_test)
```

The expected output will be as follows:

```
157/157 [==============================] - 1s 5ms/step - loss: 0.9912 - accuracy: 0.1346
[0.9911884665489197, 0.13459999859333038]
```

Figure 5.51: Performance of the model on the test set

The preceding output shows the model achieved an accuracy score of **0.13** on the test set, which is extremely low, while it got an accuracy of **0.62** on the training set. This model is struggling to learn the relevant pattern to correctly predict the different genres of movies. You could try different architectures with different numbers of hidden layers and units on your own. You can also try different learning rates and optimizers. As the scores are very different on the training and test sets, the model is overfitting and has simply learned patterns relevant to just the training set.

CHAPTER 6: REGULARIZATION AND HYPERPARAMETER TUNING

ACTIVITY 6.01: PREDICTING INCOME WITH L1 AND L2 REGULARIZERS

Solution:

1. Open a new Jupyter notebook.

2. Import the pandas library and use **pd** as the alias:

```
import pandas as pd
```

3. Create a list called **usecols** containing the column names **AAGE**, **ADTIND**, **ADTOCC**, **SEOTR**, **WKSWORK**, and **PTOTVAL**:

```
usecols = ['AAGE','ADTIND','ADTOCC','SEOTR','WKSWORK', 'PTOTVAL']
```

4. Create a variable called **train_url** that contains the URL to the training set:

```
train_url = 'https://raw.githubusercontent.com/PacktWorkshops'\
            '/The-TensorFlow-Workshop/master/Chapter06'\
            '/dataset/census-income-train.csv'
```

5. Load the training dataset into a DataFrame, **train_data**, using the **read_csv()** method. Provide the URL to the CSV file and the **usecols** list to the **usecols** parameter. Print the first five rows using the **head()** method:

```
train_data = pd.read_csv(train_url, usecols=usecols)
train_data.head()
```

The expected output will be as follows:

	AAGE	ADTIND	ADTOCC	SEOTR	WKSWORK	PTOTVAL
0	73	0	0	0	0	1700.09
1	58	4	34	0	52	1053.55
2	18	0	0	0	0	991.95
3	9	0	0	0	0	1758.14
4	10	0	0	0	0	1069.16

Figure 6.23: First five rows of the training set

6. Extract the target variable (**PTOTVAL**) using the **pop ()** method and save it in a variable called **train_target**:

```
train_target = train_data.pop('PTOTVAL')
```

7. Create a variable called **test_url** that contains the URL to the test set:

```
test_url = 'https://github.com/PacktWorkshops'\
           '/The-TensorFlow-Workshop/blob/master/Chapter06'\
           '/dataset/census-income-test.csv?raw=true'
```

8. Load the test dataset into a DataFrame, **X_test**, using the **read_csv ()** method. Provide the URL to the CSV file and the **usecols** list to the **usecols** parameter. Print the first five rows using the **head ()** method:

```
test_data = pd.read_csv(test_url, usecols=usecols)
test_data.head()
```

The expected output will be as follows:

	AAGE	ADTIND	ADTOCC	SEOTR	WKSWORK	PTOTVAL
0	38	6	36	0	12	1032.38
1	44	37	12	0	26	1462.33
2	2	0	0	0	0	1601.75
3	35	29	3	2	52	1866.88
4	49	4	34	0	50	1394.54

Figure 6.24: First five rows of the test set

9. Extract the target variable (**PTOTVAL**) using the **pop ()** method and save it in a variable called **test_target**:

```
test_target = test_data.pop('PTOTVAL')
```

10. Import the TensorFlow library and use **tf** as the alias. Then, import the **Dense** class from **tensorflow.keras.layers**:

```
import tensorflow as tf
from tensorflow.keras.layers import Dense
```

11. Set the seed as **8** using **tf.random.set_seed()** to get reproducible results:

```
tf.random.set_seed(8)
```

12. Instantiate a sequential model using **tf.keras.Sequential()** and store it in a variable called **model**:

```
model = tf.keras.Sequential()
```

13. Import the **Dense** class from **tensorflow.keras.layers**:

```
from tensorflow.keras.layers import Dense
```

14. Create a fully connected layer of **1048** units with **Dense()** and specify ReLu as the activation function and the input shape as **(5,)**, which corresponds to the number of features from the dataset. Save it in a variable called **fc1**:

```
fc1 = Dense(1048, input_shape=(5,), activation='relu')
```

15. Create three fully connected layers of **512**, **128**, and **64** units with **Dense()** and specify ReLu as the activation function. Save them in three variables, called **fc2**, **fc3**, and **fc4**, respectively:

```
fc2 = Dense(512, activation='relu')
fc3 = Dense(128, activation='relu')
fc4 = Dense(64, activation='relu')
```

16. Create a fully connected layer of three units (corresponding to the number of classes) with **Dense()** and specify softmax as the activation function. Save it in a variable called **fc5**:

```
fc5 = Dense(3, activation='softmax')
```

17. Create a fully connected layer of a single unit with **Dense()**. Save it in a variable called **fc5**:

```
fc5 = Dense(1)
```

18. Sequentially add all five fully connected layers to the model using the **add()** method:

```
model.add(fc1)
model.add(fc2)
model.add(fc3)
model.add(fc4)
model.add(fc5)
```

19. Print the summary of the model:

```
model.summary()
```

You will get the following output:

```
Model: "sequential"

_____
 Layer (type)              Output Shape            Param #
============================================================
 dense (Dense)             (None, 1048)            6288

 dense_1 (Dense)           (None, 512)             537088

 dense_2 (Dense)           (None, 128)             65664

 dense_3 (Dense)           (None, 64)              8256

 dense_4 (Dense)           (None, 1)               65


============================================================
Total params: 617,361
Trainable params: 617,361
Non-trainable params: 0
_____
```

Figure 6.25: Summary of the model architecture

20. Instantiate **Adam()** from **tf.keras.optimizers** with **0.05** as the learning rate and save it in a variable called **optimizer**:

```
optimizer = tf.keras.optimizers.Adam(0.05)
```

21. Compile the model, specify the optimizer, and set **mse** as the loss and metric to be displayed:

```
model.compile(optimizer=optimizer, loss='mse', metrics=['mse'])
```

22. Start the model training process using the **fit()** method for five epochs and split the data into a validation set with 20% of the data:

```
model.fit(train_data, train_target, epochs=5, \
          validation_split=0.2)
```

The expected output will be as follows:

```
Epoch 1/5
4989/4989 [==============================] - 41s 8ms/step - loss: 1068044.1250 - mse: 1068044.1250 - val_loss: 962788.6875 - va
l_mse: 962788.6875
Epoch 2/5
4989/4989 [==============================] - 39s 8ms/step - loss: 1006452.6250 - mse: 1006452.6250 - val_loss: 958969.7500 - va
l_mse: 958969.7500
Epoch 3/5
4989/4989 [==============================] - 39s 8ms/step - loss: 1004209.7500 - mse: 1004209.7500 - val_loss: 966022.2500 - va
l_mse: 966022.2500
Epoch 4/5
4989/4989 [==============================] - 39s 8ms/step - loss: 1003696.9375 - mse: 1003696.9375 - val_loss: 958914.8125 - va
l_mse: 958914.8125
Epoch 5/5
4989/4989 [==============================] - 40s 8ms/step - loss: 1003325.2500 - mse: 1003325.2500 - val_loss: 1022399.5000 - v
al_mse: 1022399.5000

<keras.callbacks.History at 0x7f6e26f22d50>
```

Figure 6.26: Logs of the training process

The preceding output shows the model is overfitting. It achieved an MSE score of **1005740** on the training set and only **1070237** on the validation set. Now, train another model with L1 and L2 regularization.

23. Create five fully connected layers similar to the previous models and specify both L1 and L2 regularizers for the **kernel_regularizer** parameters. Use the value **0.001** for the regularizer factor. Save them into five variables, called **reg_fc1**, **reg_fc2**, **reg_fc3**, **reg_fc4**, and **reg_fc5**:

```
reg_fc1 = Dense(1048, input_shape=(5,), activation='relu', \
                kernel_regularizer=tf.keras.regularizers\
                              .l1_l2(l1=0.001, l2=0.001))
reg_fc2 = Dense(512, activation='relu', \
                kernel_regularizer=tf.keras.regularizers\
                              .l1_l2(l1=0.001, l2=0.001))
reg_fc3 = Dense(128, activation='relu', \
                kernel_regularizer=tf.keras.regularizers\
                              .l1_l2(l1=0.001, l2=0.001))
reg_fc4 = Dense(64, activation='relu', \
                kernel_regularizer=tf.keras.regularizers\
                              .l1_l2(l1=0.001, l2=0.001))
reg_fc5 = Dense(1, activation='relu')
```

24. Instantiate a sequential model using **tf.keras.Sequential()**, store it in a variable called **model2**, and add all five fully connected layers sequentially to the model using the **add()** method:

```
model2 = tf.keras.Sequential()
model2.add(reg_fc1)
```

```
model2.add(reg_fc2)
model2.add(reg_fc3)
model2.add(reg_fc4)
model2.add(reg_fc5)
```

25. Print the summary of the model:

```
model2.summary()
```

The output will be as follows:

```
Model: "sequential_1"
```

Layer (type)	Output Shape	Param #
dense_5 (Dense)	(None, 1048)	6288
dense_6 (Dense)	(None, 512)	537088
dense_7 (Dense)	(None, 128)	65664
dense_8 (Dense)	(None, 64)	8256
dense_9 (Dense)	(None, 1)	65

```
Total params: 617,361
Trainable params: 617,361
Non-trainable params: 0
```

Figure 6.27: Summary of the model architecture

26. Compile the model using the **compile()** method, specify the optimizer, and set **mse** as the loss and metric to be displayed:

```
optimizer = tf.keras.optimizers.Adam(0.1)
model2.compile(optimizer=optimizer, loss='mse', metrics=['mse'])
```

27. Start the model training process using the **fit()** method for five epochs and split the data into a validation set with 20% of the data:

```
model2.fit(train_data, train_target, epochs=5, \
           validation_split=0.2)
```

The output will be as follows:

```
Epoch 1/5
4989/4989 [==============================] - 61s 12ms/step - loss: 13579405.0000 - mse: 13579149.0000 - val_loss: 3970202.7500
- val_mse: 3969996.2500
Epoch 2/5
4989/4989 [==============================] - 62s 12ms/step - loss: 4028319.2500 - mse: 4028123.2500 - val_loss: 3970189.5000 -
val_mse: 3969996.2500
Epoch 3/5
4989/4989 [==============================] - 60s 12ms/step - loss: 4028304.7500 - mse: 4028129.5000 - val_loss: 3970171.0000 -
val_mse: 3969996.2500
Epoch 4/5
4989/4989 [==============================] - 61s 12ms/step - loss: 4028274.5000 - mse: 4028119.5000 - val_loss: 3970124.2500 -
val_mse: 3969996.2500
Epoch 5/5
4989/4989 [==============================] - 61s 12ms/step - loss: 4028207.0000 - mse: 4028115.5000 - val_loss: 3970040.5000 -
val_mse: 3969996.2500

<keras.callbacks.History at 0x7f6e2267d3d0>
```

Figure 6.28: Logs of the training process

With the addition of L1 and L2 regularization, the model has similar accuracy scores between the training (**4028182**) and test (**3970020**) sets. Therefore, the model is not overfitting much.

ACTIVITY 6.02: PREDICTING INCOME WITH BAYESIAN OPTIMIZATION FROM KERAS TUNER

Solution:

1. Open a new Jupyter notebook.

2. Import the pandas library and use **pd** as the alias:

```
import pandas as pd
```

3. Create a list called **usecols** containing the following column names: **AAGE**, **ADTIND**, **ADTOCC**, **SEOTR**, **WKSWORK**, and **PTOTVAL**:

```
usecols = ['AAGE','ADTIND','ADTOCC','SEOTR','WKSWORK', 'PTOTVAL']
```

4. Create a variable called **train_url** that contains the URL to the training set:

```
train_url = 'https://raw.githubusercontent.com/PacktWorkshops'\
            '/The-TensorFlow-Workshop/master/Chapter06'\
            '/dataset/census-income-train.csv'
```

5. Load the training dataset into a DataFrame called **train_data** using the **read_csv()** method, and provide the URL to the CSV file and the **usecols** list to the **usecols** parameter. Print the first five rows using the **head()** method:

```
train_data = pd.read_csv(train_url, usecols=usecols)
train_data.head()
```

6. You will get the following output:

	AAGE	ADTIND	ADTOCC	SEOTR	WKSWORK	PTOTVAL
0	73	0	0	0	0	1700.09
1	58	4	34	0	52	1053.55
2	18	0	0	0	0	991.95
3	9	0	0	0	0	1758.14
4	10	0	0	0	0	1069.16

Figure 6.29: First five rows of the training set

7. Extract the target variable (**PTOTVAL**) using the **pop ()** method, and save it in a variable called **train_target**:

```
train_target = train_data.pop('PTOTVAL')
```

8. Create a variable called **test_url** that contains the URL to the test set:

```
test_url = 'https://github.com/PacktWorkshops'\
           '/The-TensorFlow-Workshop/blob/master/Chapter06'\
           '/dataset/census-income-test.csv?raw=true'
```

9. Load the test dataset into a DataFrame called **X_test** using the **read_csv ()** method and provide the URL to the CSV file and the **usecols** list to the **usecols** parameter. Print the first five rows using the **head ()** method:

```
test_data = pd.read_csv(test_url, usecols=usecols)
test_data.head()
```

The output will be the following:

	AAGE	ADTIND	ADTOCC	SEOTR	WKSWORK	PTOTVAL
0	38	6	36	0	12	1032.38
1	44	37	12	0	26	1462.33
2	2	0	0	0	0	1601.75
3	35	29	3	2	52	1866.88
4	49	4	34	0	50	1394.54

Figure 6.30: First five rows of the test set

10. Extract the target variable (**PTOTVAL**) using the **pop ()** method, and save it in a variable called **test_target**:

```
test_target = test_data.pop('PTOTVAL')
```

11. Import the TensorFlow library and use **tf** as the alias. Then, import the **Dense** class from **tensorflow.keras.layers**:

```
import tensorflow as tf
from tensorflow.keras.layers import Dense
```

12. Set the seed as **8** using **tf.random.set_seed()** to get reproducible results:

```
tf.random.set_seed(8)
```

13. Define a function called **model_builder** to create a sequential model with the same architecture as *Activity 6.01, Predicting Income with L1 and L2 Regularizers*. But this time, provide a hyperparameter, **hp.Choice**, for the learning rate, **hp.Int** for the number of units for the input layer, and **hp.Choice** for L2 regularization:

```
def model_builder(hp):
model = tf.keras.Sequential()

hp_l2 = hp.Choice('l2', values = [0.1, 0.01, 0.001])
hp_units = hp.Int('units', min_value=128, max_value=512, step=64)

reg_fc1 = Dense(hp_units, input_shape=(5,), activation='relu', \
                kernel_regularizer=tf.keras.regularizers\
                                   .l2(l=hp_l2))
reg_fc2 = Dense(512, activation='relu', \
                kernel_regularizer=tf.keras.regularizers\
                                   .l2(l=hp_l2))
reg_fc3 = Dense(128, activation='relu', \
                kernel_regularizer=tf.keras.regularizers\
                                   .l2(l=hp_l2))
reg_fc4 = Dense(128, activation='relu', \
                kernel_regularizer=tf.keras.regularizers\
                                   .l2(l=hp_l2))
reg_fc5 = Dense(1)
model.add(reg_fc1)
model.add(reg_fc2)
model.add(reg_fc3)
model.add(reg_fc4)
model.add(reg_fc5)
hp_learning_rate = hp.Choice('learning_rate', \
                             values = [0.01, 0.001])
```

```
optimizer = tf.keras.optimizers.Adam(hp_learning_rate)
model.compile(optimizer=optimizer, loss='mse', metrics=['mse'])
return model
```

14. Install the **keras-tuner** package and then import it and assign it the **kt** alias:

```
!pip install keras-tuner
import kerastuner as kt
```

15. Instantiate a **BayesianOptimization** tuner, and assign **val_mse** to **objective** and **10** to **max_trials**:

```
tuner = kt.BayesianOptimization(model_builder, \
                                objective = 'val_mse', \
                                max_trials = 10)
```

16. Launch the hyperparameter search with **search()** on the training and test sets:

```
tuner.search(train_data, train_target, \
             validation_data=(test_data, test_target))
```

17. Extract the best hyperparameter combination (index **0**) with **get_best_hyperparameters()** and save it in a variable called **best_hps**:

```
best_hps = tuner.get_best_hyperparameters()[0]
```

18. Extract the best value for the number of units for the input layer, save it in a variable called **best_units**, and print its value:

```
best_units = best_hps.get('units')
best_units
```

You will get the following output:

```
128
```

The best value for the number of units of the input layer found by Hyperband is **128**.

19. Extract the best value for the learning rate, save it in a variable called **best_lr**, and print its value:

```
best_lr = best_hps.get('learning_rate')
best_lr
```

The best value for the learning rate hyperparameter found by Hyperband is **0.001**:

```
0.001
```

20. Extract the best value for the L2 regularization, save it in a variable called **best_l2**, and print its value:

```
best_l2 = best_hps.get('l2')
best_l2
```

21. The best value for the learning rate hyperparameter found by Hyperband is **0.001**:

```
0.001
```

22. Start the model training process using the **fit()** method for five epochs and use the test set for **validation_data**:

```
model = tuner.hypermodel.build(best_hps)
model.fit(X_train, y_train, epochs=5, \
          validation_data=(X_test, y_test))
```

You should get an output similar to the following:

```
Epoch 1/5
6236/6236 [==============================] - 27s 4ms/step - loss: 1057466.0000 - mse: 1057465.1250 - val_loss: 1018290.7500 - v
al_mse: 1018289.8750
Epoch 2/5
6236/6236 [==============================] - 27s 4ms/step - loss: 1004625.5625 - mse: 1004624.7500 - val_loss: 999515.1250 - va
l_mse: 999514.6250
Epoch 3/5
6236/6236 [==============================] - 28s 5ms/step - loss: 1000156.8750 - mse: 1000156.5000 - val_loss: 995369.2500 - va
l_mse: 995368.3750
Epoch 4/5
6236/6236 [==============================] - 26s 4ms/step - loss: 997135.5000 - mse: 997134.9375 - val_loss: 1004317.0000 - val
_mse: 1004316.2500
Epoch 5/5
6236/6236 [==============================] - 27s 4ms/step - loss: 994731.3125 - mse: 994730.7500 - val_loss: 989958.5000 - val_
mse: 989957.7500

<keras.callbacks.History at 0x7f8c2be79b50>
```

Figure 6.31: Logs of the training process

With Bayesian optimization, you found the best combination of hyperparameters for the number of units for the input layer (**128**), learning rate (**0.001**), and L2 regularization (**0.001**). With these hyperparameters, the final model achieved an MSE score of **994174** on the training set and **989335** on the test set. This is a great improvement from *Activity 6.01, Predicting Income with L1 and L2 Regularizers*, and the model is not overfitting much.

CHAPTER 7: CONVOLUTIONAL NEURAL NETWORKS

ACTIVITY 7.01: BUILDING A CNN WITH MORE ANN LAYERS

Solution:

There are several possible ways to arrive at a solution for this activity. The following steps describe one of these methods and are similar to those used on the **CIFAR-10** dataset earlier in the chapter:

1. Start a new Jupyter notebook.

2. Import the TensorFlow library:

```
import tensorflow as tf
```

3. Import the additional libraries needed:

```
import numpy as np
import matplotlib.pyplot as plt
import tensorflow as tf
import tensorflow_datasets as tfds
from tensorflow.keras.layers import Input, Conv2D, Dense, Flatten, \
    Dropout, Activation, Rescaling
from tensorflow.keras.models import Model
from sklearn.metrics import confusion_matrix, ConfusionMatrixDisplay
```

4. Load the **CIFAR-100** dataset directly from **tensorflow_datasets** and view its properties:

```
(c100_train_dataset, c100_test_dataset), \
dataset_info = tfds.load('cifar100',\
                         split = ['train', 'test'],\
                         data_dir = 'content/Cifar100/',\
                         shuffle_files = True,\
                         as_supervised = True,\
                         with_info = True)
assert isinstance(c100_train_dataset, tf.data.Dataset)

image_shape = dataset_info.features["image"].shape
print(f'Shape of Images in the Dataset: \t{image_shape}')
num_classes = dataset_info.features["label"].num_classes
print(f'Number of Classes in the Dataset: \t{num_classes}')
```

```
names_of_classes = dataset_info.features["label"].names
print(f'Names of Classes in the Dataset: \t{names_of_classes}\n')

print(f'Total examples in Train Dataset: \
    \t{len(c100_train_dataset)}')
print(f'Total examples in Test Dataset: \
    \t{len(c100_test_dataset)}')
```

This will give the following output:

```
Shape of Images in the Dataset:        (32, 32, 3)
Number of Classes in the Dataset:      100
Names of Classes in the Dataset:       ['apple', 'aquarium_fish', 'baby', 'bear',

Total examples in Train Dataset:       50000
Total examples in Test Dataset:        10000
```

Figure 7.42: Properties of the CIFAR-100 dataset

5. Use a rescaling layer to rescale images. Then, build a test and train data pipeline by rescaling, caching, shuffling, batching, and prefetching the images:

```
normalization_layer = Rescaling(1./255)

c100_train_dataset = c100_train_dataset.map\
                    (lambda x, y: (normalization_layer(x), y), \
                     num_parallel_calls = \
                     tf.data.experimental.AUTOTUNE)
c100_train_dataset = c100_train_dataset.cache()
c100_train_dataset = c100_train_dataset.shuffle\
                    (len(c100_train_dataset))
c100_train_dataset = c100_train_dataset.batch(32)
c100_train_dataset = c100_train_dataset.prefetch(tf.data.
experimental.AUTOTUNE)

c100_test_dataset = c100_test_dataset.map\
                    (lambda x, y: (normalization_layer(x), y), \
                     num_parallel_calls = \
                     tf.data.experimental.AUTOTUNE)
c100_test_dataset = c100_test_dataset.cache()
c100_test_dataset = c100_test_dataset.batch(128)
c100_test_dataset = \
c100_test_dataset.prefetch(tf.data.experimental.AUTOTUNE)
```

6. Build the model using the functional API:

```
input_layer = Input(shape=image_shape)
x = Conv2D(filters = 32, kernel_size = \
            (3, 3), strides=2)(input_layer)
x = Activation('relu')(x)


x = Conv2D(filters = 64, kernel_size = (3, 3), strides=2)(x)
x = Activation('relu')(x)


x = Conv2D(filters = 128, kernel_size = (3, 3), strides=2)(x)
x = Activation('relu')(x)


x = Flatten()(x)
x = Dropout(rate = 0.5)(x)


x = Dense(units = 1024)(x)
x = Activation('relu')(x)
x = Dropout(rate = 0.2)(x)


x = Dense(units = num_classes)(x)
output = Activation('softmax')(x)


c100_classification_model = Model(input_layer, output)
```

7. Compile and fit the model:

```
c100_classification_model.compile(\
    optimizer='adam', \
    loss='sparse_categorical_crossentropy', \
    metrics = ['accuracy'], loss_weights = None, \
    weighted_metrics = None, run_eagerly = None, \
    steps_per_execution = None
)


history = c100_classification_model.fit\
        (c100_train_dataset, \
        validation_data=c100_test_dataset, \
        epochs=15)
```

The output will look like the following image:

```
Epoch 1/15
1563/1563 [==============================] - 13s 4ms/step - loss: 6.3395 - accuracy: 0.0375 - val_loss: 4.0599 - val_accuracy: 0.0707
Epoch 2/15
1563/1563 [==============================] - 5s 3ms/step - loss: 3.9944 - accuracy: 0.0829 - val_loss: 3.7596 - val_accuracy: 0.1258
Epoch 3/15
1563/1563 [==============================] - 5s 3ms/step - loss: 3.7720 - accuracy: 0.1189 - val_loss: 3.5724 - val_accuracy: 0.1675
Epoch 4/15
1563/1563 [==============================] - 4s 3ms/step - loss: 3.6113 - accuracy: 0.1491 - val_loss: 3.4189 - val_accuracy: 0.1909
Epoch 5/15
1563/1563 [==============================] - 4s 3ms/step - loss: 3.4475 - accuracy: 0.1772 - val_loss: 3.2566 - val_accuracy: 0.2149
Epoch 6/15
1563/1563 [==============================] - 4s 3ms/step - loss: 3.3013 - accuracy: 0.2020 - val_loss: 3.1660 - val_accuracy: 0.2416
Epoch 7/15
1563/1563 [==============================] - 4s 3ms/step - loss: 3.1800 - accuracy: 0.2207 - val_loss: 3.0630 - val_accuracy: 0.2589
Epoch 8/15
1563/1563 [==============================] - 4s 3ms/step - loss: 3.0794 - accuracy: 0.2403 - val_loss: 2.9940 - val_accuracy: 0.2709
Epoch 9/15
1563/1563 [==============================] - 4s 3ms/step - loss: 2.9897 - accuracy: 0.2561 - val_loss: 2.9154 - val_accuracy: 0.2869
Epoch 10/15
1563/1563 [==============================] - 5s 3ms/step - loss: 2.9122 - accuracy: 0.2728 - val_loss: 2.8597 - val_accuracy: 0.3027
Epoch 11/15
1563/1563 [==============================] - 5s 3ms/step - loss: 2.8356 - accuracy: 0.2852 - val_loss: 2.8523 - val_accuracy: 0.2980
Epoch 12/15
1563/1563 [==============================] - 4s 3ms/step - loss: 2.7869 - accuracy: 0.2950 - val_loss: 2.7875 - val_accuracy: 0.3176
Epoch 13/15
1563/1563 [==============================] - 4s 3ms/step - loss: 2.7299 - accuracy: 0.3097 - val_loss: 2.7684 - val_accuracy: 0.3194
Epoch 14/15
1563/1563 [==============================] - 5s 3ms/step - loss: 2.6782 - accuracy: 0.3192 - val_loss: 2.7655 - val_accuracy: 0.3211
Epoch 15/15
1563/1563 [==============================] - 5s 3ms/step - loss: 2.6242 - accuracy: 0.3305 - val_loss: 2.7470 - val_accuracy: 0.3285
```

Figure 7.43: Model fit

8. Plot the loss and accuracy by using the following code:

```python
def plot_trend_by_epoch(tr_values, val_values, title):
    epoch_number = range(len(tr_values))
    plt.plot(epoch_number, tr_values, 'r')
    plt.plot(epoch_number, val_values, 'b')
    plt.title(title)
    plt.xlabel('epochs')
    plt.legend(['Training '+title, 'Validation '+title])
    plt.figure()

hist_dict = history.history

tr_loss, val_loss = hist_dict['loss'], \
                    hist_dict['val_loss']
plot_trend_by_epoch(tr_loss, val_loss, "Loss")

tr_accuracy, val_accuracy = hist_dict['accuracy'], \
                            hist_dict['val_accuracy']
plot_trend_by_epoch(tr_accuracy, val_accuracy, "Accuracy")
```

Loss plot would look like the following:

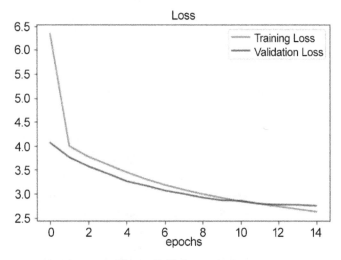

Figure 7.44: Loss plot

Accuracy plot would look like the following:

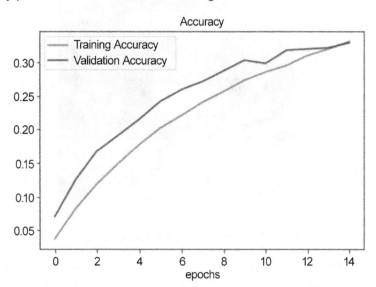

Figure 7.45: Accuracy plot

9. Display a misclassified example. Use the following code:

```
test_labels = []
test_images = []
for image, label in tfds.as_numpy(c100_test_dataset.unbatch()):
    test_images.append(image)
```

```
    test_labels.append(label)
test_labels = np.array(test_labels)

predictions = c100_classification_model.predict\
            (c100_test_dataset).argmax(axis=1)

incorrect_predictions = np.where(predictions != test_labels)[0]
index = np.random.choice(incorrect_predictions)

plt.imshow(test_images[index])
print(f'True label: {names_of_classes[test_labels[index]]}')
print(f'Predicted label: {names_of_classes[predictions[index]]}')
```

This will produce the following output:

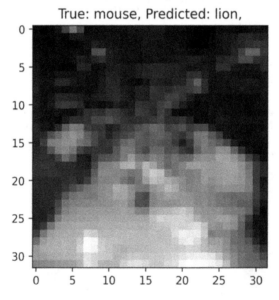

Figure 7.46: Wrong classification example

The output shows an example of a wrong classification: the prediction was lion, and the true value was mouse. In this activity, the number of classes was 100, which makes it significantly more difficult than in *Exercise 7.05, Building a CNN*, in which there were only 10 classes. Nevertheless, you can see that after 15 epochs, the accuracy continued to increase, and loss continued to decrease even on the validation dataset. You could then expect better model performance if you were to let the model train for more epochs.

CHAPTER 8: PRE-TRAINED NETWORKS

ACTIVITY 8.01: FRUIT CLASSIFICATION WITH FINE-TUNING

Solution:

1. Open a new Jupyter notebook.

2. Import the TensorFlow library as **tf**:

```
import tensorflow as tf
```

3. Create a variable called **file_url** containing a link to the dataset:

```
file_url = 'https://github.com/PacktWorkshops/'\
           'The-TensorFlow-Workshop/blob/master'\
           '/Chapter08/dataset/fruits360.zip'
```

4. Download the dataset using **tf.keras.get_file** with **'fruits360.zip'**, **origin=file_url**, and **extract=True** as parameters, and save the result to a variable called **zip_dir**:

```
zip_dir = tf.keras.utils.get_file('fruits360.zip', \
                                   origin=file_url, extract=True)
```

5. Import the **pathlib** library:

```
import pathlib
```

6. Create a variable called **path** containing the full path to the **fruits360_filtered** directory using **pathlib.Path(zip_dir).parent**:

```
path = pathlib.Path(zip_dir).parent / 'fruits360_filtered'
```

7. Create two variables called **train_dir** and **validation_dir** that take the full path to the train (**Training**) and validation (**Test**) folders, respectively:

```
train_dir = path / 'Training'
validation_dir = path / 'Test'
```

8. Create two variables called **total_train** and **total_val** that get the number of images for the training and validation sets:

```
total_train = 11398
total_val = 4752
```

9. Import **ImageDataGenerator** from
   tensorflow.keras.preprocessing:

```
from tensorflow.keras.preprocessing.image
    import ImageDataGenerator
```

10. Create an **ImageDataGenerator** model called **train_img_gen** with
    data augmentation:

```
train_img_gen = ImageDataGenerator(rescale=1./255, \
                            rotation_range=40, \
                            width_shift_range=0.1, \
                            height_shift_range=0.1, \
                            shear_range=0.2, \
                            zoom_range=0.2, \
                            horizontal_flip=True, \
                            fill_mode='nearest'))
```

11. Create an **ImageDataGenerator** mode called **val_img_gen** with rescaling
    by dividing by **255**:

```
val_img_gen = ImageDataGenerator(rescale=1./255)
```

12. Create four variables called **batch_size**, **img_height**, **img_width**, and
    channel that take the values **32**, **224**, **224**, and **3**, respectively:

```
Batch_size = 32
img_height = 224
img_width = 224
channel = 3
```

13. Create a data generator called **train_data_gen** using
    flow_from_directory() and specify the batch size, training folder, and
    target size:

```
train_data_gen = train_image_generator.flow_from_directory\
            (batch_size=batch_size, directory=train_dir, \
             target_size=(img_height, img_width))
```

14. Create a data generator called **val_data_gen** using
    flow_from_directory() and specify the batch size, validation folder, and
    target size:

```
val_data_gen = validation_image_generator.flow_from_directory\
                (batch_size=batch_size, directory=validation_dir,\ ,
                target_size=(img_height, img_width))
```

15. Import **numpy** as **np**, **tensorflow** as **tf**, and **layers** from
    tensorflow.keras:

```
import numpy as np
import tensorflow as tf
from tensorflow.keras import layers
```

16. Set **8** as the seed for **numpy** and **tensorflow**:

```
np.random.seed(8)
tf.random.set_seed(8)
```

17. Import **NASNetMobile** from **tensorflow.keras.applications**:

```
from tensorflow.keras.applications
import NASNetMobile
```

18. Instantiate a **NASNetMobile** model into a variable called **base_model**:

```
base_model = NASNetMobile(input_shape=(img_height, img_width, \
                                    channel), \
                        weights='imagenet', include_top=False)
```

19. Print a summary of this **NASNetMobile** model:

```
base_model.summary()
```

The expected output is as follows:

```
Model: "NASNet"
```

Layer (type)	Output Shape	Param #	Connected to
input_3 (InputLayer)	[(None, 224, 224, 3)	0	
stem_conv1 (Conv2D)	(None, 111, 111, 32)	864	input_3[0][0]
stem_bn1 (BatchNormalization)	(None, 111, 111, 32)	128	stem_conv1[0][0]
activation_376 (Activation)	(None, 111, 111, 32)	0	stem_bn1[0][0]
reduction_conv_1_stem_1 (Conv2D	(None, 111, 111, 11)	352	activation_376[0][0]
reduction_bn_1_stem_1 (BatchNor	(None, 111, 111, 11)	44	reduction_conv_1_stem_1[0][0]
activation_377 (Activation)	(None, 111, 111, 11)	0	reduction_bn_1_stem_1[0][0]

Figure 8.8: Summary of the model

20. Create a new model using **tf.keras.Sequential()** by adding the base model to the **Flatten** and **Dense** layers. Save this model to a variable called **model**:

```
model = tf.keras.Sequential([base_model,\
                            layers.Flatten(),\
                            layers.Dense(500, \
                                        activation='relu'), \
                            layers.Dense(120, \
                                        activation='softmax')])
```

21. Instantiate a **tf.keras.optimizers.Adam()** class with **0.001** as the learning rate and save it to a variable called **optimizer**:

```
optimizer = tf.keras.optimizers.Adam(0.001)
```

22. Compile the neural network using the **compile()** method with **categorical_crossentropy** as the loss function, an Adam optimizer with a learning rate of **0.001**, and **accuracy** as the metric to be displayed:

```
model.compile(loss='categorical_crossentropy', \
             optimizer=optimizer, metrics=['accuracy'])
```

23. Fit the neural networks with **fit()** method. This model may take a few minutes to train:

```
model.fit(train_data_gen,
          steps_per_epoch=len(features_train) // batch_size,\
          epochs=5,\
          validation_data=val_data_gen,\
          validation_steps=len(features_test) // batch_size\
    )
```

The expected output is as follows:

```
Epoch 1/5
712/712 [==============================] - 148s 207ms/step - loss: 2.7766 - accuracy: 0.4375 - val_loss: 3.0142 - val_accuracy: 0.4840
Epoch 2/5
712/712 [==============================] - 144s 202ms/step - loss: 0.5059 - accuracy: 0.8475 - val_loss: 3.0363 - val_accuracy: 0.5149
Epoch 3/5
712/712 [==============================] - 142s 200ms/step - loss: 0.2538 - accuracy: 0.9220 - val_loss: 0.9776 - val_accuracy: 0.7919
Epoch 4/5
712/712 [==============================] - 142s 200ms/step - loss: 0.2049 - accuracy: 0.9380 - val_loss: 0.8568 - val_accuracy: 0.8523
Epoch 5/5
712/712 [==============================] - 142s 199ms/step - loss: 0.1554 - accuracy: 0.9549 - val_loss: 0.8855 - val_accuracy: 0.8264
<tensorflow.python.keras.callbacks.History at 0x7fa90f22c860>
```

Figure 8.9: Epochs of the trained model

In this activity, you used fine-tuning to customize a **NASNetMobile** model pre-trained on ImageNet on a dataset containing images of fruit. You froze the first 700 layers of this model and trained only the last few on five epochs. You achieved an accuracy score of **0.9549** for the training set and **0.8264** for the test set.

ACTIVITY 8.02: TRANSFER LEARNING WITH TENSORFLOW HUB

Solution:

1. Open a new Jupyter notebook.

2. Import the TensorFlow library:

```
import tensorflow as tf
```

3. Create a variable called **file_url** containing a link to the dataset:

```
file_url = 'https://storage.googleapis.com'\
           '/mledu-datasets/cats_and_dogs_filtered.zip'
```

4. Download the dataset using **tf.keras.get_file** with **cats_and_dogs.zip**, **origin=file_url**, and **extract=True** as parameters and save the result to a variable called **zip_dir**:

```
zip_dir = tf.keras.utils.get_file('cats_and_dogs.zip', \
                                 origin=file_url, extract=True)
```

5. Import the **pathlib** library:

```
import pathlib
```

6. Create a variable called **path** containing the full path to the **cats_and_dogs_filtered** directory using **pathlib.Path(zip_dir).parent**:

```
path = pathlib.Path(zip_dir).parent / 'cats_and_dogs_filtered'
```

7. Create two variables called **train_dir** and **validation_dir** that take the full path to the **train** and **validation** folders:

```
train_dir = path / 'train'
validation_dir = path / 'validation'
```

8. Create two variables called **total_train** and **total_val** that will get the number of images for the training and validation sets (**2000** and **1000**, respectively):

```
total_train = 2000
total_val = 1000
```

9. Import **ImageDataGenerator** from **tensorflow.keras.preprocessing**:

```
from tensorflow.keras.preprocessing.image
import ImageDataGenerator
```

10. Instantiate two **ImageDataGenerator** classes and call them **train_image_generator** and **validation_image_generator**. These will rescale images by dividing by **255**:

```
train_image_generator = ImageDataGenerator(rescale=1./255)
validation_image_generator = ImageDataGenerator(rescale=1./255)
```

11. Create three variables called **batch_size**, **img_height**, and **img_width** that take the values **32**, **224**, and **224**, respectively:

```
batch_size = 32
img_height = 224
img_width = 224
```

12. Create a data generator called **train_data_gen** using **flow_from_directory()** and specify the batch size, the path to the training folder, target size, and mode of the class:

```
train_data_gen = train_image_generator.flow_from_directory\
                (batch_size=batch_size, directory=train_dir, \
                 shuffle=True, target_size=(img_height, \
                                            img_width), \
                 class_mode='binary')
```

13. Create a data generator called **val_data_gen** using **flow_from_directory()** and specify the batch size, paths to the validation folder, target size, and mode of the class:

```
val_data_gen = validation_image_generator.flow_from_directory\
                (batch_size=batch_size, \
                 directory=validation_dir, \
                 target_size=(img_height, img_width), \
                 class_mode='binary')
```

14. Import **numpy** as **np**, **tensorflow** as **tf**, and **layers** from **tensorflow.keras**:

```
import numpy as np
import tensorflow as tf
from tensorflow.keras import layers
```

15. Set **8** (this is totally arbitrary) as **seed** for numpy and tensorflow:

```
np.random.seed(8)
tf.random.set_seed(8)
```

16. Import **tensorflow_hub**, as shown here:

```
import tensorflow_hub as hub
```

17. Load the EfficientNet B0 feature vector from TensorFlow Hub:

```
MODULE_HANDLE = 'https://tfhub.dev/google/efficientnet/b0'\
                '/feature-vector/1'
module = hub.load(MODULE_HANDLE)
```

18. Create a new model that combines the EfficientNet B0 module with two new top layers, with **500** and **1** as units, and ReLu and sigmoid as the activation functions:

```
model = tf.keras.Sequential\
        ([hub.KerasLayer(MODULE_HANDLE,\
                        input_shape=(224, 224, 3)),
          layers.Dense(500, activation='relu'),
          layers.Dense(1, activation='sigmoid')])
```

19. Compile this model by providing **binary_crossentropy** as the **loss** function, an Adam optimizer with a learning rate of **0.001**, and **accuracy** as the metric to be displayed:

```
model.compile(loss='binary_crossentropy', \
              optimizer=tf.keras.optimizers.Adam(0.001), \
              metrics=['accuracy'])
```

20. Fit the model and provide the train and validation data generators. Run it for five epochs:

```
model.fit(train_data_gen, \
          steps_per_epoch = total_train // batch_size, \
          epochs=5, \
          validation_data=val_data_gen, \
          validation_steps=total_val // batch_size)
```

The expected output will be as follows:

```
Epoch 1/5
62/62 [==============================] - 28s 240ms/step - loss: 0.0554 - accuracy: 0.9776 - val_loss: 0.0247 - val_accuracy: 0.9899
Epoch 2/5
62/62 [==============================] - 14s 223ms/step - loss: 0.0116 - accuracy: 0.9959 - val_loss: 0.0253 - val_accuracy: 0.9919
Epoch 3/5
62/62 [==============================] - 14s 221ms/step - loss: 0.0032 - accuracy: 0.9995 - val_loss: 0.0245 - val_accuracy: 0.9909
Epoch 4/5
62/62 [==============================] - 14s 222ms/step - loss: 0.0011 - accuracy: 1.0000 - val_loss: 0.0265 - val_accuracy: 0.9919
Epoch 5/5
62/62 [==============================] - 14s 222ms/step - loss: 5.1736e-04 - accuracy: 1.0000 - val_loss: 0.0243 - val_accuracy: 0.9929
<keras.callbacks.History at 0x7f04bc0a6b10>
```

Figure 8.10: Model training output

In this activity, you achieved a very high accuracy score (with **1** and **0.99** for the training and test sets, respectively), using transfer learning from TensorFlow Hub. You used the **EfficientNet B0** feature vector combined with two custom final layers, and your final model is almost perfectly predicting images of cats and dogs.

CHAPTER 9: RECURRENT NEURAL NETWORKS

ACTIVITY 9.01: BUILDING AN RNN WITH MULTIPLE LSTM LAYERS TO PREDICT POWER CONSUMPTION

Solution:

Perform the following steps to complete this activity.

1. Open a new Jupyter or Colab notebook.

2. Import the libraries needed. Use **numpy**, **pandas**, **datetime**, and **MinMaxScaler** to scale the dataset between zero and one:

```
import numpy as np
import pandas as pd
import datetime
from sklearn.preprocessing import MinMaxScaler
```

3. Use the **read_csv()** function to read in your CSV file and store your dataset in a pandas DataFrame, **data**:

```
data = pd.read_csv("household_power_consumption.csv")
```

4. Create a new column, **Datetime**, by combining **Date** and **Time** columns using the following code:

```
data['Date'] = pd.to_datetime(data['Date'], format="%d/%m/%Y")
data['Datetime'] = data['Date'].dt.strftime('%Y-%m-%d') + ' ' \
                   +  data['Time']
data['Datetime'] = pd.to_datetime(data['Datetime'])
```

5. Sort the DataFrame in ascending order using the **Datetime** column:

```
data = data.sort_values(['Datetime'])
```

6. Create a list called **num_cols** containing the columns that have numeric values – **Global_active_power**, **Global_reactive_power**, **Voltage**, **Global_intensity**, **Sub_metering_1**, **Sub_metering_2**, and **Sub_metering_3**:

```
num_cols = ['Global_active_power', 'Global_reactive_power', \
            'Voltage', 'Global_intensity', 'Sub_metering_1', \
            'Sub_metering_2', 'Sub_metering_3']
```

7. Convert all columns listed in **num_cols** to a numeric datatype:

```
for col in num_cols:
    data[col] = pd.to_numeric(data[col], errors='coerce')
```

8. Call the **head()** function on your data to take a look at the first five rows of your DataFrame:

```
data.head()
```

You should get the following output:

	Date	Time	Global_active_power	Global_reactive_power	Voltage	Global_intensity	Sub_metering_1	Sub_metering_2	Sub_metering_3	Datetime
0	2007-01-01	0:00:00	2.580	0.136	241.97	10.6	0.0	0.0	0.0	2007-01-01 00:00:00
1	2007-01-01	0:01:00	2.552	0.100	241.75	10.4	0.0	0.0	0.0	2007-01-01 00:01:00
2	2007-01-01	0:02:00	2.550	0.100	241.64	10.4	0.0	0.0	0.0	2007-01-01 00:02:00
3	2007-01-01	0:03:00	2.550	0.100	241.71	10.4	0.0	0.0	0.0	2007-01-01 00:03:00
4	2007-01-01	0:04:00	2.554	0.100	241.98	10.4	0.0	0.0	0.0	2007-01-01 00:04:00

Figure 9.40: First five rows of the DataFrame

9. Call **tail()** on your data to take a look at the last five rows of your DataFrame:

```
data.tail()
```

You should get the following output:

	Date	Time	Global_active_power	Global_reactive_power	Voltage	Global_intensity	Sub_metering_1
260635	2007-06-30	23:55:00	2.880	0.360	239.01	12.0	0.0
260636	2007-06-30	23:56:00	2.892	0.358	238.86	12.2	0.0
260637	2007-06-30	23:57:00	2.882	0.280	239.05	12.0	0.0
260638	2007-06-30	23:58:00	2.660	0.290	238.98	11.2	0.0
260639	2007-06-30	23:59:00	2.548	0.354	239.25	10.6	0.0

Figure 9.41: Last five rows of the DataFrame

10. Iterate through columns in **num_cols** and fill in missing values with the average using the following code:

```
for col in num_cols:
    data[col].fillna(data[col].mean(), inplace=True)
```

11. Use **drop()** to remove **Date**, **Time**, **Global_reactive_power**, and **Datetime** columns from your DataFrame and save the results in a variable called **df**:

```
df = data.drop(['Date', 'Time', 'Global_reactive_power', 'Datetime'], \
                axis = 1)
```

12. Create a scaler from **MinMaxScaler** to your DataFrame to numbers between zero and one. Use **fit_transform** to fit the model to the data and then transform the data according to the fitted model:

```
scaler = MinMaxScaler()
scaled_data = scaler.fit_transform(df)
scaled_data
```

You should get the following output:

```
array([[0.23592747, 0.67445255, 0.22173913, 0.        , 0.        ,
         0.        ],
       [0.23328296, 0.66642336, 0.2173913 , 0.        , 0.        ,
         0.        ],
       [0.23309407, 0.66240876, 0.2173913 , 0.        , 0.        ,
         0.        ],
       ...,
       [0.26445032, 0.56788321, 0.25217391, 0.        , 0.        ,
         0.9        ],
       [0.24348319, 0.56532847, 0.23478261, 0.        , 0.        ,
         0.9        ],
       [0.23290518, 0.57518248, 0.22173913, 0.        , 0.01282051,
         0.85        ]])
```

Figure 9.42: Standardized training data

The preceding screenshot shows the data has been standardized. Values sit between 0 and 1 now.

13. Create two empty lists called **X** and **y** that will be used to store features and target variables:

```
X = []
y = []
```

14. Create a training dataset that has the previous 60 minutes' power consumption so that you can predict the value for the next minute. Use a **for** loop to create data in 60 time steps:

```
for i in range(60, scaled_data.shape[0]):
    X.append(scaled_data [i-60:i])
    y.append(scaled_data [i, 0])
```

15. Convert **X** and **y** into NumPy arrays in preparation for training your model:

```
X, y = np.array(X), np.array(y)
```

16. Split the dataset into training and testing sets with data before and after the index **217440**, respectively:

```
X_train = X[:217440]
y_train = y[:217440]
X_test = X[217440:]
y_test = y[217440:]
```

17. You will need some additional libraries for building LSTM. Use **Sequential** to initialize the neural net, **Dense** to add a dense layer, **LSTM** to add an LSTM layer, and **Dropout** to help prevent overfitting:

```
from tensorflow.keras import Sequential
from tensorflow.keras.layers import Dense, LSTM, Dropout
```

18. Initialize your neural network. Add LSTM layers with **20**, **40**, and **80** units. Use a ReLU activation function and set **return_sequences** to **True**. The **input_shape** should be the dimensions of your training set (the number of features and days). Finally, add your dropout layer:

```
regressor = Sequential()

regressor.add(LSTM(units= 20, activation = 'relu',\
                   return_sequences = True,\
                   input_shape = (X_train.shape[1], X_train.
shape[2])))
regressor.add(Dropout(0.5))

regressor.add(LSTM(units= 40, \
                   activation = 'relu', \
                   return_sequences = True))
regressor.add(Dropout(0.5))
```

```
regressor.add(LSTM(units= 80, \
                 activation = 'relu'))
regressor.add(Dropout(0.5))


regressor.add(Dense(units = 1))
```

19. Print the architecture of the model using the **summary()** function:

```
regressor.summary()
```

The preceding command gives valuable information about the model, layers, and parameters:

```
Model: "sequential"
```

Layer (type)	Output Shape	Param #
lstm (LSTM)	(None, 60, 20)	2160
dropout (Dropout)	(None, 60, 20)	0
lstm_1 (LSTM)	(None, 60, 40)	9760
dropout_1 (Dropout)	(None, 60, 40)	0
lstm_2 (LSTM)	(None, 80)	38720
dropout_2 (Dropout)	(None, 80)	0
dense (Dense)	(None, 1)	81

```
Total params: 50,721
Trainable params: 50,721
Non-trainable params: 0
```

Figure 9.43: Model summary

20. Use the **compile()** method to configure your model for training. Select Adam as your optimizer and mean squared error to measure your loss function:

```
regressor.compile(optimizer='adam', loss = 'mean_squared_error')
```

21. Fit your model and set it to run on two epochs. Set your batch size to **32**:

```
regressor.fit(X_train, y_train, epochs=2, batch_size=32)
```

22. Save the predictions on the test set in a variable called **y_pred** using **regressor.predict(X_test)**:

```
y_pred = regressor.predict(X_test)
```

23. Take a look at the real household power consumption and your predictions for the last hour of data from your test set:

```
plt.figure(figsize=(14,5))
plt.plot(y_test[-60:], color = 'black', \
         label = "Real Power Consumption")
plt.plot(y_pred[-60:], color = 'gray', \
         label = 'Predicted Power Consumption')
plt.title('Power Consumption Prediction')
plt.xlabel('time')
plt.ylabel('Power Consumption')
plt.legend()
plt.show()
```

You should get the following output:

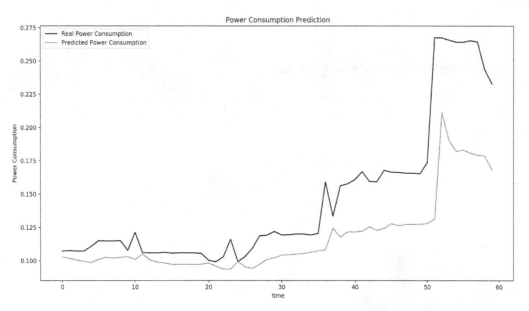

Figure 9.44: Household power consumption prediction visualization

As you can see in *Figure 9.44*, your results are pretty good. You can observe that for the most part, your predictions are close to the actual values.

ACTIVITY 9.02: BUILDING AN RNN FOR PREDICTING TWEETS' SENTIMENT

Solution:

Perform the following steps to complete this activity:

1. Open a new Jupyter or Colab notebook.

2. Import the libraries needed. Use **numpy** for computation and **pandas** to work with your dataset:

```
import numpy as np
import pandas as pd
```

3. Use the **read_csv** method to read in your CSV file and store your dataset in a pandas DataFrame, **data**:

```
data = pd.read_csv("https://raw.githubusercontent.com"\
                   "/PacktWorkshops/The-TensorFlow-Workshop"\
                   "/master/Chapter09/Datasets/tweets.csv")
```

4. Call the **head()** method on your data to take a look at the first five rows of your DataFrame:

```
data.head()
```

You should get the following output:

	tweet_id	airline_sentiment	airline_sentiment_confidence	negativereason	negativereason_confidence	airline
0	570306133677760513	neutral	1.0000	NaN	NaN	Virgin America
1	570301130888122368	positive	0.3486	NaN	0.0000	Virgin America
2	570301083672813571	neutral	0.6837	NaN	NaN	Virgin America
3	570301031407624196	negative	1.0000	Bad Flight	0.7033	Virgin America
4	570300817074462722	negative	1.0000	Can't Tell	1.0000	Virgin America

Figure 9.45: First five rows of the DataFrame

In the preceding screenshot, you can see the different sentiments stored in the **airline_sentiment** column.

5. Call **tail()** on your data to take a look at the last five rows of your DataFrame:

```
data.tail()
```

You should get the following output:

	tweet_id	airline_sentiment	airline_sentiment_confidence	negativereason	negativereason_confidence
14635	569587686496825344	positive	0.3487	NaN	0.0000
14636	569587371693355008	negative	1.0000	Customer Service Issue	1.0000
14637	569587242672398336	neutral	1.0000	NaN	NaN
14638	569587188687634433	negative	1.0000	Customer Service Issue	0.6659
14639	569587140490866689	neutral	0.6771	NaN	0.0000

Figure 9.46: Last five rows of the DataFrame

6. Create a new DataFrame called **df** that will have only **text** as features and **airline_sentiment** as the target variable:

```
df = data[['text','airline_sentiment']]
```

7. Subset **df** by removing all rows where **airline_sentiment** is equal to **neutral** by using the following command:

```
df = df[df['airline_sentiment'] != 'neutral']
```

8. Transform the **airline_sentiment** column to a numeric type by replacing **negative** with **0** and **positive** with **1**. Save the result to a variable, **y**:

```
y = df['airline_sentiment'].map({'negative':0, 'positive':1}).values
```

9. Create a variable, **X**, that will contain the data from the text column in **df**:

```
X = df['text']
```

10. Import **Tokenizer** from **tensorflow.keras.preprocessing.text** and **pad_sequences** from **tensorflow.keras.preprocessing.sequence**:

```
from tensorflow.keras.preprocessing.text import Tokenizer
from tensorflow.keras.preprocessing.sequence \
    import pad_sequences
```

11. Instantiate a **Tokenizer()** class with **num_words** equal to **10000**. This will keep only the first 10,000 most frequent words. Save it into a variable, **tokenizer**:

```
tokenizer = Tokenizer(num_words=10000)
```

12. Fit **tokenizer** on the data **X**:

```
tokenizer.fit_on_texts(X)
```

13. Print the vocabulary from **tokenizer**:

```
tokenizer.word_index
```

You should get output like the following:

```
{'to': 1,
 'the': 2,
 'i': 3,
 'a': 4,
 'united': 5,
 'you': 6,
 'for': 7,
 'flight': 8,
 'and': 9,
 'on': 10,
 'my': 11,
 'usairways': 12,
 'americanair': 13,
 'is': 14,
```

Figure 9.47: Vocabulary defined by tokenizer

From the output vocabulary, you can see the word **to** has been assigned the index **1**, **the** is assigned **2**, and so on. You can use it to map the raw text into a numerical version of it.

14. Create the **vocab_size** variable, to contain the length of the tokenizer vocabulary plus an additional character that will be used for unknown words:

```
vocab_size = len(tokenizer.word_index) + 1
```

15. Transform the raw text from **X** to an encoded version using the vocabulary from **tokenizer**. Save the result in a variable called **encoded_tweets**:

```
encoded_tweets = tokenizer.texts_to_sequences(X)
```

16. Pad **encoded_tweets** with **0** at the end for a maximum of 280 characters. Save the result in a variable called **padded_tweets**:

```
padded_tweets = pad_sequences(encoded_tweets, maxlen=280,
padding='post')
```

17. Print the shape of **padded_tweets**:

```
padded_tweets.shape
```

You should get the following result:

```
(11541, 280)
```

18. As you can see, prepared tweets now all have the same length, that is, 280 characters.

19. Randomly permute the indices of **padded_tweets**. Save the result in the **indices** variable:

```
indices = np.random.permutation(padded_tweets.shape[0])
```

20. Create two variables, **train_idx** and **test_idx**, to contain the first 10,000 indices and the remaining ones respectively:

```
train_idx = indices[:10000]
test_idx = indices[10000:]
```

21. Using **padded_tweets** and **y**, split the data into training and testing sets. Save them into four different variables called **X_train**, **X_test**, **y_train**, and **y_test**:

```
X_train = padded_tweets[train_idx,]
X_test = padded_tweets[test_idx,]
y_train = y[train_idx,]
y_test = y[test_idx,]
```

22. You will need some additional libraries to build your model. Import **Sequential**, **Dense**, **LSTM**, **Dropout**, and **Embedding** using the following code:

```
from tensorflow.keras import Sequential
from tensorflow.keras.layers import Dense, LSTM, Dropout, Embedding
```

23. Initialize your neural network. Add an embedding layer by providing the length of the vocabulary, the length of the embedding layer, and the input length. Add two LSTM layers with **50** and **100** units. Use a ReLU activation function and set **return_sequences** to **True**. Then, add a dropout layer for each LSTM with a dropout of 20%. Finally, add a fully-connected layer with sigmoid as the final activation function:

```
model = Sequential()

model.add(Embedding(vocab_size, embedding_vector_length, input_
length=280))
model.add(LSTM(units= 50, activation = 'relu', return_sequences =
True))
model.add(Dropout(0.2))

model.add(LSTM(100, activation = 'relu'))
model.add(Dropout(0.2))

model.add(Dense(1, activation='sigmoid'))
```

24. Check the summary of the model using the **summary()** function:

```
model.summary()
```

You should get the following output:

```
Model: "sequential"

_____
Layer (type)              Output Shape              Param #
===============================================================
embedding (Embedding)     (None, 280, 300)          3974400

lstm (LSTM)               (None, 280, 50)           70200

dropout (Dropout)         (None, 280, 50)           0

lstm_1 (LSTM)             (None, 100)               60400

dropout_1 (Dropout)       (None, 100)               0

dense (Dense)             (None, 1)                 101
===============================================================
Total params: 4,105,101
Trainable params: 4,105,101
Non-trainable params: 0
```

Figure 9.48: Model summary

25. Use the **compile()** method to configure your model for training. Select **adam** as your optimizer, **binary_crossentropy** to measure your loss function, and **accuracy** as the metric to be displayed:

```
model.compile(optimizer='adam', loss='binary_crossentropy',
metrics=['accuracy'])
```

26. Fit your model and set it to run on two epochs. Set your batch size to **32**:

```
model.fit(X_train, y_train, epochs=2, batch_size=32)
```

You should get the following output:

```
Epoch 1/2
313/313 [==============================] - 116s 365ms/step - loss: 0.5219 - accuracy: 0.7931
Epoch 2/2
313/313 [==============================] - 116s 370ms/step - loss: 0.5112 - accuracy: 0.7947
<keras.callbacks.History at 0x1ea121ff8c8>
```

Figure 9. 49: Training the model

As you can see in *Figure 9.49*, your model achieved an accuracy of **0.7978** on the training set with minimal data preparation. You can try to improve this by removing stop words or extremely frequent words such as **the** and **a** that don't really help to assess the sentiment of a tweet and see if you can achieve the same performance on the testing set. You can deduce that the model can correctly predict almost 80% of the sentiments for the tweets in the training data.

CHAPTER 10: CUSTOM TENSORFLOW COMPONENTS

ACTIVITY 10.01: BUILDING A MODEL WITH CUSTOM LAYERS AND A CUSTOM LOSS FUNCTION

Solution:

To get started, open a new Colab or Jupyter Notebook. If you are using Google Colab, you will need to download the dataset into your Google Drive first:

1. Open a new Jupyter notebook or Google Colab notebook.

2. If you are using Google Colab, you can upload your dataset locally with the following code. Otherwise, go to *step 4*. Click on **Choose Files** to navigate to the CSV file and click **Open**. Save the file as **uploaded**. Then, go to the folder where you saved the dataset:

```
from google.colab import files
uploaded = files.upload()
```

3. Unzip the dataset in the current folder:

```
!unzip \*.zip
```

4. Create a variable, **directory**, that contains the path to the dataset:

```
directory = "/content/gdrive/My Drive/Datasets/pneumonia-or-healthy/"
```

5. Import all the required libraries:

```
import numpy as np
import pandas as pd
import pathlib
import os
import matplotlib.pyplot as plt
from keras.models import Sequential
from keras import optimizers
from tensorflow.keras.preprocessing.image import ImageDataGenerator
import tensorflow as tf
from tensorflow.keras.layers import Input, Conv2D, ReLU, \
    BatchNormalization,Add, AveragePooling2D, Flatten, Dense
from tensorflow.keras.models import Model
```

6. Create a variable, **path**, that contains the full path to the data using **pathlib.Path**:

```
path = pathlib.Path(directory)
```

7. Create two variables, called **train_dir** and **validation_dir**, that take the full paths to the train and validation folders, respectively:

```
train_dir = path / 'training_set'
validation_dir = path / 'test_set'
```

8. Create four variables, called **train_table_dir**, **train_glass_dir**, **validation_table_dir**, and **validation_glass_dir**, that take the full paths to the glass and table folders for the train and validation sets, respectively:

```
train_table_dir = train_dir / 'table'
train_glass_dir = train_dir /'glass'
validation_table_dir = validation_dir / 'table'
validation_glass_dir = validation_dir / 'glass'
```

9. Create four variables that will contain the number of images of glasses and tables for the training and validation sets:

```
num_train_table = len([f for f in os.listdir(train_table_dir)if \
                  os.path.isfile(os.path.join\
                                 (train_table_dir, f))])
num_train_glass = len([f for f in os.listdir(train_glass_dir)if \
                  os.path.isfile(os.path.join\
                                 (train_glass_dir, f))])
num_validation_table = len([f for f in os.listdir\
                       (validation_table_dir)if
os.path.isfile(os.path.join(validation_table_dir, f))])
num_validation_glass = len([f for f in os.listdir\
                       (validation_glass_dir)if \
                       os.path.isfile\
                       (os.path.join\
                       (validation_glass_dir, f))])
```

10. Display a bar chart with the total number of images of glasses and tables:

```
plt.bar(['table', 'glass'], \
        [num_train_table + num_validation_table, \
         num_train_glass + num_validation_glass], \
        align='center', \
```

```
        alpha=0.5)
plt.show()
```

You should get the following output:

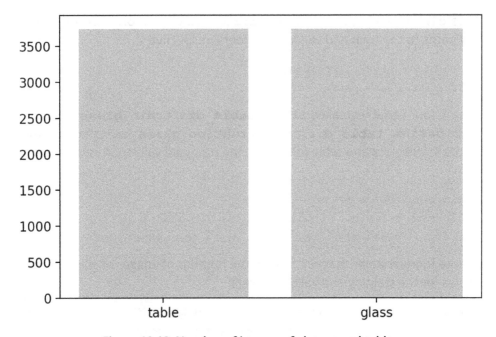

Figure 10.12: Number of images of glasses and tables

The preceding chart shows you the dataset is well balanced. There are almost as many images of glasses as tables, around 3,500 images each.

11. Create two variables, called **total_train** and **total_val**, that will get the number of images for the training and validation sets, respectively:

```
total_train = len(os.listdir(train_table_dir)) + \
              len(os.listdir(validation_table_dir))
total_val = len(os.listdir(train_glass_dir)) + \
            len(os.listdir(validation_glass_dir))
```

12. Import the **ImageDataGenerator** class:

```
from tensorflow.keras.preprocessing.image \
    import ImageDataGenerator
```

13. Instantiate two **ImageDataGenerator** classes, **train_image_generator** and **validation_image_generator**, that will rescale the images by dividing by 255:

```
train_image_generator = ImageDataGenerator(rescale=1./255)
validation_image_generator = ImageDataGenerator(rescale=1./255)
```

14. Create three variables, called **batch_size**, **img_height**, and **img_width**, that take the values **32**, **100**, and **100**, respectively:

```
batch_size = 32
img_height = 100
img_width = 100
```

15. Create a data generator called **train_data_gen** using **flow_from_directory()** method and specify the batch size, the path to the training folder, the value of the **shuffle** parameter, the size of the target, and the class mode:

```
train_data_gen = train_image_generator.flow_from_directory\
                    (batch_size=batch_size, directory=train_dir, \
                     shuffle=True, \
                     target_size=(img_height, img_width), \
                     class_mode='binary')
```

16. Create a data generator called **val_data_gen** using **flow_from_directory()** method and specify the batch size, the path to the validation folder, the size of the target, and the class mode:

```
val_data_gen = validation_image_generator.flow_from_directory\
                    (batch_size=batch_size, directory=validation_dir,\
                     target_size=(img_height, img_width), \
                     class_mode='binary')
```

17. Create your custom loss function. Use **def** and choose a name for your custom loss, **custom_loss_function**, in this case. Then, add your two arguments, **y_true** and **y_pred**. Now, create a variable, **squared_difference**, to store the square of **y_true** minus **y_pred**. Finally, return the calculated loss using your **tf.reduce_mean** from **squared_difference**:

```
def custom_loss_function(y_true, y_pred):
    squared_difference = tf.square(float(y_true) - float(y_pred))
    return tf.reduce_mean(squared_difference, axis=-1)
```

18. Build a function that takes your input as a tensor and adds ReLU and batch normalization to it:

```
def relu_batchnorm_layer(input):
    return BatchNormalization()(ReLU()(input))
```

19. Create a function to build the residual block. You will need to take a tensor as your input and pass it to two Conv2D layers. Next, add the input to the output, followed by ReLU and batch normalization.

    Since you used an **Add** layer for the skip connection in your **residual_block**, you need to make sure that its inputs are always of the same shape. The **downsample** parameter is used to specify the strides of the first Conv2D layer. It specifies **strides=2** if **True** and **strides=1** if **False**. When **strides=1**, the output (**int_output**) is the same size as the input. But when **strides=2**, the dimensions of **int_ouput** are halved. To take this into account, add a Conv2D layer with **kernel_size=1** to the skip connection:

```
def residual_block(input, downsample: bool, filters: int, \
                   kernel_size: int = 3):
    int_output = Conv2D(filters=filters, kernel_size=kernel_size,
                        strides= (1 if not downsample else 2),
                        padding="same")(input)
    int_output = relu_batchnorm_layer(int_output)
    int_output = Conv2D(filters=filters, kernel_size=kernel_size,
                        padding="same")(int_output)

    if downsample:
        int_output2 = Conv2D(filters=filters, kernel_size=1, strides=2,
                             padding="same")(input)
        output = Add()([int_output2, int_output])
    else:
        output = Add()([input, int_output])

    output = relu_batchnorm_layer(output)
    return output
```

20. Now, use the **keras.layers.Input()** layer to define the input layer of your model. Here, your shape is 100 pixels by 100 pixels and has three colors (RGB). Then, create your model with your custom architecture. Finally, reference your input and output tensors with **model = Model (inputs, outputs)**:

```
inputs = Input(shape=(100, 100, 3))
num_filters = 32

t = BatchNormalization()(inputs)
t = Conv2D(kernel_size=3,
            strides=1,
            filters=32,
            padding="same")(t)
t = relu_batchnorm_layer(t)

num_blocks_list = [1, 3, 5, 6, 1]
for i in range(len(num_blocks_list)):
    num_blocks = num_blocks_list[i]
    for j in range(num_blocks):
        t = residual_block(t, downsample=(j==0 and i!=0), filters=num_
filters)
    num_filters *= 2

t = AveragePooling2D(4)(t)
t = Flatten()(t)
outputs = Dense(1, activation='sigmoid')(t)

model = Model(inputs, outputs)
```

21. Get a summary of your model:

```
model.summary()
```

The summary will be shown on running the preceding command:

```
Model: "model"

Layer (type)                    Output Shape          Param #    Connected to
==================================================================================
input_1 (InputLayer)            [(None, 100, 100, 3)  0

batch_normalization (BatchNorma (None, 100, 100, 3)   12         input_1[0][0]

conv2d (Conv2D)                 (None, 100, 100, 32)  896        batch_normalization[0][0]

re_lu (ReLU)                    (None, 100, 100, 32)  0          conv2d[0][0]

batch_normalization_1 (BatchNor (None, 100, 100, 32)  128        re_lu[0][0]

conv2d_1 (Conv2D)               (None, 100, 100, 32)  9248       batch_normalization_1[0][0]

re_lu_1 (ReLU)                  (None, 100, 100, 32)  0          conv2d_1[0][0]
```

Figure 10.13: Model summary

22. Compile this model by providing your custom loss function, using Adam as the optimizer and accuracy as the metric to be displayed:

```
model.compile(
        optimizer='adam',
        loss=custom_loss_function,
        metrics=['accuracy']
)
```

23. Fit the model and provide the train and validation data generators, the number of epochs, the steps per epoch, and the validation steps:

```
history = model.fit(
    Train_data_gen,
    steps_per_epoch=total_train // batch_size,
    epochs=5,
    validation_data=val_data_gen,
    validation_steps=total_val // batch_size
)
```

You should get the following output:

```
Epoch 1/5
116/116 [==============================] - ETA: 0s - loss: 0.2266 - accuracy: 0.7170WARNING:tensorflow:Your input ran out of da
ta; interrupting training. Make sure that your dataset or generator can generate at least `steps_per_epoch * epochs` batches (i
n this case, 116 batches). You may need to use the repeat() function when building your dataset.
116/116 [==============================] - 747s 6s/step - loss: 0.2266 - accuracy: 0.7170 - val_loss: 0.3582 - val_accuracy: 0.
5251
Epoch 2/5
116/116 [==============================] - 669s 6s/step - loss: 0.1059 - accuracy: 0.8584
Epoch 3/5
116/116 [==============================] - 712s 6s/step - loss: 0.0855 - accuracy: 0.8887
Epoch 4/5
116/116 [==============================] - 779s 7s/step - loss: 0.0826 - accuracy: 0.8871
Epoch 5/5
116/116 [==============================] - 704s 6s/step - loss: 0.0741 - accuracy: 0.9014
```

Figure 10.14: Screenshot of the training progress

The preceding screenshot shows the information displayed by TensorFlow during the training of your model. You can see the accuracy achieved on the training and validation sets for each epoch. On the fifth epoch, the model is **85.9%** accurate on the training set and **88.5%** on the validation set.

24. Plot your training and validation accuracy:

```
plt.plot(history.history['accuracy'])
plt.plot(history.history['val_accuracy'])
plt.title('Training Accuracy vs Validation Accuracy')
plt.ylabel('Accuracy')
plt.xlabel('Epoch')
plt.legend(['Train', 'Validation'], loc='upper left')
plt.show()
```

You should get the following output:

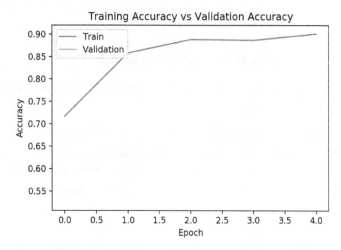

Figure 10.15: Training and validation accuracy

The preceding chart shows the accuracy scores for the training and validation sets for each epoch.

25. Plot your training and validation loss:

```
plt.plot(history.history['loss'])
plt.plot(history.history['val_loss'])
plt.title('Training Loss vs Validation Loss')
plt.ylabel('Loss')
plt.xlabel('Epoch')
plt.legend(['Train', 'Validation'], loc='upper left')
plt.show()
```

You should get the following output:

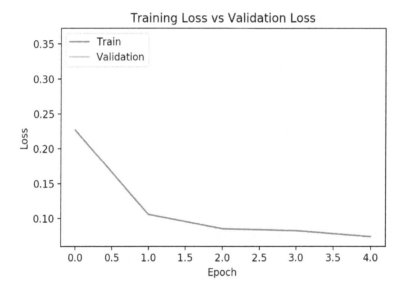

Figure 10.16: Training and validation loss

The preceding chart shows the loss scores for the training and validation sets for each epoch.

With this activity, you have successfully built a custom MSE loss function and a custom residual block layer and trained this custom deep learning model on the glass versus table dataset. You now know how to go beyond the default classes offered by TensorFlow and build your own custom deep learning models.

CHAPTER 11: GENERATIVE MODELS

ACTIVITY 11.01: GENERATING IMAGES USING GANS

Solution:

Perform the following steps to complete this activity:

1. Load Google Colab and Google Drive:

```
try:
    from google.colab import drive
    drive.mount('/content/drive', force_remount=True)
    COLAB = True
    print("Note: using Google CoLab")
    %tensorflow_version 2.x
except:
    print("Note: not using Google CoLab")
    COLAB = False
```

Your output should look something like this:

```
Mounted at /content/drive
Note: using Google CoLab
```

2. Import the libraries that you will be using:

```
import tensorflow as tf
from tensorflow.keras.models import Sequential, Model, load_model
from tensorflow.keras.layers import InputLayer, Reshape, Dropout, Dense
from tensorflow.keras.layers import Flatten, BatchNormalization
from tensorflow.keras.layers import UpSampling2D, Conv2D
from tensorflow.keras.layers import Activation, ZeroPadding2D
from tensorflow.keras.optimizers import Adam
from tensorflow.keras.layers import LeakyReLU
import zipfile
import matplotlib.pyplot as plt
import numpy as np
from PIL import Image
from tqdm import tqdm
import os
import time
from skimage.io import imread
```

3. Create a function to format a time string to track your time usage:

```
def time_string(sec_elapsed):
    hour = int(sec_elapsed / (60 * 60))
    minute = int((sec_elapsed % (60 * 60)) / 60)
    second = sec_elapsed % 60
    return "{}:{:>02}:{:>05.2f}".format(hour, minute, second)
```

4. Set the generation resolution to **3**. Also, set **img_rows** and **img_cols** to **5** and **img_margin** to **16** so that your preview images will be a **5x5** array (25 images) with a 16-pixel margin. Set **seed_vector** equal to **200**, **data_path** to where you stored your image dataset, and **epochs** to **500**. Finally, print the parameters:

```
gen_res = 3
gen_square = 32 * gen_res
img_chan = 3
img_rows = 5
img_cols = 5
img_margin = 16
seed_vector = 200
data_path = 'banana-or-orange/training_set/'
epochs = 500
num_batch = 32
num_buffer = 60000


print(f"Will generate a resolution of {gen_res}.")
print(f"Will generate {gen_square}px square images.")
print(f"Will generate {img_chan} image channels.")
print(f"Will generate {img_rows} preview rows.")
print(f"Will generate {img_cols} preview columns.")
print(f"Our preview margin equals {img_margin}.")
print(f"Our data path is: {data_path}.")
print(f"Our number of epochs are: {epochs}.")
print(f"Will generate a batch size of {num_batch}.")
print(f"Will generate a buffer size of {num_buffer}.")
```

Your output should look something like this:

```
Will generate a resolution of 3.
Will generate 96px square images.
Will generate 3 image channels.
Will generate 5 preview rows.
Will generate 5 preview columns.
Our preview margin equals 16.
Our data path is: banana-or-orange/training_set/.
Our number of epochs are: 500.
Will generate a batch size of 32.
Will generate a buffer size of 60000.
```

Figure 11.30: Output showing the parameters

5. If a NumPy preprocessed file exists from prior execution, then load it into memory; otherwise, preprocess the data and save the image binary:

```
training_binary_path = os.path.join(data_path,
        f'training_data_{gen_square}_{gen_square}.npy')


print(f"Looking for file: {training_binary_path}")


if not os.path.isfile(training_binary_path):
    start = time.time()
    print("Loading training images…")

    train_data = []
    images_path = os.path.join(data_path,'banana')
    for filename in tqdm(os.listdir(images_path)):
        path = os.path.join(images_path,filename)
        images = Image.open(path).resize((gen_square,
                                          gen_square),\
                                         Image.ANTIALIAS)
        train_data.append(np.asarray(images))
    train_data = np.reshape(train_data, (-1,gen_square,
             gen_square,img_chan))
    train_data = train_data.astype(np.float32)
    train_data = train_data / 127-5 - 1.

    print("Saving training image binary...")
    np.save(training_binary_path,train_data)
    elapsed = time.time()-start
    print (f'Image preprocess time: {time_string(elapsed)}')
```

```
else:
    print("Loading training data...")
    train_data = np.load(training_binary_path)
```

6. Batch and shuffle the data. Use the **tensorflow.data.Dataset** object library to use its functions to shuffle the dataset and create batches:

```
train_dataset = tf.data.Dataset.from_tensor_slices(train_data) \
                        .shuffle(num_buffer).batch(num_batch)
```

7. Build the generator for the DCGAN:

```
def create_dc_generator(seed_size, channels):
    model = Sequential()

    model.add(Dense(4*4*256,activation="relu",input_dim=seed_size))
    model.add(Reshape((4,4,256)))

    model.add(UpSampling2D())
    model.add(Conv2D(256,kernel_size=3,padding="same"))
    model.add(BatchNormalization(momentum=0.8))
    model.add(Activation("relu"))

    model.add(UpSampling2D())
    model.add(Conv2D(256,kernel_size=3,padding="same"))
    model.add(BatchNormalization(momentum=0.8))
    model.add(Activation("relu"))

    # Output resolution, additional upsampling
    model.add(UpSampling2D())
    model.add(Conv2D(128,kernel_size=3,padding="same"))
    model.add(BatchNormalization(momentum=0.8))
    model.add(Activation("relu"))

    if gen_res>1:
        model.add(UpSampling2D(size=(gen_res,gen_res)))
        model.add(Conv2D(128,kernel_size=3,padding="same"))
        model.add(BatchNormalization(momentum=0.8))
        model.add(Activation("relu"))

    # Final CNN layer
    model.add(Conv2D(channels,kernel_size=3,padding="same"))
```

```
        model.add(Activation("tanh"))

        return model
```

8. Build the discriminator for the DCGAN:

```
    def create_dc_discriminator(image_shape):
        model = Sequential()

        model.add(Conv2D(32, kernel_size=3, strides=2, \
                        input_shape=image_shape,
                        padding="same"))
        model.add(LeakyReLU(alpha=0.2))

        model.add(Dropout(0.25))
        model.add(Conv2D(64, kernel_size=3, strides=2, padding="same"))
        model.add(ZeroPadding2D(padding=((0,1),(0,1))))
        model.add(BatchNormalization(momentum=0.8))
        model.add(LeakyReLU(alpha=0.2))

        model.add(Dropout(0.25))
        model.add(Conv2D(128, kernel_size=3, strides=2, padding="same"))
        model.add(BatchNormalization(momentum=0.8))
        model.add(LeakyReLU(alpha=0.2))

        model.add(Dropout(0.25))
        model.add(Conv2D(256, kernel_size=3, strides=1, padding="same"))
        model.add(BatchNormalization(momentum=0.8))
        model.add(LeakyReLU(alpha=0.2))

        model.add(Dropout(0.25))
        model.add(Conv2D(512, kernel_size=3, strides=1, padding="same"))
        model.add(BatchNormalization(momentum=0.8))
        model.add(LeakyReLU(alpha=0.2))

        model.add(Dropout(0.25))
        model.add(Flatten())
        model.add(Dense(1, activation='sigmoid'))

        return model
```

9. Build the generator for the vanilla GAN:

```
def create_generator(seed_size, channels):
    model = Sequential()

    model.add(Dense(96*96*3,activation="tanh",input_dim=seed_size))
    model.add(Reshape((96,96,3)))

    return model
```

10. Build the discriminator for the vanilla GAN:

```
def create_discriminator(img_size):
    model = Sequential()
    model.add(InputLayer(input_shape=img_size))
    model.add(Dense(1024, activation="tanh"))
    model.add(Flatten())
    model.add(Dense(1, activation='sigmoid'))

    return model
```

11. Create a function to generate and save images that can be used to view progress during the model's training:

```
def save_images(generator, cnt, noise, prefix=None):
    img_array = np.full((
        img_margin + (img_rows * (gen_square+img_margin)),
        img_margin + (img_cols * (gen_square+img_margin)), 3),
        255, dtype=np.uint8)

    gen_imgs = generator.predict(noise)

    gen_imgs = 0.5 * gen_imgs + 0.5

    img_count = 0
    for row in range(img_rows):
        for col in range(img_cols):
            r = row * (gen_square+16) + img_margin
            c = col * (gen_square+16) + img_margin
            img_array[r:r+gen_square,c:c+gen_square] \
                = gen_imgs[img_count] * 255
```

```
        img_count += 1


    output_path = os.path.join(data_path, 'output')
    if not os.path.exists(output_path):
        os.makedirs(output_path)


    filename = os.path.join(output_path, f"train{prefix}-{cnt}.png")
    im = Image.fromarray(img_array)
    im.save(filename)
```

12. Initialize the generator for the DCGAN and view the output:

```
dc_generator = create_dc_generator(seed_vector, img_chan)

noise = tf.random.normal([1, seed_vector])
gen_img = dc_generator(noise, training=False)

plt.imshow(gen_img[0, :, :, 0])
```

Your output should look something like this:

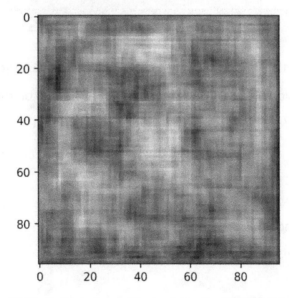

Figure 11.31: Output showing noise from the DCGAN generator

13. Initialize the generator for the vanilla GAN and view the output:

```
generator = create_generator(seed_vector, img_chan)
gen_van_img = generator(noise, training=False)
plt.imshow(gen_van_img[0, :, :, 0])
```

You should get the following output:

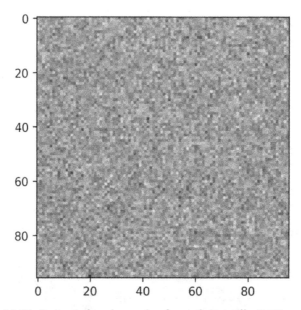

Figure 11.32: Output showing noise from the vanilla GAN generator

14. Print the decision of the DCGAN discriminator evaluated on the seed image:

```
img_shape = (gen_square,gen_square,img_chan)

discriminator = create_discriminator(img_shape)
decision = discriminator(gen_img)
print (decision)
```

Your output should look something like this:

```
tf.Tensor([[0.4994658]], shape=(1,1), dtype=float32)
```

15. Print the decision of the vanilla GAN evaluated on the seed image:

```
discriminator = create_discriminator(img_shape)
decision = discriminator(gen_img)
print(decision)
```

Your output should look something like this:

```
tf.Tensor([[0.5055983]], shape=(1,1), dtype=float32)
```

16. Create your loss functions. Since the output of both the discriminator and generator networks is different, you can define two separate loss functions for them. Moreover, they need to be trained separately in independent passes through the networks. Both GANs can utilize the same loss functions for their discriminators and generators. You can use **tf.keras.losses.BinaryCrossentropy** for **cross_entropy**. This calculates the loss between true and predicted labels. Then, define the **discrim_loss** function from **real_output** and **fake_output** using **tf.ones** and **tf.zeros** to calculate **total_loss**:

```
cross_entropy = tf.keras.losses.BinaryCrossentropy()

def discrim_loss(real_output, fake_output):
    real_loss = cross_entropy(tf.ones_like(real_output), real_output)
    fake_loss = cross_entropy(tf.zeros_like(fake_output), fake_output)
    total_loss = real_loss + fake_loss
    return total_loss


def gen_loss(fake_output):
    return cross_entropy(tf.ones_like(fake_output), fake_output)
```

17. Create two Adam optimizers, one for the generator and one for the discriminator. Use the same learning rate and momentum for each:

```
gen_optimizer = tf.keras.optimizers.Adam(1.5e-4,0.5)
disc_optimizer = tf.keras.optimizers.Adam(1.5e-4,0.5)
```

Here, you have your individual training step. It's very important that you only modify one network's weights at a time. With **tf.GradientTape()**, you can train the discriminator and generator at the same time, but separately from one another. This is how TensorFlow does automatic differentiation. It calculates the derivatives. You'll see that it creates two "tapes" – **gen_tape** and **disc_tape**. Think of these as recordings of the calculations for each.

18. Create **real_output** and **fake_output** for the discriminator. Use this for the generator loss (**g_loss**). Then, calculate the discriminator loss (**d_loss**) and the gradients of both the generator and discriminator with **gradients_of_generator** and **gradients_of_discriminator** and apply them. Encapsulate these steps within a function, passing in the generator, discriminator, and images, and returning the generator loss (**g_loss**) and discriminator loss (**d_loss**):

```
@tf.function
def train_step(generator, discriminator, images):
    seed = tf.random.normal([num_batch, seed_vector])

    with tf.GradientTape() as gen_tape, \
        tf.GradientTape() as disc_tape:
         gen_imgs = generator(seed, training=True)

         real_output = discriminator(images, training=True)
         fake_output = discriminator(gen_imgs, training=True)

         g_loss = gen_loss(fake_output)
         d_loss = discrim_loss(real_output, fake_output)


         gradients_of_generator = gen_tape.gradient(\
             g_loss, generator.trainable_variables)
         gradients_of_discriminator = disc_tape.gradient(\
             d_loss, discriminator.trainable_variables)

         gen_optimizer.apply_gradients(zip(
             gradients_of_generator, generator.trainable_variables))
         disc_optimizer.apply_gradients(zip(
             gradients_of_discriminator,
             discriminator.trainable_variables))
    return g_loss,d_loss
```

19. Create a number of fixed seeds with **fixed_seeds** equal to the number of images to display so that you can track the same images. This allows you to see how individual seeds evolve over time, tracking your time with **for epoch in range**. Now, loop through each batch with **for image_batch in dataset**. Continue to track your loss for both the generator and discriminator with **generator_loss** and **discriminator_loss**. Now, you have a nice display of all this information as it trains:

```
def train(generator, discriminator, dataset, epochs, prefix=None):
    fixed_seed = np.random.normal(0, 1, (img_rows * img_cols,
                                          seed_vector))

    start = time.time()

    for epoch in range(epochs):
        epoch_start = time.time()

        g_loss_list = []
        d_loss_list = []

        for image_batch in dataset:
            t = train_step(image_batch)
            g_loss_list.append(t[0])
            d_loss_list.append(t[1])

        generator_loss = sum(g_loss_list) / len(g_loss_list)
        discriminator_loss = sum(d_loss_list) / len(d_loss_list)

        epoch_elapsed = time.time() - epoch_start
        if (epoch + 1) % 100 == 0:
            print (f'Epoch {epoch+1}, gen loss={generator_loss},
            disc loss={discriminator_loss},'\
                    f' {time_string(epoch_elapsed)}')
        save_images(epoch,fixed_seed)

    elapsed = time.time()-start
    print (f'Training time: {time_string(elapsed)}')
```

20. Train the DCGAN model on your training dataset:

```
train(dc_generator, dc_discriminator, train_dataset, \
      epochs, prefix='-dc-gan')
```

Your output should look something like this:

```
Epoch 100, gen loss=3.098963975906372,disc loss=0.953665018081665, 0:00:14.92
Epoch 200, gen loss=2.757673501968384,disc loss=0.5467668175697327, 0:00:14.62
Epoch 300, gen loss=3.607388734817505,disc loss=0.4326052963733673, 0:00:14.70
Epoch 400, gen loss=4.211923599243164,disc loss=0.32031702995300293, 0:00:14.70
Epoch 500, gen loss=4.57510232925415,disc loss=0.3236366808414459, 0:00:14.69
Training time: 2:05:32.68
```

Figure 11.33: Output during training of the DCGAN model

The output shows the loss for the generator and discriminator at each epoch.

21. Train the vanilla model on your training dataset:

```
train(generator, discriminator, train_dataset, epochs, \
      prefix='-vanilla')
```

Your output should look something like this:

```
Epoch 100, gen loss=348.518798828125,disc loss=3.244476742948345e-26, 0:00:17.40
Epoch 200, gen loss=294.2938537597656,disc loss=4.122039078874912e-25, 0:00:21.63
Epoch 300, gen loss=270.9067077636719,disc loss=2.2674438840869547e-15, 0:00:16.34
Epoch 400, gen loss=218.98326110839844,disc loss=2.304341698060536e-15, 0:00:17.51
Epoch 500, gen loss=199.6638641357422,disc loss=5.25980567545048e-07, 0:00:18.92
Training time: 2:24:26.21
```

Figure 11.34: Output during training of the vanilla GAN model

22. View your images generated by the DCGAN model after the 100th epoch:

```
a = imread('banana-or-orange/training_set/output'\
           '/train-dc-gan-99.png')
plt.imshow(a)
```

You will get output like the following:

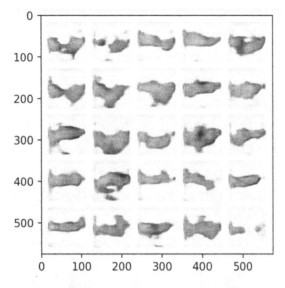

Figure 11.35: Output images from the DCGAN model after 100 epochs

23. View your images generated by the DCGAN model after the 500th epoch:

```
a = imread('/ banana-or-orange/training_set'\
          '/output/train-dc-gan-499.png')
plt.imshow(a)
```

You will get output like the following:

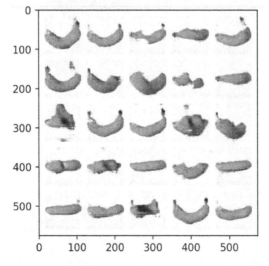

Figure 11.36: Output images from the DCGAN model after 500 epochs

24. View your images generated by the vanilla GAN model after the 100th epoch:

```
a = imread('banana-or-orange/training_set'\
           '/output/train-vanilla-99.png')
plt.imshow(a)
```

You will get output like the following:

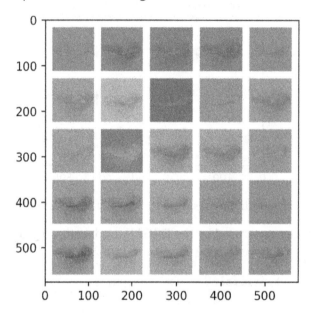

Figure 11.37: Output images from the vanilla GAN model after 100 epochs

25. View your images generated by the vanilla GAN model after the 500th epoch:

```
a = imread('/ banana-or-orange/training_set'\
           '/output/train-vanilla-499.png')
plt.imshow(a)
```

You will get output like the following:

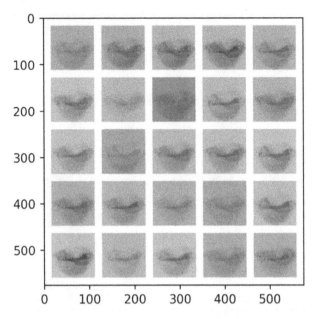

Figure 11.38: Output images from the vanilla GAN model after 500 epochs

The output shows the images generated by the vanilla GAN after 500 epochs. You can see that they are very different from those generated by the DCGAN.

You've just completed the last activity of the book. You created your own images with a DCGAN and compared them to a vanilla GAN model. As you can see from *Figure 11.36* and *Figure 11.38*, the results are very different from those of the DCGAN model, which were clearly recognizable as banana-like with different variations and orientations. With that model, though some images were more banana-like than others, all still exhibit at least some identifiable characteristics of bananas, such as color, shape, and presence of the black tip. The results from the vanilla GAN model, however, look more like pixel averages of the training dataset, which is overall not a good representation of real-life bananas. All images seem to have the same orientation, which may be another indicator that the results are more of a pixel average of the training data.

Matthew Moocarme

Anthony So

Anthony Maddalone

HEY!

We're Matthew Moocarme, Anthony So, and Anthony Maddalone, the authors of this book. We really hope you enjoyed reading our book and found it useful for learning TensorFlow.

It would really help us (and other potential readers!) if you could leave a review on Amazon sharing your thoughts on *The TensorFlow Workshop*.

Go to the link https://packt.link/r/1800205252.

OR

Scan the QR code to leave your review.

Your review will help us to understand what's worked well in this book and what could be improved upon for future editions, so it really is appreciated.

Best wishes,

Matthew Moocarme, Anthony So, and Anthony Maddalone

INDEX

P

package: 58, 67, 85, 112-113, 126, 203, 223-224, 226, 228, 230-232, 237, 253-254, 297, 357, 367, 381

pandas: 55-57, 60-63, 66, 91, 116, 134-135, 139, 144, 149, 155, 166, 181, 190, 196, 200, 208, 216, 227, 231, 236, 315, 350, 400

pneumonia: 380-381

pydata: 56, 66

pyplot: 69, 84, 253-254, 266, 270, 273-274, 276, 315, 368, 382, 417, 434

python: 5-7, 9, 12-15, 24-27, 34, 36, 42, 47, 55, 80, 94, 109-110, 112-113, 120, 124, 126, 294

R

regressor: 221, 236, 319-321, 323, 334-337, 341, 396

rmsprop: 42, 130, 136, 140, 145, 150, 155, 253-254, 258

S

sigmoid: 43, 45, 101, 127, 147-148, 150, 154-155, 161-163, 165, 170, 186, 198, 200-201, 241, 258, 269, 299, 306, 331-332, 371, 376, 385, 395, 412, 418, 428-429, 438

skimage: 434, 448

sklearn: 59, 135, 140, 227, 232, 270, 276, 315

softmax: 186-187, 190, 193, 198, 201, 210, 212, 217, 225, 228, 233, 241, 272, 278, 295, 301, 303, 305-306, 353-354, 376, 404

T

tensor: 1-2, 8-17, 19-26, 28-30, 32-33, 36-38, 40, 45-46, 50, 53, 77-81, 83, 85, 87-89, 94, 98, 101-102, 115, 117-118, 126, 129, 133, 175, 251, 375, 384, 423, 436, 440

tensordot: 81-82, 85

tfimage: 260-261

tf-version: 305

tuners: 223-224, 237

U

usecols: 221

V

verbose: 266, 354-355, 405-406, 415-416, 420

W

waveform: 84

workflow: 93, 94-95, 120

X

xlabel: 58-59, 84, 86, 273, 280, 324, 342

Y

ylabel: 58-59, 84, 86, 324, 342